Jixiang Yan

Optical Electronics

Also of interest

Semiconductor Spintronics
Thomas Schäpers, 2016
ISBN 978-3-11-036167-4, e-ISBN (PDF) 978-3-11-042544-4,
e-ISBN (EPUB) 978-3-11-043758-4

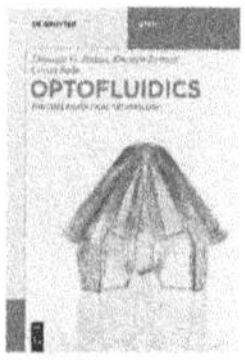

Optofluidics. Process Analytical Technology
Dominik G. Rabus, Cinzia Sada, Karsten Rebner, 2018
ISBN 978-3-11-054614-9, e-ISBN (PDF) 978-3-11-054615-6,
e-ISBN (EPUB) 978-3-11-054622-4

Elastic Light Scattering Spectrometry
Cheng Zhi Huang, Jian Ling, Jian Wang, 2018
ISBN 978-3-11-057310-7, e-ISBN (PDF) 978-3-11-057313-8,
e-ISBN (EPUB) 978-3-11-057324-4

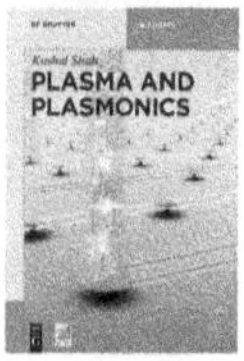

Plasma and Plasmonics
Kushal Shah, 2018
ISBN 978-3-11-056994-0, e-ISBN (PDF) 978-3-11-057003-8,
e-ISBN (EPUB) 978-3-11-057016-8

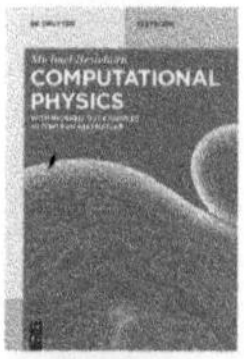

Computational Physics. With Worked Out Examples in FORTRAN and MATLAB
Michael Bestehorn, 2018
ISBN 978-3-11-051513-8, e-ISBN (PDF) 978-3-11-051514-5,
e-ISBN (EPUB) 978-3-11-051521-3

Optical Imaging and Photography.
Introduction to Science and Technology of Optics, Sensors and Systems
Ulrich Teubner, Hans Josef Brückner, 2018
ISBN 978-3-11-047293-6, e-ISBN (PDF) 978-3-11-047294-3,
e-ISBN (EPUB) 978-3-11-047295-0

Jixiang Yan

Optical Electronics

An Introduction

DE GRUYTER

清華大学出版社
TSINGHUA UNIVERSITY PRESS

Physics and Astronomy Classification Scheme 2010
Primary: physics; Secondary: radio electronics

Author
Prof. Jixiang Yan
Beijing Institute of Technology
School of optoelectronics
5 Zhongguancun South Street
100081 Beijing
People's Republic of China
yanjx@bit.edu.cn

ISBN 978-3-11-050049-3
e-ISBN (PDF) 978-3-11-050060-8
e-ISBN (EPUB) 978-3-11-049800-4

Library of Congress Control Number: 2018953198

Bibliographic information published by the Deutsche Nationalbibliothek
The Deutsche Nationalbibliothek lists this publication in the Deutsche Nationalbibliografie; detailed bibliographic data are available on the Internet at http://dnb.dnb.de.

Typesetting: Integra Software Services Pvt. Ltd.
Printing and binding: CPI books GmbH, Leck

www.degruyter.com

Preface

This textbook introduces the main principles involved in study and practice of optical electronics. Chapter 1 introduces development of light nature. Other subjects such as optical radiation and radiation sources, optical transmission and transmission media, optical detection and detectors, and optical imaging and the imaging systems are treaded in Chapters 2, 5, 6 and 7, respectively. Chapters 3 and 4 focus on several currently popular solid-state lasers. Finally, the nonlinear optical effects produced by the interaction between light and matter are treaded in Chapter 8. The textbook is primarily for senior and graduate students in optical and electrical engineering.

This book was originally written in Chinese by Jixiang Yan and was published by Tsinghua University Press, Beijing, China. The English version is translated by Professors Jixiang Yan and Xiaofang Zhang of Beijing Institute of Technology, and Professor Tianbao Xie of Linfield College in the United States. I would like to apologize to any of my colleagues whose work have not been acknowledged or not adequately represented in this book.

Thanks to Zhongye Ji, Hongxi Ren, Zexia Zhang, Chuhan Wu, Zhili Zheng, Yanna Han, Ruida Lang and Lu Zhang for their assistance in processing diagrams, formulas and terms. I am also grateful to editor Shan Zeng for her assistance in preparing the work for printing.

A special thanks to Mr. Wei Wang and Ms. Man Wu for their true and useful proofreading of this book. The second proofreading was completed by Dr. Yan Yan.

Jixiang Yan
Professor, Beijing Institute of Technology
China

https://doi.org/10.1515/9783110500608-201

Contents

1 The development of nature of light

Light is one of the major source of energy through which humans seek information. Statistics shows that of all the information that humans receive, 90% comes from light radiation, reflection and scattering. Since the mid- or late seventeenth century, the nature of light has been studied; there are two different views on these studies. One view was by Newton who believed the theory of particle nature of light. Another one was suggested by Huygens who believed the theory of wave nature of light. Two centuries later, Maxwell established the theory of electromagnetic wave nature of light in addition to relying on the earlier theory of light quantum. This led to the theory of "wave–particle duality." These theories brought a significant change in the perceptions that described the nature of light. Thus, it can be concluded that the concept of dual property of light did not end the studies on the exploration of nature of light. This topic will continue to be studied as an important subject of science and hopefully we can answer the questions on the nature of light much better. Therefore, we do not want to name this chapter as "the nature of light"; instead, it will be named as "development of nature of light." This is because the theories we discuss here are just the up-to-date theories.

This chapter will provide an introduction to the theories of the nature of light and in the following chapters the history and development process of light will be discussed. First, we will introduce the theories of wave nature and particle nature of light, followed by a discussion on the electromagnetic theory of light. Later, we will discuss about the earlier quantum theories of light and "dual nature of light." We will conclude with modern light quantum theory.

1.1 Earlier theories

As we mentioned earlier, there are two earlier theories based on the nature of light: Newton's particle theory and Huygens wave theory. This section will provide a brief introduction. Let us begin with the classical concepts of particles and wave.

1.1.1 Classical concepts of particle and wave

A typical particle should have the following basic properties:
(1) It should have inertial mass.
(2) It will be accelerated when exerted by an external force.
(3) It has momentum.

https://doi.org/10.1515/9783110500608-001

Wave has a property of transferring energy from one place to another. Electromagnetic waves can travel in both vacuum and medium, whereas mechanical waves require a medium to propogate.

1.1.2 Particle theory of light

The major postulates of Newton's light particle theory formed in 1670 are as follows:
(1) Light is composed of a large number of particles. All these particles have inertial mass and travel in a straight line at very high speed.
(2) When this particle beam meets a larger object, it will be stopped.
(3) When these particles rebound from surfaces, they obey the law of reflection.
(4) When the particles enter into a denser medium, the particle beam would refract toward the normal at the surface of the boundary of two media, and therefore it would speed up.

Newton's theory can successfully explain the phenomena such as light travels in a straight line, and behind a large object, there is a shade. It also explained the refraction law. To explain the refraction law, an imagination force needs to be added. This force can change the speed of light when light travels through the boundary. This theory arose some problems. First, the existence of such an imaginative force lacked evidence and support. In addition, it led to a conclusion that the speed of light is directly proportional to the index of the medium. This means that the light will travel faster in a denser medium. However, the process of measuring the speed of light was not made until the beginning of nineteenth century. In 1862, Foucalt proved in his experiment that this conclusion was totally wrong.

1.1.3 Wave theory of light

Ten years later after Newton proved his particle theory of light, Huygens suggested the wave theory of light. The basic concepts of Huygens theory are as follows:
(1) Light is composed of many wavelets; each point present on these wavelets is a source of secondary wavelet.
(2) Plane wave still produce plane wave and spherical wave still produces spherical wave.
(3) Light wave is a longitudinal wave just like a sound wave.
(4) When reflected from a surface, it obeys the law of reflection.
(5) Light travels from low index medium to high index medium; this process will slow down and refract toward the normal surface of the boundary.
(6) Light can produce interference and diffraction pattern.

So, similar to Newton's particle theory of light, Huygens's wave theory of light can explain that light travels in a straight line and follows the law of reflection. The latter can explain the law of reflection successfully without any assumption. Further, Huygens's theory predicted that light should have both interference and diffraction. Since the technologies had not advanced enough to show these experimentally, later, Young's double slit interference experiment in 1807 and Poisson's small hole experiment in 1820 proved that light had both interference and diffraction characters. These experiments strongly supported the wave theory of light.

Huygens thought light is both a longitudinal wave and a mechanical wave that can travel only in an elastic medium. Fresnel proposed that the light wave was transverse according to his light polarization experiment in 1820. In 1860, Maxwell established the electromagnetic theory; his theory took the light wave theory to a another level. This theory can explain all other experiments at that time; it also completely rejected the assumption that light is a mechanic wave. We will discuss Maxwell theory in detail now.

1.2 Electromagnetic theory of light

1.2.1 Electromagnetic induction law

Suppose that a charge q is moving at speed of $\boldsymbol{v}$ ($\boldsymbol{v} \ll c$, so relativity effect is neglected); at point $\boldsymbol{r}$ the electric field generated by q is as follows:

$$\boldsymbol{E} = \frac{q}{4\pi\varepsilon_0 r^3}\boldsymbol{r} \tag{1.1}$$

The B field at $\boldsymbol{r}$ is as follows:

$$\boldsymbol{B} = \frac{\mu_0 q(\boldsymbol{v}\times\boldsymbol{r})}{4\pi r^3} \tag{1.2}$$

where ε_0 is the permittivity of vacuum and μ_0 is the permeability of vacuum.

Combining equations (1.1) and (1.2), we then have

$$\boldsymbol{B} = \mu_0\varepsilon_0(\boldsymbol{v}\times\boldsymbol{E}) \tag{1.3}$$

Biot–Savart also derived a similar expression as equation (1.2) for magnetic field generated by a section of current. Suppose that there is a section of current-carrying wire, where the current is I and for an arbitrary small current element $\boldsymbol{Idl}$, $\boldsymbol{dl}$ has the direction of current, and B is the field generated by this small element at distance $\boldsymbol{r}$ away from $\boldsymbol{Idl}$ is given as follows:

$$d\boldsymbol{B} = \frac{\mu_0}{4\pi}\frac{\boldsymbol{Idl}\times\boldsymbol{r}}{r^3} \tag{1.4}$$

In 1820, Ampere found from experiments that just as reaction exists between point charges, there also exists reactions between current elements. This was later called Ampere's law. Ampere's law states that the force acting on a current element $\boldsymbol{Idl}$ in a magnetic field **B** is given as follows:

$$d\boldsymbol{F} = I d\boldsymbol{l} \times \boldsymbol{B} \tag{1.5}$$

After Oested discovered the magnetic effect of a current, many scientists thought that as the current can produce magnetic field, can magnetic field in turn generate electric field? Based on many experiments, Faraday published an important paper on November 24, 1831. In this paper, Faraday summarized that there are five cases in which current can be generated: (1) alternating current, (2) moving charge or moving steady current, (3) changing magnetic field, (4) moving magnet and (5) conduct wire moving in a magnetic field. He called these phenomena as magnetic induction and emphasized that the induction current did not depend on the steady current or the magnetic field, but only depend on their change. A general expression of induction law (differential form) is as follows:

$$\nabla \times \boldsymbol{E} = -\frac{\partial \boldsymbol{B}}{\partial t}$$

This is one of Maxwell's equations. This function indicates that even if no conductor exists, changing the magnetic field can produce whirlpool-like induction electric field in the space.

1.2.2 Maxwell's electromagnetic theory

Based on Maxwell's systematic study and the work summarized by Oersted, Biot, Savart, Ampere and Faraday, Maxwell proposed a hypothesis of curl electric field and displacement current and finally established a set of equations for description of the motion of electromagnetic field. This set of equations are called Maxwell's equations. The differentiation form of these equations is as follows:

$$\nabla \cdot \boldsymbol{D} = \rho_{eo} \tag{1.6}$$

$$\nabla \times \boldsymbol{E} = -\frac{\partial \boldsymbol{B}}{\partial t} \tag{1.7}$$

$$\nabla \cdot \boldsymbol{B} = 0 \tag{1.8}$$

$$\nabla \times \boldsymbol{H} = \boldsymbol{j}_0 + \frac{\partial \boldsymbol{D}}{\partial t} \tag{1.9}$$

where ρ_{eo} is free charge density, $\boldsymbol{j}_0$ is current density, $\boldsymbol{D}$ is displacement current, $\frac{\partial \boldsymbol{D}}{\partial t}$ is displacement current density and $\boldsymbol{H}$ is the magnetic field strength.

Maxwell's equations hold good for both vacuum and media. While handling problems in media, the expressions of electric and magnetic fields in media should be applied. For isotropic media, the fields can be expressed as follows:

$$\boldsymbol{D} = \varepsilon_r \varepsilon_0 \boldsymbol{E} = \varepsilon \boldsymbol{E} \tag{1.10}$$

$$\boldsymbol{B} = \mu_r \mu_0 \boldsymbol{H} = \mu \boldsymbol{H} \tag{1.11}$$

$$\boldsymbol{j} = \sigma \boldsymbol{E} \tag{1.12}$$

where ε_r and μ_r are relative permittivity and relative permeability, respectively, and σ is conductivity. In vacuum, $\varepsilon_r = \mu_r = 1$. When we study electric and magnetic field, we are usually more interested in $\boldsymbol{E}$ component. For a monochromatic E and M wave with a frequency ω traveling in a vacuum, equations (1.6)–(1.12) will lead to the following Helmholtz's equation:

$$\nabla^2 \boldsymbol{E} + k^2 \boldsymbol{E} = 0 \tag{1.13}$$

$$\nabla \cdot \boldsymbol{E} = 0 \tag{1.14}$$

and

$$\boldsymbol{B} = -\frac{i}{\omega} \nabla \times \boldsymbol{E} \tag{1.15}$$

k in equation (1.13) is called wave number, which is as follows:

$$k = \omega\sqrt{\mu\varepsilon} = \frac{2\pi}{\lambda}$$

For a plane E and M wave, E can be expressed as follows:

$$\boldsymbol{E}(\boldsymbol{r}, t) = \boldsymbol{E}_0 exp[i(\boldsymbol{k} \cdot \boldsymbol{r} - \omega t)] \tag{1.16}$$

where $\boldsymbol{k}$ is a vector that is present in the direction of propagation of E and M wave. Substituting equation (1.16) into equation (1.14) gives the following equation:

$$\boldsymbol{k} \cdot \boldsymbol{E} = 0 \tag{1.17}$$

Similarly,

$$\boldsymbol{k} \cdot \boldsymbol{B} = 0 \tag{1.18}$$

From equation (1.15), we can get

$$\boldsymbol{B} = \frac{1}{\omega}(\boldsymbol{k} \times \boldsymbol{E}) \tag{1.19}$$

Equations (1.17)–(1.19) show that vectors $\boldsymbol{E}$, $\boldsymbol{B}$ and $\boldsymbol{k}$ are orthogonal to each other. Thus, Maxwell's electromagnetic theory shows electromagnetic wave is a transverse wave, and vectors $\boldsymbol{E}$ and $\boldsymbol{B}$ are perpendicular to the wave vector $\boldsymbol{k}$ and are also perpendicular to each other. $\boldsymbol{E} \times \boldsymbol{B}$ is in the direction of $\boldsymbol{k}$. The propagation speed of electromagnetic wave is as follows:

$$v = \frac{1}{\sqrt{\mu\varepsilon}} = \frac{1}{\sqrt{\varepsilon_r\varepsilon_0\mu_r\mu_0}} = \frac{v_0}{\sqrt{\varepsilon_r\mu_r}} \tag{1.20}$$

where $v_0 = \frac{1}{\sqrt{\varepsilon_0\mu_0}}$ is the propagation speed of electromagnetic wave in vacuum.

1.2.3 Electromagnetic theory of light

On December 8, 1860, Maxwell read his summation paper "A Dynamic Theory of the Electromagnetic Field" to the Royal Society. In this paper, Maxwell gave the calculated propagation speed of electromagnetic wave. The date was agreed with Foucalt's measured speed in 1862. Thus, Maxwell drew a conclusion that "light is an electromagnetic wave it follows the electromagnetic law propagates via field." In simple words, light is a electromagnetic wave whose wavelength falls in a certain range. Maxwell hereby established the light electromagnetic theory. It led human understanding of the nature of light that led to a big step.

The accepted values of ε_0 and μ_0 are as follows: $\varepsilon_0 = 8.854187817 \times 10^{-12}\,\mathrm{F{\cdot}m^{-1}}$ and $\mu_0 = 12.566370614 \times 10^{-7}\,\mathrm{N{\cdot}A^{-1}}$. The associated speed of light in vacuum is as follows:

$$c = \frac{1}{\sqrt{\varepsilon_0\mu_0}} = 2.99792458\,\mathrm{m{\cdot}s^{-1}}$$

In any medium with permittivity ε_r and permeability μ_r, the speed of light is given as follows:

$$u = \frac{c}{\sqrt{\varepsilon_r\mu_r}} \tag{1.21}$$

as both ε_r and μ_r are bigger than 1, so u is smaller than c. Their ratio is

$$\frac{c}{u} = n = \sqrt{\varepsilon_r\mu_r} \tag{1.22}$$

This is called the index of refraction of a medium.

For visible light, material magnetization is neglected, so $\mu_r \approx 1$, then

$$n = \frac{c}{u} \approx \sqrt{\varepsilon_r} \tag{1.23}$$

This also signifies that the speed of light varies with different media.

Note that

$$u = 2\pi\omega\lambda \tag{1.24}$$

where ω is the angular frequency of light that does not change as the medium changes. Therefore, the wavelength λ will be different in different medium.

Since light is an electromagnetic wave, it should have all the characteristics of electromagnetic wave introduced in Section 1.2.2. These include the following: light should be a transverse wave, and its electric field vector $\boldsymbol{E}$ and magnetic strength vector $\boldsymbol{H}$ are perpendicular to each other. In addition, both $\boldsymbol{E}$ and $\boldsymbol{H}$ are perpendicular to the direction of propogation $\boldsymbol{k}$. Light can travel in a medium as well as in vacuum.

As light is an electromagnetic wave, its process of propogation should associate with the electromagnetic energy transmission similar to the behavior of any other electromagnetic wave. Usually light energy is expressed as intensity I. Light intensity is defined as the average energy value in a cycle on a unit area, which is perpendicular to the direction of propogation of light. The main component interacts through human eyes and the film is electric field $\boldsymbol{E}$. Light intensity I is proportional to $|\boldsymbol{E}|^2 = \boldsymbol{E_o}^2$. In wave optics, relative intensity is more important, so light intensity I is defined as follows:

$$I = \boldsymbol{E_o}^2 \tag{1.25}$$

1.3 Superposition and interference of light

The electromagnetic theory successfully explained phenomena such as interference and diffraction. Detailed description of these phenomena has been provided in most of the textbooks on physical optics. Here we briefly introduce the fundamental theories of these phenomena: (1) law of independent propagation and (2) superposition principle of light. We will also discuss the interference conditions of two light waves.

1.3.1 The law of independent propagation of light waves

When several light waves travel in the space, they do not interrupt with each other. In other words, how each wave propagates in the space has nothing to do with other waves existing in the same space. This is the so-called law of independent propagation of light.

It is obvious that light propagates independently in free space. However, whether the law of the independent propagation of light in a medium still holds good will depend on the properties of the medium and the characteristics of incident light. Two situations will lead to a failure of the law of independent propagation of light:

(1) Under the influence of light field with certain intensity and frequency, the property of absorption and transmission of the medium will change. These changes will affect other forms of light transmitting in the same medium.
(2) Under the reaction of a very strong light wave, the medium can be polarized and can produce polarization vector. The latter can radiate new electromagnetic

waves. Under certain conditions, when two different light waves incident into a medium, a third light wave with different frequency may appear. This phenomenon is called “nonlinear optics” effect. To create nonlinear optics effect, the incident wave must have very high intensity. That is why before laser appeared, nonlinear optics phenomenon was hardly observed. After laser was created, many different nonlinear optic effects have been gradually noticed and understood and applied in many fields. Nonlinear optics or high-intensity light optics have become an important research area in modern optics. We will discuss these concepts in Chapter 8. In this book, we assume that all media are linear and light waves obey the law of independent propagation unless otherwise specified.

1.3.2 Light wave superposition principle

If independent propagation principle of light waves is true, when two or more light waves existed simultaneously in an area, then the light vibration of each point in this area will be the summation of vibration caused by each light wave. Suppose that two light waves are as follows:

$$\boldsymbol{E}_1(\boldsymbol{r}_1, t) = \boldsymbol{E}_{10}(\boldsymbol{r}_1) exp\{-i[\omega_1 t - \varphi_1(\boldsymbol{r}_1)]\}$$

and

$$\boldsymbol{E}_2(\boldsymbol{r}_2, \boldsymbol{t}) = \boldsymbol{E}_{20}(\boldsymbol{r}_2) exp\{-i[\omega_2 t - \varphi_2(\boldsymbol{r}_2)]\} \tag{1.26}$$

Summation of these two waves is as follows:

$$\mathbf{E} = \boldsymbol{E}_1(\boldsymbol{r}_1, \boldsymbol{t}) + \boldsymbol{E}_2(\boldsymbol{r}_2, \boldsymbol{t}) = \boldsymbol{E}_{10}(\boldsymbol{r}_1) exp\{-i[\omega_1 t - \varphi_1(\boldsymbol{r}_1)]\} + \boldsymbol{E}_{20}(\boldsymbol{r}_2) exp\{-i[\omega_2 t - \varphi_2(\boldsymbol{r}_2)]\} \tag{1.27}$$

The principle of superposition of light wave works only if the law of independent propagation of light wave is true. Therefore, if two or more light waves satisfy superposition principle, light media have to meet proper conditions and the intensity of each light wave cannot be too high. As different medium appears nonlinearly optical corresponding to different light intensities, the so-called high intensity light wave varies with different medium.

1.3.3 Interference conditions of light waves

In the section 1.3.2, superposition of amplitudes of light field has been discussed. While most light receivers do not respond to the amplitude but to the intensity of light waves, this section will discuss about the relationship between intensity of different waves in superposition. The conditions of light interference will also be explained.

The interference of light wave is a phenomenon in which light intensity is redistributed when two or more waves undergo superposition.

For simplicity without loss of generality, let us consider two scalar light waves, each with single frequency. At time t, two scalar electric fields are as follows:

$$\boldsymbol{E}_1(P,t) = \boldsymbol{E}_{10}(P)exp\{-i[\omega_1 t - \varphi_1(P)]\}$$

and

$$\boldsymbol{E}_2(P,t) = \boldsymbol{E}_{20}(P)exp\{-i[\omega_2 t - \varphi_2(P)]\} \tag{1.28}$$

The combination intensity is

$$I = [E_1(P,t) + E_2(P,t)] \cdot [E_1(P,t) + E_2(P,t)]^* \tag{1.29}$$

"*" stands for complex conjugate (CC).

Substituting equation (1.28) into (1.29), we obtain the following:

$$\begin{aligned} I &= I_1 + I_2 + (E_{10}E_{20}exp\{-i[(\omega_1-\omega_2)t-(\varphi_1-\varphi_2)]\} + c.c.) \\ &= I_1 + I_2 + 2\sqrt{I_1I_2}cos[(\omega_1-\omega_2)t-(\varphi_1-\varphi_2)] = I_1 + I_2 + 2\sqrt{I_1I_2}cos\delta \end{aligned} \tag{1.30}$$

CC here represents the complex conjugate of the former term. $I_1 = E_{10}^2$ and $I_2 = E_{20}^2$ are intensities of these two waves at point P, respectively. $\delta = (\omega_1-\omega_2)t-(\varphi_1-\varphi_2)$ is the total phase difference of these two waves at point P.

Equation (1.30) shows two waves traveling independently and their altitudes obey the principle of superposition, but the intensity does not satisfy the same principle. The summation of each wave's intensity $I_1 + I_2$ differs from the total intensity by a factor of $2\sqrt{I_1I_2}cos\delta$. This term is called interference term. Theoretically speaking, the interference term always exists. This means that the redistribution of intensity cannot be avoided. In other words, light waves always interfere with each other. However, the theoretical interference cannot always be displayed. In practice, only redistribution of intensity can be observed or recorded when we think the two waves are interfering. We will discuss interference conditions of light waves based on this justification.

First, we check the condition of frequency. Frequency of visible light is in the order of 10^{14} Hz. Human eyes and any other type of detectors cannot detect any intensity with such rapid change. All we can do is to measure or record the average value of intensity. When $\varphi_1 - \varphi_2 \neq 0$, the average value of cosδ is zero. The interference term disappears. So for two waves that can make interference, the first condition is that the frequencies of these two waves are the same.

The second term in δ is the initial phase difference. If it is not stable, the time average of cosδ will be zero too. Therefore, the second condition for the interference of two waves can be observed and recorded as a stable phase difference between these two waves. For two simple harmonic (SH) scalar waves, as we pointed here, if they satisfy the aforementioned conditions, their interference can be observed and recorded. While for two vector SH waves, their vibration directions have to be taken into consideration. If two vibration directions are parallel to each other, their superposition is similar as for scalar waves. If their direction of vibration is perpendicular to each other, then from the summation of vector, we know that there is no interference.

$$I = I_1 + I_2$$

In a general case, if there is an angle between vibration directions of two waves, the parallel components of these two waves can produce interference.

Summarizing our discussion, the conditions for two waves can generate the following interference:
(1) They have the same frequency
(2) They have stable phase difference
(3) They have vibration directions that are not perpendicular to each other

More explicitly, the above-mentioned conditions are for interference that are being able to observed and recorded. Moreover, the degree of stability of frequency and phase difference vary with the characters of detectors and the environmental conditions of recording.

1.4 Further discussion of coherence

In this section, we will discuss the space and time coherence. First, let us talk how to utilize complex variables to express field.

1.4.1 Complex variable expression of polychromatic field

For the sake of simplicity, consider a line polarized electromagnetic wave; it can be expressed as $V^{(r)}(\boldsymbol{r}, t)$ Its Fourier integrated form is given as follows:

$$V(\boldsymbol{r}, t) = \frac{1}{2\pi} \int_{-\infty}^{+\infty} V(\boldsymbol{r}, \omega) exp(-i\omega t) d\omega \tag{1.31}$$

$$V(\boldsymbol{r}, \omega) = \int_{-\infty}^{+\infty} V(\boldsymbol{r}, t) exp(i\omega t) dt \tag{1.32}$$

Because $V^{(r)}$ is real, equation (1.32) shows $V(\boldsymbol{r}, -\omega) = V^*(\boldsymbol{r}, \omega)$; this means all the information is included in the positive spectrum. Therefore, we can use $(\boldsymbol{r}, t)$ as follows:

$$V(\boldsymbol{r}, t) = \frac{1}{2\pi} \int_{0}^{+\infty} V(\boldsymbol{r}, \omega) exp(-i\omega t) d\omega \tag{1.33}$$

The complex variable $V(\mathbf{r}, t)$ is used to represent $V^{(r)}$; it is associated with $V^{(r)}$ as a complex analytic signal. Obviously, there exists a one-to-one relationship between these two functions. In fact, given V, from equations (1.31) and (1.33), we can get the following equation:

$$V^{(r)} = 2Re(V) \tag{1.34}$$

On the other hand, once $V^{(r)}$ has been defined, from equation (1.32), we can get $V(\mathbf{r}, \omega)$; substituting it into equation (1.33), we can find V.

It is more convenient to use complex signals to express electromagnetic field. For instance, if a signal is monochromatic, it can be expressed as follows:

$$V^{(r)} = Asin\omega t$$

or

$$V = Aexp(-i\omega t)/2$$

The advantage of using the complex signal is well known. A common situation in practice is the spectrum of analytic signal has a significant value only when its spectrum is within a very small interval $\Delta\omega$ compared to the average frequency $<\omega>$. This means that this signal can be treated as a quiz-monochromatic wave. Under this circumstance, we can have the following equation:

$$V(t) = A(t)exp\{i[\Psi(t) - \langle\omega\rangle t]\} \tag{1.35}$$

where $A(t)$ and $\Psi(t)$ are slow changing functions; this means

$$\left[\frac{dA}{Adt}, \frac{d\Psi}{dt}\right] \ll \langle\omega\rangle \tag{1.36}$$

It was noticed that if we use quantum theory to describe electromagnetic field, complex signals will have deeper meanings. It was also noticed that V and V^* almost correspond to the operator of photon creation and annihilation (will be discussed later). Finally, other quantities can be expressed as functions of analytic signals. For example, the strength of a light beam can be defined as follows:

$$I(\mathbf{r}, t) = V(\mathbf{r}, t) \cdot V^*(\mathbf{r}, t) \tag{1.37}$$

In fact, this definition means I is not directly proportional to $\left(V^{(r)}\right)^2$. If light wave is monochromatic, it can be proved from equation (1.35) that $I(\mathbf{r}, t)$ equals to the average value of $\left(V^{(r)}\right)^2/2$ in a few optical periods.[1]

---

1 From equation (1.35), $V^{(r)} = 2A(t)\cos[\psi(t) - \langle\omega\rangle t]$, $\left(V^{(r)}\right)^2/2 = 2A^2(t)\cos^2[\psi(t) - \langle\omega\rangle t]$ $= A^2(t)\{1 + \cos 2[\psi(t) - \langle\omega\rangle t]\}$. When carrying the integration for sometime, the last term is zero. Meanwhile, $A^2(t)$ changes little over time so that it can be treated as a constant, so we have $\frac{1}{T}\int_0^T \left(V^{(r)}\right)^2/2dt \approx A^2(t)\frac{1}{T}\int_0^T 1 \cdot dt = A^2(t) = V \cdot V^* = I$.

1.4.2 Degree of spatial and temporal coherence

In order to describe the properties of a light beam, we introduce a classical correlation function; for now the function is restricted as first order.

For a point $\boldsymbol{r}_1$, a first-order correlation function $\Gamma^{(1)}$ can be defined as follows:

$$\Gamma^{(1)}(\boldsymbol{r}_1, \boldsymbol{r}_1, \tau) = \lim_{T\to\infty} \frac{1}{2T} \int_{-T}^{T} V(\boldsymbol{r}_1, t+\tau) V^*(\boldsymbol{r}_1, t) dt \tag{1.38}$$

It is $V(\boldsymbol{r}, t)$'s autocorrelation function; in other words, it is the average value of product $V(\boldsymbol{r}, t+\tau)V^*(\boldsymbol{r}, t)$. Thus, it can be simplified as follows:

$$\Gamma(\boldsymbol{r}_1, \boldsymbol{r}_1, \tau) = \langle V(\boldsymbol{r}_1, t+\tau) \cdot V^*(\boldsymbol{r}_1, t)\rangle \tag{1.39}$$

We can also define a normalized function $\gamma^{(1)}(\boldsymbol{r}_1, \boldsymbol{r}_1, \tau)$

$$\gamma^{(1)} = \frac{\Gamma^{(1)}}{\langle V(\boldsymbol{r}_1, t) \cdot V^*(\boldsymbol{r}_1, t)\rangle} = \frac{\Gamma^{(1)}}{\langle I(\boldsymbol{r}_1, t)\rangle} \tag{1.40}$$

From Schwarz inequality, we can know that

$$\left|\gamma^{(1)}(\boldsymbol{r}_1, \boldsymbol{r}_1, \tau)\right| \le 1$$

From equation (1.38) we also can see that

$$\gamma^{(1)}(\boldsymbol{r}_1, \boldsymbol{r}_1, -\tau) = \gamma^{(1)*}(\boldsymbol{r}_1, \boldsymbol{r}_1, \tau)$$

Function $\gamma^{(1)}(\boldsymbol{r}_1, \boldsymbol{r}_1, \tau)$ is called complex degree of temporal coherence. Its mode $|\gamma^{(1)}|$ is called the degree of temporal coherence. Obviously, $\Gamma^{(1)}$ (and so is $\gamma^{(1)}$) is a measure of the degree of correlation of two analytic signals that arrive at the same point $\boldsymbol{r}_1$ with a time difference τ. In case of lacking temporal coherence, from equations (1.38) and (1.40), we know that when $\gamma^{(1)} = 0$, we have $\tau > 0(\tau = 0, \gamma^{(1)} = 1)$. In this situation of complete temporal correlation (for instance, a sine wave with constant amplitude $V = A(\boldsymbol{r}_1)exp(-i\omega t)$, $|\gamma^{(1)}| = 1$ is true for any given τ. So $|\gamma^{(1)}|$ is a function with values between 0 and 1. It gives the degree of temporal coherence. Generally, the function $|\gamma^{(1)}(\tau)|$ has the form shown in Fig. 1.1 (again, we have $|\gamma^{(1)}(-\tau)| = |\gamma^{(1)}(\tau)|$.

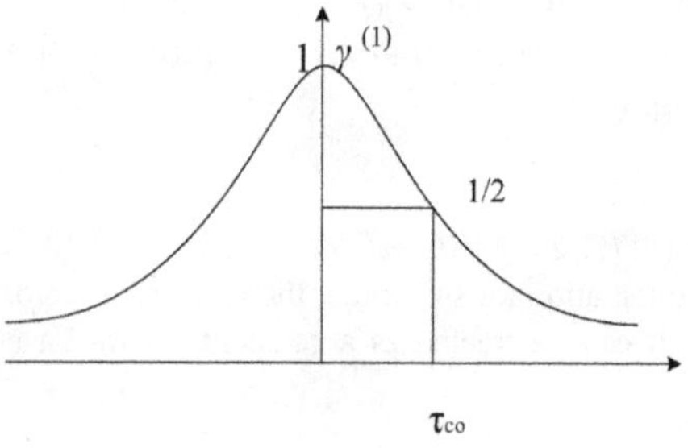

Fig. 1.1: A possible behavior of degree of temporal correlation.

We can define a characteristic time interval τ_{co} (called correlation time) as a time interval, for example, $|\gamma^{(1)}(\tau)| = 1/2$. For complete coherent waves, $\tau_{co} = \infty$, whereas for complete incoherent wave, $\tau_{co} = 0$. We can also define a temporal correlation length L_c and $L_c = c\tau_c$.

Following the same method, we can also define two first-order correlation functions at the same time but for different location $\boldsymbol{r}_1$ and $\boldsymbol{r}_2$.

$$\Gamma^{(1)}(\boldsymbol{r}_1,\boldsymbol{r}_1,0) = \lim_{T\to\infty}\frac{1}{2T}\int_{-T}^{T} V(\boldsymbol{r}_1,t)\cdot V^*(\boldsymbol{r}_2,t)dt = \langle V(\boldsymbol{r}_1,t), V^*(\boldsymbol{r}_2,t)\rangle \tag{1.41}$$

The corresponding normalization function $\gamma^{(1)}(\boldsymbol{r}_1,\boldsymbol{r}_1,0)$ is as follows:

$$\gamma^{(1)} = \frac{\Gamma^{(1)}(\boldsymbol{r}_1,\boldsymbol{r}_2,0)}{\left[\Gamma^{(1)}(\boldsymbol{r}_1,\boldsymbol{r}_1,0)\Gamma^{(1)}(\boldsymbol{r}_2,\boldsymbol{r}_2,0)\right]^{1/2}} \tag{1.42}$$

Once again the Schwarz inequality leads to $|\gamma^{(1)}| \le 1$. $\boldsymbol{R}^{(1)}(\boldsymbol{r}_1,\boldsymbol{r}_2,0)$; it is called the complex spatial correlation. Its mode is called the degree of spatial correlation. From earlier discussion, we know if fixing $\boldsymbol{r}_1$, $\gamma^{(1)}$ as a function of $\boldsymbol{r}_2$, decreases from 1 (when $\boldsymbol{r}_1=\boldsymbol{r}_2$) to zero as $|\boldsymbol{r}_1-\boldsymbol{r}_2|$ increases. Therefore, for a certain character area, $\gamma^{(1)}$ will be bigger for a special value (say ½). This area is on a wave plane and around a point P_1 defined by vector $\boldsymbol{r}_1$. It is called the coherent area of P_1.

The correlation time can be defined as the half width at half maximum of the curve.

The concepts of spatial and temporal correlation can be combined with a cross-correlation function; the latter is defined as follows:

$$\Gamma^{(1)}(\boldsymbol{r}_1,\boldsymbol{r}_2,\tau) = \langle V(\boldsymbol{r}_1,t+\tau), V^*(\boldsymbol{r}_2,t)\rangle \tag{1.43}$$

It can be normalized as follows:

$$\gamma^{(1)}(\boldsymbol{r}_1,\boldsymbol{r}_2,\tau) = \frac{\Gamma^{(1)}(\boldsymbol{r}_1,\boldsymbol{r}_2,\tau)}{\left[\Gamma^{(1)}(\boldsymbol{r}_1,\boldsymbol{r}_1,0)\Gamma^{(1)}(\boldsymbol{r}_2,\boldsymbol{r}_2,0)\right]^{1/2}} \tag{1.44}$$

This function is a measure of correlation between two points on a wave plane and is called complex correlation degree. For a quiz-monochromatic wave, combining equations (1.35) and (1.44), we can get the following equation:

$$\gamma^{(1)}(\tau) = |\gamma^{(1)}(\tau)|exp\{i[\Psi(\tau) - \langle\omega\rangle t]\} \tag{1.45}$$

where $|\gamma^{(1)}|$ and $\Psi(\tau)$ are slow changing functions, which means

$$\left[\frac{d|\gamma^{(1)}(\tau)|}{|\gamma^{(1)}|d\tau}, \frac{d\Psi}{d\tau}\right] \ll \langle\omega\rangle \tag{1.46}$$

1.4.3 Measurement of correlation of spatial and temporal

A very simple way to measure spatial coherence between two points on a wave plane is by using Young's interferometer (Fig. 1.2). This device contains two screens. There are two small holes in the first screen located at x_1 and x_2. Light passes through these two holes and forms an interference pattern on the second screen. More precisely, at any time t, the interference at point P is the superposition of two waves radiated from x_1 and x_2 at $(t - (L_1/c))$ and $[t - (L_2/c)]$, respectively. While measuring interference fringes (during the film exposure time), the higher the correlation degree of analytic signals $V(x_1, t - (L_1/c))$ and $V(x_2, t_2 - (L_2/c))$, the clearer will be the resolution fringes.

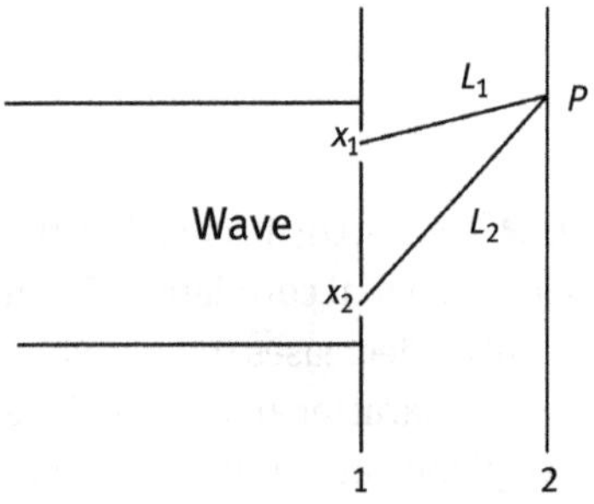

Fig. 1.2: Young's interferometer for measuring spatial coherence of electromagnetic wave.

If point P is chosen so that $L_1 = L_2$, then the visibility of fringes around point P can be a measure of spatial coherence of t_1 and t_2. More specifically, we define clarity of fringe at point P as follows:

$$V_{(P)} = \frac{I_{max} - I_{min}}{I_{max} + I_{min}} \tag{1.47}$$

where I_{max} and I_{min} are the maximum and minimum intensities of fringes near P. If two holes 1 and 2 generate same illuminance at P and waves have ideal spatial coherence, then $I_{min} = 0$ and $V_{(\mathrm{P})} = 1$.

When signals of t_1 and t_2 are completely incoherent, the fringe disappears (e.g., $I_{max} = I_{min}$) and $V_{(P)} = 0$. From the discussion in the previous section, $V_{(P)}$ must be related to the mode of function $\gamma^{(1)}(x_1, x_2, 0)$. Generally, for any point P on the screen, $V_{(P)}$ is expected to be related to the mode of function $\gamma^{(1)}(x_1, x_2, \tau)$, where $\tau = (L_1 - L_2)/c$. At the end of this section, we will prove that if two holes generate same illuminance at point P, then

$$V_{(P)}(\tau) = \left|\gamma^{(1)}(x_1, x_2, \tau)\right| \tag{1.48}$$

Thus, measuring fringes' visibility at a point P, such as at $L_1 = L_2$, we can find spatial coherence between x_1 and x_2.

Michelson interferometer (Fig. 1.3) provides a simple method to measure temporal coherence. Let us assume that P is the point at which temporal coherence is to be measured. A small hole at P and a lens with P as one of its focus points change the incident wave into a plane wave; this plane wave is projected onto a partially reflecting mirror S_1 (reflection rate R= 50%). This mirror splits the wave into two beams A and B, which are reflected back from mirrors S_2 and S_3 (R=1) to form beam C. Because of the interference between A and B, whether illuminance in direction C is brighter or dimmer depends on whether $2(L_3 - L_2)$ is an even number of times of half the wavelength or an odd number of times of half the wavelength. Obviously, as far as $(L_3 - L_2)$ does not take too long to make A and B incoherent, interference can always be observed. For partial interference wave, intensity I_C of beam C as a function of $2(L_3 - L_2)$ will have the behavior as shown in Fig. 1.3(b). In this case, just as equation (1.47), we can again define a fringe clarity. Here, I_{max} and I_{min} are as shown in Fig. 1.3(b). As in Young's interferometer, it can be proved that

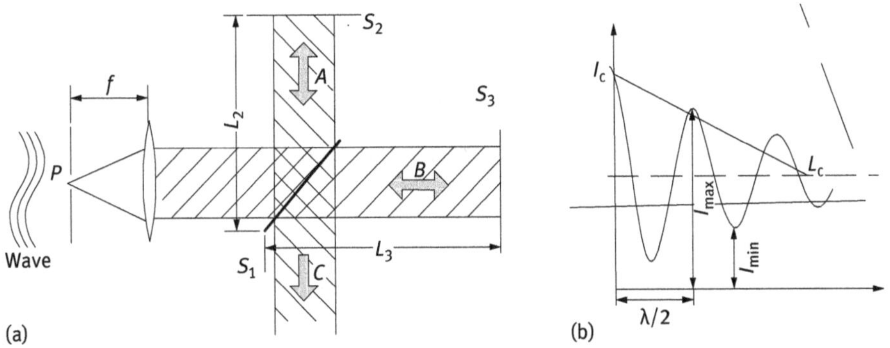

Fig. 1.3: (a) Michelson interferometer, used for measuring electromagnetic temporal correlation; (b) intensity of beam C as a function of L_3L_2.

$$V_{(P)}(\tau) = \left|\gamma^{(1)}(P, P, \tau)\right| \tag{1.49}$$

Here, $\tau = 2(L_3 - L_2)/c$. Therefore, under this circumstance, measuring fringe clarity can give the temporal coherence at P. Once we know $V_{(P)}(\tau)$, we can find coherence time τ_{co}, and coherence length L_c=$c_o\tau_{co}$. L_c equals to $(L_3 - L_2)$ when clarity reduces to $V_{(P)}(\tau) = 1/2$.

We will end this section by proving equation (1.48); this can be an exercise for applying analytic signals. Similar theory can be used to prove equation (1.49). Assume $V(t')$ is an analytic signal at point P at time t. Since it is a superposition of signals from two holes, therefore $V(t')$ can be written as follows:

$$\mathrm{V} = k_1 V(x_1, t' - t_1) + k_2 V(x_2, t' - t_2) \tag{1.50}$$

where $t_1 = L_1/c$, $t_2 = L_2/c$. Coefficients k_1 and k_2 are inversely proportional to L_1 and L_2, and relay on the sizes of holes and the angle between incident beam and diffraction beams from points x_1 and x_2. As the phase of diffraction wave is only a quarter of incident wave, so we have the following equation:

$$k_1 = |k_1| exp(-i\pi/2) \tag{1.51a}$$

$$k_2 = |k_2| exp(-i\pi/2) \tag{1.51b}$$

Let $t = t' - t_2$ and $\tau = t_2 - t_1$, equation (1.50) becomes

$$V = k_1 V(x_1, t+\tau) + k_2 V(x_2, t) \tag{1.52}$$

And the intensity at P is

$$I = V * V^*$$

That is,

$$I = |k_1|^2 |V(x_1, t+\tau)|^2 + |k_2|^2 |V(x_2, t)|^2 + \left[k_1 V(x_1, t+\tau) k_2^* V^*(x_2, t) + c.c.\right]$$

$$= I_1 + I_2 + 2Re\left[k_1 k_2^* V(x_1, t+\tau) V^*(x_2, t)\right]$$

$$I = V \cdot V^* = I_1(t+\tau) + I_2(t) + 2Re\left[k_1 k_2^* V(x_1, t+\tau) V^*(x_2, t)\right] \tag{1.53}$$

where I_1 and I_2 are intensities at P of beams from x_1 or x_2, respectively. The expressions of I_1 and I_2 are as follows:

$$I_1 = |k_1|^2 |V(x_1, t+\tau)|^2 = |k_1|^2 I(x_1, t+\tau) \tag{1.54a}$$

$$I_2 = |k_2|^2 |V(x_2, t)|^2 = |k_2|^2 I(x_2, t) \tag{1.54b}$$

$I(x_1, t+\tau)$ and $I(x_2, t)$ are intensities at points x_1 and x_2, respectively. Taking average of time in equation (1.53) and combining equations (1.53) and (1.51), we get the following equation:

$$\langle I \rangle = \langle I_1 \rangle + \langle I_2 \rangle + 2|k_1||k_2| Re\left[\Gamma^{(1)}(x_1, x_2, \tau)\right] \tag{1.55}$$

From equation (1.44),

$$\Gamma^{(1)} = \boldsymbol{r}^{(1)} [\langle I(x_1, t+\tau) \rangle \langle I(x_2, t) \rangle]^{1/2} \tag{1.56}$$

Substituting equation (1.56) into equation (1.55) and using equation (1.54), we obtain the following equation:

$$\langle I \rangle = \langle I_1 \rangle + \langle I_2 \rangle + 2(\langle I_1 \rangle \langle I_2 \rangle)^{1/2} Re\left[\gamma^{(1)}(x_1, x_2, \tau)\right]$$

$$= \langle I_1 \rangle + \langle I_2 \rangle + 2(\langle I_1 \rangle \langle I_2 \rangle)^{1/2} \left|\gamma^{(1)}\right| cos[\Psi(\tau) - \langle \omega \rangle \tau] \tag{1.57}$$

Equation (1.45) has been applied to the last step. As $\left|\gamma^{(1)}\right|$ and $\Psi(\tau)$ change slowly, the intensity $\langle I\rangle$ varies with P that comes from the rapid change of cosine $\langle\omega\rangle\tau$. So about P, we have the following equation:

$$I_{max} = \langle I_1\rangle + \langle I_2\rangle + 2(\langle I_1\rangle\langle I_2\rangle)^{1/2}\left|\gamma^{(1)}\right| \tag{1.58a}$$

$$I_{min} = \langle I_1\rangle + \langle I_2\rangle - 2(\langle I_1\rangle\langle I_2\rangle)^{1/2}\left|\gamma^{(1)}\right| \tag{1.58b}$$

So, from equation (1.47), we obtain the following:

$$V_{(P)} = \frac{2(\langle I_1\rangle\langle I_2\rangle)^{1/2}}{\langle I_1\rangle + \langle I_2\rangle}\left|\gamma^{(1)}(x_1, x_2, \tau)\right| \tag{1.59}$$

In the case of $\langle I_1\rangle = \langle I_2\rangle$, equation (1.59) reduces to equation (1.48).

1.5 Early quantum theory of light and wave–particle duality

As we mentioned earlier in this chapter, in the late nineteenth century, theory and experiments had proved that light is some kind of electromagnetic wave. Most people at that time thought that the knowledge of the nature of light was already complete and perfect; there was very little (if there were any) to add into this theory.

But unfortunately, or we would rather like to say fortunately, wave theory of light could not explain a number of experiments that appeared later. This promoted a new theory – the quantum theory of light. In this section, we will first review how quantum theory of light rose and its applications in explaining some of the experiments. While quantum theory of light does not mean that we should abandon wave theory of light, in fact, light has wave–particle duality; we will discuss this at the end of this section.

1.5.1 Concepts of radiation and energy quanta

1 Earlier theory of radiation

Black body radiation is the earliest experiment that shook the classical physics. According to the thermodynamic theory, every object can radiate electromagnetic waves in certain wavelength range. The radiated energy and its energy distribution versus wavelength depends on the temperature of the radiator. Usually, we define total electromagnetic wave energy emitted from an unit area of an object in an unit time as the radiant exitance $R(T)$; it is the function of the object's temperature T. We call monochromatic radiant exitance $R(\lambda, T)$as the emitted electromagnetic wave energy falling between λ and $\lambda + d\lambda$ as follows:

$$R(\lambda, T) = \frac{dR(T)}{d\lambda} \tag{1.60}$$

Reverse to radiation, any object has a certain capacity of absorbing electromagnetic wave that is projected on its surface. The ratio of absorbed electromagnetic energy between λ and $\lambda + d\lambda$ to incident wave's wavelength λ is called monochromatic absorption ratio. It is also a function of the object's temperature, and is denoted as $\alpha(\lambda, T)$. If for any temperature and any incident electromagnetic wavelength, an object has

$$\alpha(\lambda, T) = 1$$

then this object is then called a black body. If in a given unit time, the total energy radiated from a black body exactly equals to the total energy absorbed, we call this black body as in a thermal equilibrium state. If an object is in thermal equilibrium sate, it will have a well-defined temperature T; therefore, it also has a defined monochromic radiant exitance $\boldsymbol{R}_0(\lambda, T)$.

At the end of the nineteenth century, people measured radiant exitance with respect to wavelength (Fig. 1.4) from various experiments. However, how to derive an analytic expression of $\boldsymbol{R}_0(\lambda, T)$ from theory has become an important question.

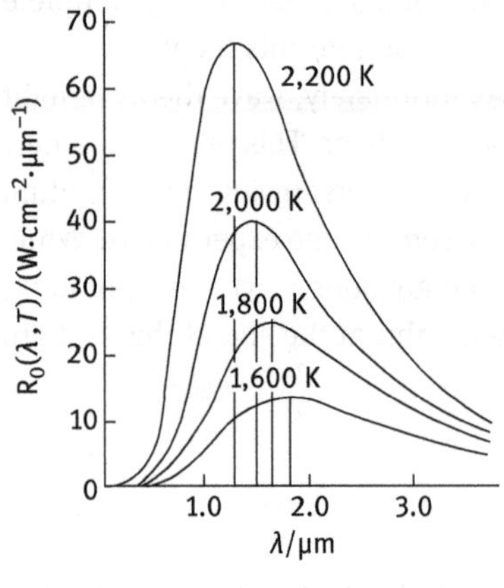

Fig. 1.4: Distribution of radiant exitance of black body with wavelength.

The German physicist Wilhelm Wien was the first person who gave an analytic expression of $\boldsymbol{R}_0(\lambda, T)$. In 1896, on the basis of the thermodynamic theory under a certain assumption, Wien derived a semiempirical formula as follows:

$$\boldsymbol{R}_0(\lambda, T) = c_1 \lambda^{-5} e^{c_2/\lambda T} \tag{1.61}$$

where c_1 and c_2 are two parameters to be decided by the experiment. Equation (1.61) very well fitted into the experimental data in the shorter wavelength range, but it failed to explain the radiation of longer wavelength. Four years later, Rayleigh and

Jeans applied equipartition of energy from statistical physics to electromagnetic radiation and derived the Rayleigh–Jeans formula:

$$\boldsymbol{R}_0(\lambda, T) = 2\pi ck\lambda^{-4} \tag{1.62}$$

Here

$$k = 1.380658 \times 10^{-2} JK^{-1}$$

This is called Boltzmann's constant and C is the speed of light in vacuum.

Unlike Wien's formula, the Rayleigh–Jeans formula very well fitted into the actual curve in a very long wavelength range but way too much against the data in a short wavelength range.

2. Planck's quantum concept

In 1900, Planck summarized the Wien and Rayleigh–Jeans formulas; he proposed the following equation:

$$\boldsymbol{R}_0(\lambda, T) = c_1\lambda^{-5}\frac{1}{e^{c_2/\lambda T} - 1} \tag{1.63}$$

Equation (1.63) gave results that could fit all the experimental data. However, the constants c_1 and c_2 had to be determined by experiment, so this was also an empirical formula. In order to derive equation (1.63) and determine the constants, later in the same year, Planck proposed a hypothesis. In this hypothesis, Planck assumed that radiators were composed of charged harmonic oscillators; these oscillators can only exist in some special states. In these states, their energies were integral multiples of smallest energy ε. When a radiator radiates radiation, the energy carried by emitted electromagnetic waves was also integral multiples of ε. This smallest energy ε is called energy quanta. For an oscillator with frequency v, the corresponding energy is as follows:

$$\varepsilon = hv \tag{1.64}$$

where $h = 6.6260755 \times 10^{-34}\ J{\cdot}s$ is called Planck's constant.

On the basis of the above-mentioned assumption, Planck derived constants in equation (1.63) that is as follows: $c_1 = 2\pi hc^2$ and $c_2 = hc/k$. Finally, Planck derived the following equation:

$$\boldsymbol{R}_0(\lambda, T) = 2\pi hc^2\lambda^{-5}\frac{1}{e^{hc/k\lambda T} - 1} \tag{1.65}$$

This is Planck's black body radiation formula. The results calculated from this formula very well fitted into the experimental data, both in short and in long wavelength ranges (Fig. 1.5).

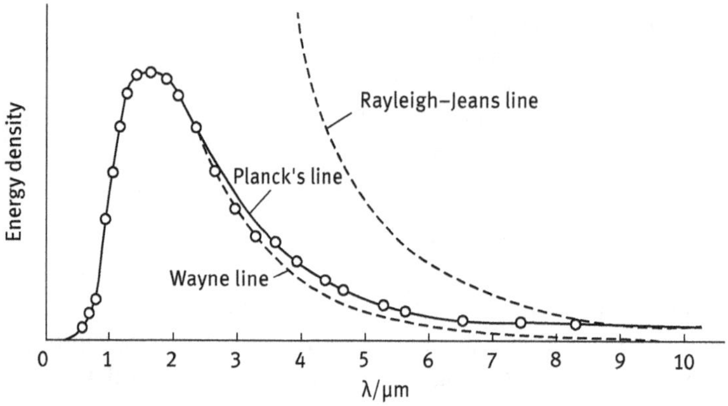

Fig. 1.5: Comparison of theoretical data and experimental data of black body radiation. Theoretical values; Experimental data.

Planck's quanta assumption explains the experimental results of black body perfectly; most importantly, he proposed an idea that classical theory almost cannot prove. This led to a complete new theory. However, as was mentioned earlier, the field of physics at that time utilized more of classical physics because this theory was thought to be perfect. Planck's new theory did not get enough attention as it should have been. Not until late on more and more experimental phenomena made classical physics facing difficulty to explain them, especially when Einstein successfully interpreted photoelectric effect, was the new theory widely accepted.

1.5.2 Photoelectric effect and the concept of optical quanta

1. Photoelectric effect

When light with optimum short wavelength falls on the surface of a metal, the surface emits electrons with certain kinetic energy. This phenomenon is called photoelectric effect. This effect was first discovered by Heinrich Hertz in 1887.

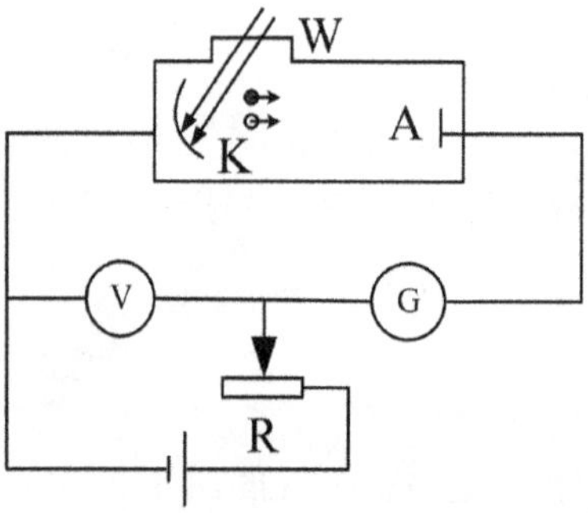

Fig. 1.6: A sketch of photoelectron experimental setup.
K—cathode; *A*—anode; *W*—quartz window; *V*—voltmeter; *G*—galvanometer

A simple set up used for studying photoelectric effect is illustrated in Fig. 1.6. The main part of the device is a vacuum tube in which there is a metal plate K that works as a cathode and another metal plate that works as an anode. K and A are connected to the negative and positive terminal of a battery, respectively. Cathode K is made of a material whose photoelectric effect was studied. The variable resistor R is used to change the voltage between A and K. This voltage is measured by a voltmeter. G is used for measuring photoelectric current.

The experiment starts with a certain wavelength falling on cathode K through the quartz window W. When the photoelectric effect occurs, electrons emitted from K are collected by anode A, and then form a current in the circuit; this current can be read by G. Changing the voltage V can give information about photoelectrons' kinetic energy. The main results from this experiment are as follows:

(1) Relationship between photoelectron effect and the frequency of incident light
For certain metals, when photoelectron effect occurs depends totally on the frequency of the incident light. It has nothing to do with the intensity of the light. This means that for different metals, there exists a threshold frequency ν_0; if $\nu < \nu_0$, irrespective of the intensity of the incident light, there is no photoelectric effect.[2] On the other hand, when $\nu > \nu_0$, there is always some electrons that are emitted. ν_0 is called the stopping frequency or red limit frequency.

(2) Relationship between maximum initial kinetic energy of photoelectron and the frequency of incident light
When photoelectron effect occurs, the maximum initial speed of photoelectron is v_M, so the maximum kinetic energy is $\frac{1}{2}mv_M^2/(1.6 \times 10^{19} J)$, which is also a quantity that is only related to the frequency of incident light and has nothing to do with the light intensity either. It can be proved that this kinetic energy is

$$\frac{1}{2}m_e v_m^2 = eh'(\nu - \nu_0) \tag{1.66}$$

where e is the electron charge, m_e is the electron mass and h' is a universal constant.

Figure 1.7 shows the maximum electron kinetic energy of electron emitted from sodium metal that is changing linearly with the frequency ν of incident light in a photoelectron effect.

(1) Relationship between photoelectric current and intensity of incident light
The number of electrons (so is the photoelectric current) generated from photoelectron effect is directly proportional to the intensity of incident light.
(2) Time behavior of photoelectron effect.

2 Photoelectric effect was observed more than 70 years before laser was invented; there was no strong light source that can generate multiphoton process.

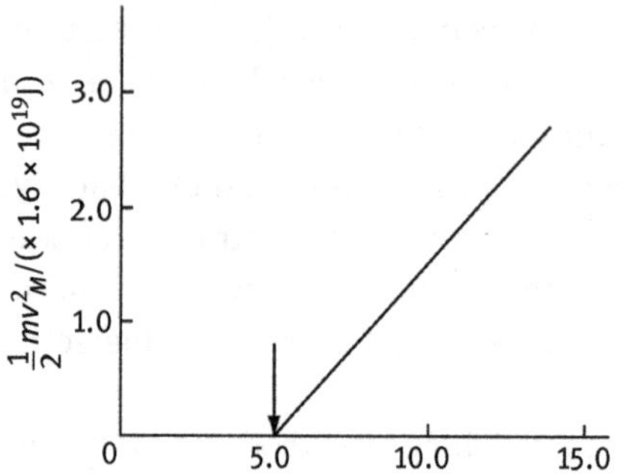

Fig. 1.7: Relationship between maximum photoelectron kinetic energy and applied voltage.

Experiments showed when the frequency of incident light $\nu > \nu_0$, irrespective of the intensity, the time from which light hits the surface of the metal to the photoelectron is released in the order of nanosecond. As the accuracy of measurement was limited in the past, it was believed that the photoelectric effect occurred instantly.

This result once again brought classical physics into a difficult situation. According to the classical wave theory, under illuminating light, whether electrons can escape or not from the inside of the metal and the initial mechanical energy have nothing to do with the frequency of the light but it is related with the intensity of incident light. Furthermore, it takes time for electrons to gain and accumulate energy; this means there is a finite time delay between light falling on the surface of the metal till photoelectrons are generated. Evidently, these are expectations against experimental results. All these helped Einstein, who became a great physicist at the young age of 30 and is known for expanding the light quanta theory, and explained the photoelectron effect successfully.

2. Light quantum theory of Einstein

Planck's quantum theory successfully explains the black body radiation. It only assumed that radiation energy is quantized; his theory did not apply to problems that are related to absorption and propagation of electromagnetic waves. In 1905, to explain the photoelectron effect based on Planck's theory, Einstein proposed that the electromagnetic wave has particle properties not only when it is emitted, but also when it is absorbed and propagated. These particles were called light quanta.[3] The relationship of photon energy ε and electromagnetic wave frequency ν is given by equation (1.64).

Photoelectron effect can be explained perfectly by photon theory:

(1) For electrons to escape from the surface of the metal, they have to overcome work function W. After an electron absorbs a photon with frequency ν and

3 In 1926, people accepted G. N. Lewis's suggestion to call it photon.

gains energy $h\nu$, whether it can escape from a metal surface will depend on the value of $h\nu$ and W; if $h\nu > W$, then electrons can escape, otherwise they cannot, and

$$\nu_0 = W/h \tag{1.67}$$

It is known as a stopping frequency or red limit frequency.

(2) When $h\nu > W$, from the law of conservation of energy, electrons can escape from the surface of the metal and will also have certain kinetic energy:

$$\frac{1}{2}m_e v_M^2 = h\upsilon - W = h(\nu - \nu_0) \tag{1.68}$$

Obviously, equations (1.68) and (1.66) are equivalent; by comparing both equations, we find that the universal constant h' and Planck's constant h has the following relation: $h' = h/e$.

(3) As intensity of incident light increases, the total number of photons that reach the surface of the metal increases; this results in a proportional increase of the total number of photoelectrons. This explains the increase in the photocurrent with the increase in the intensity.

(4) When $\nu \geq \nu_0$, as long as an electron absorbs a photon, it will obtain enough energy to escape from the surface of the metal. Here accumulation of energy is not necessary; therefore, the photoelectron effect occurs almost instantly.

According to Einstein's mass–energy formula, photon with energy $h\nu$ will have the following mass:

$$m_p = h\upsilon/c^2 \tag{1.69}$$

and momentum:

$$p = m_p c = h\upsilon/c = h/\lambda \tag{1.70}$$

Equations (1.70) and (1.64) together are called the Planck–Einstein relation.

Till now the nature of light particles is never doubted. However, it is so contradictory with the well-developed classical theories that made it to be widely accepted only after Millikan finished further experiments to verify it.

1.5.3 Compton scattering and further approval of particle property of light

Compton scattering is another convictive evidence of particle nature of light. As early as in 1904, some experiments showed that the wavelength of γ rays increased after it is scattered from an object. In 1921, Compton studied γ ray scattering. He measured the wavelengths of scattering waves in different directions and found the scattering spectra; in the scattering spectra there was both increase in the wavelength as well as unchanged wavelength of the waves. This phenomenon is called the Compton effect.

According to the classical electromagnetic theory, when electromagnetic waves act on an object, it will cause the charged particles inside the object to undergo forced oscillation. The oscillating charged particles then radiate secondary radiations in different directions with the same frequency of the incident waves. Therefore, scattered waves should have no changes in the wavelength. Therefore, classical theory cannot explain the red shift in scattering waves.

According to the light quanta theory, scattering of light results from the collision of photons of incoming waves and electrons inside the object. If scattered electrons have strong bond with atoms, then the collision occurs between photons and the atoms. As photons have very little mass compared to atoms, based on the collision theory, photon almost does not lose energy during the collision (this is similar to a small elastic ball scattered from a solid wall without loss of much energy), so that the wavelength of the scattered waves remain same. However, if photon collides with free electrons or electrons with small bonds, photons will transfer some energy to those electrons; therefore, the scattering waves will have longer wavelength than that of the incident light.

Light quanta theory not only can explain Compton scattering quantitatively, but also can calculate the change in wavelength with respect to the scattering angle qualitatively. For doing so, as shown in Fig. 1.8, a single photon collides with a free electron. Let the photon move toward the electron as shown in the figure; if a photon has a frequency ν_0, then photon has energy and momentum $h\nu_0$ and $\frac{h\nu_0}{c}k_0$, respectively where k_0 is the unit vector in the direction of photon.

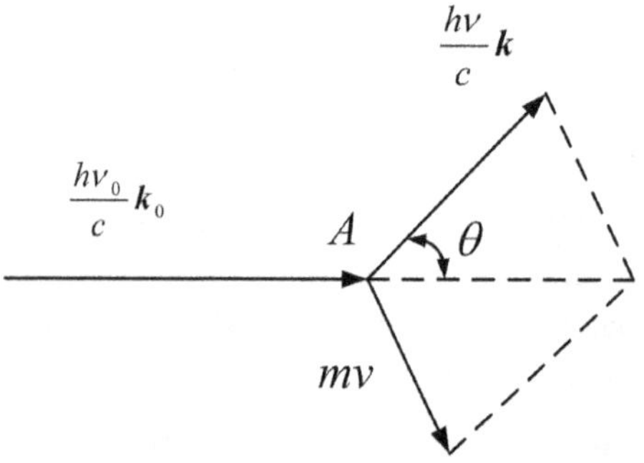

Fig. 1.8: Collision between photon and free electron.

If electron is at a rest position before collision and mass is m_0, at the moment of collision, the initial energy and momentum are m_0c^2 and 0, respectively.

After collision, at a direction θ with respect to the incident light, the measured frequency of scattering wave is ν_0. If the unit vector of scattering direction is $\boldsymbol{k}$, then the energy and momentum of scattering photons are $h\nu$ and $\frac{h\nu}{c}\boldsymbol{k}$, respectively. If the velocity of electron after collision is v and mass is m_0, then according to relativity,

$$m = \frac{m_0}{\sqrt{1 - v^2/c^2}} \tag{1.71}$$

From the conservation of energy before and after collision:

$$h\nu_0 = h\nu + m_0c^2\left[\frac{1}{\sqrt{1 - v^2/c^2}} - 1\right] \tag{1.72}$$

and

$$\frac{k\nu_0}{c}\boldsymbol{k}_0 = \frac{k\nu}{c}\boldsymbol{k} + \frac{m_0\boldsymbol{v}}{\sqrt{1 - v^2/c^2}}$$

Solving equation (1.72) we get the following:

$$\Delta\nu = \nu_0 - \nu = \frac{h\nu\nu_0}{m_0c^2}(1 - cos\theta) = 2\frac{h}{m_0c^2}\nu\nu_0 sin^2\frac{\theta}{2} \tag{1.73}$$

or

$$\Delta\lambda = \lambda - \lambda_0 = \frac{h}{m_0c^2}(1 - cos\theta) = 2\frac{h}{m_0c^2}sin^2\frac{\theta}{2}$$

Equation (1.73) was first theoretically derived by Compton and later was experimentally verified by Compton and Wu Youxun. This once again proved the assumption that light quanta has momentum is correct. Meanwhile, equation (1.73) also provided a method to calculate Planck's constant. Measuring the wavelength change $\Delta\lambda$ in a given direction θ and applying equation (1.73), we can obtain h. The example 1 is a simple example of how to use equation (1.73) to find h.

Example 1 If in a Compton scattering experiment, wavelength change is $1.95 \times 10^{-4} nm$ at the direction of $\theta = \frac{\pi}{6}$, estimate Planck's constant h.

Solution: From equation (1.73),

$$h = \frac{\Delta\lambda}{2}m_0c\frac{1}{1/4} = 2\Delta\lambda m_0c$$

Substituting $\Delta\lambda$ and constant m_0 and c, we obtain $h \approx 6.626 \times 10^{-34} J \cdot s$.

1.5.4 Particle–wave duality of light

In earlier discussion, we used irrefutable facts to disclose the quantum nature of light. However, this does not mean that the particle theory completely disapproved the wave theory of light and the wave theory was abandoned. This is because the wave nature of light was verified through compelling evidence such as interference and diffraction. Therefore, two theories reached some kind of agreement: When discussing about the propagation of light, we use electromagnetic wave theory to describe it; while talking about light–matter interaction, we utilize quantum to deal with. However, things were not completely satisfying; in some phenomena, light has both wave and particle behaviors. This seems that the contradiction was not solved until quantum electrodynamic developed in the 1930s of the twentieth century. Quantum electrodynamics is the theoretical foundation of quantum optics. We will briefly discuss it in Section 1.6. In this section, we will base on quiz-classical concept to briefly introduce the theory of elemental light quantum.

1. Electromagnetic theory of light

Although the phenomena of light wave had been observed and acknowledged widely, but a complete electromagnetic theory was established by Maxwell in the nineteenth century. In 1864, based on the previous theoretical and experimental research, Maxwell established electromagnetic equations, for example, the Maxwell equations. Wave functions of electric field E and magnetic strength H can be derived from these equations. It was proved that electromagnetic field travels in the form of waves at a propagation speed of C. This strongly proved that light is an electromagnetic wave.

The electric field and the magnetic strength are related to each other through wave equations. In the process of electromagnetic wave propagation, both electric and magnetic fields are perpendicular to each other and both lie in planes that are perpendicular to the direction of propagation. For the sake of convenience in discussions on mathematics, an electric component is always expressed as a complex variable. The complex electric field of a scalar field that travels in the Z direction can be expressed as follows:

$$E = E_0 exp[-j(\omega t - kz)] \tag{1.74}$$

Quantities in equation (1.74) have the same meaning as discussed earlier.

According to the electromagnetic theory, the energy density of a monochromatic light in a moment is as follows:

$$\frac{1}{8\pi}(\varepsilon E^2 + \mu H^2)$$

Its average value in one cycle is given as follows:

$$\frac{1}{16\pi}(\varepsilon E^2 + \mu H^2) \tag{1.75}$$

The density of light energy flux is as follows:

$$\mathbf{S} = \frac{C}{4\pi}\boldsymbol{E} \times \boldsymbol{H} \tag{1.76}$$

This means that the light intensity is directionally proportional to the amplitude of the field.

2. Elementary quantum theory of light

Elementary quantum theory of light assumes that light is composed of photons. If light frequency is ν, then these photons have energy:

$$h\nu = hc/\lambda$$

where h is Planck's constant; the value suggested in 1986 is $6.620755 \times 10^{-34} J \cdot s$. When light acts on matter, it can only exchange energy in a value of integer number of $h\nu$.

A photon has zero rest mass, but it has motion mass; according to the relativity of mass–energy formula, photon's motion mass is given as follows:

$$m = h\nu/c^2$$

Its momentum is therefore

$$p = mc = h\nu/c = h/\lambda$$

Assume that the number density of photon is N, then the light energy density is $Nh\nu$ and energy flux density is $Nh\nu c$.

As in the above-mentioned formulas, the quantities on the left-hand side of the equations are used to describe a photon, whereas those on the right-hand side are for the wave, and so these two equations are also present in the wave–particle duality of light.

1.6 Brief introduction to modern quantum theory of light

In this section, we will use quantum theory to treat some statistical problems of photon. These include one-dimensional harmonic oscillation as in equation (1.6.2); electromagnetic quantization equation (1.6.3); Eigen states of photon number of quantized radiation field equation (1.6.4) and brief quantum theory of optics equation (1.6.5).

Dirac symbol, because it is abstract and easy to handle, is used widely in modern physics and many other areas related to quantum mechanics. In order to continue the discussion smoothly, we will introduce this notion briefly in this section.

1.6.1 Vector space and linear operator

We first introduce "bra" and "ket" vectors invented by Dirac; the physics behind the use of linear operators to express visible quantities is then discussed.

1. "Ket" and "bra" vectors

Quantum state can be expressed by a "ket" ($|\rangle$) vector. In order to distinguish different states, a letter is put inside the "ket." So for state A, its ket is$|A\rangle$. If a state is a superposition of several other states, the corresponding ket can be expressed as a linear combination of other states. For example,

$$C_1|A_1\rangle + C_2|A_2\rangle = |R\rangle \tag{1.77}$$

This means $|R\rangle$ is a superposition of $|A_1\rangle$ and $|A_2\rangle$ with associated weights. According to quantum mechanics, a state superposition with itself will be in the same state; thus,

$$C_1|A\rangle + C_2|A\rangle = (C_1 + C_2)|A\rangle$$

We can conclude that multiple ket with any nonzero complex variable will get a ket of the same state. In other words, a state is defined by the direction of a ket. All the states in a mechanical system will have one-to-one correspondence with all possible directions of this ket. Hence, in equation (1.77), to define $|R\rangle$, the most important is the ratio of the coefficients of $|A_1\rangle$ and $|A_2\rangle$ that are not the real values of C_1 and C_2. In other words, the state $|R\rangle$ depends on a complex number $|C_1/C_2|exp|j(\varphi_1 - \varphi_2)|$ or two real parameters $|C_1/C_2|$ and $(\varphi_1 - \varphi_2)$. Here φ_1 and φ_2 are arguments of C_1 and C_2, respectively.

The complex conjugate of ket $|A\rangle$ is denoted as $\langle A|$, called "bra." Ket and bra can have dot product, denoted as $\langle B|A\rangle$, which is a complex variable, and

$$\langle B|A\rangle = \overline{\langle A|B\rangle} \tag{1.78}$$

Here $\overline{\langle A|B\rangle}$ is a complex conjugate of $\langle A|B\rangle$. Let $|B\rangle = \langle A|$, equation (1.78) indicates $\langle A|A\rangle$, which is a real number; we usually assume

$$\langle A|A\rangle \geq 0$$

The equation sign holds only when $|A\rangle = 0$.

If

$$\langle A|A\rangle = 1$$

we say A is normalized. If

$$\langle A|B\rangle = 0$$

we say A and B are orthogonal.

The bra of $C|A\rangle$ is $C^*\langle A|$, $C^*$is a complex conjugate of C.

2. Linear operator

The operator is a kind of symbol; it (for instance, $\hat{\alpha}$) represents a calculation, and can change a ket to another ket as follows:

$$\hat{\alpha}|A\rangle = |B\rangle$$

If for any C_1、 C_2 and $|A_1\rangle$、 $|A_2\rangle$ always have

$$\hat{\alpha}(C_1|A_1\rangle + C_2|A_2\rangle) = C_1\hat{\alpha}|A_1\rangle + C_2\hat{\alpha}|A_2\rangle$$

$\hat{\alpha}$ is called the linear operator.

When the result of a linear operator acting on a ket is known, this operator is thought to be completely defined. If $\hat{\alpha}|A\rangle = \hat{\beta}|A\rangle$ is true for any$|A\rangle$, then $\hat{\alpha}$ and $\hat{\beta}$ are equal. Especially, if $\hat{\alpha}|A\rangle = 0$ is true for any $|A\rangle$, then $\hat{\alpha} = 0$.

The summation and product of two operators are defined as follows:

$$(\hat{\alpha} + \hat{\beta})|A\rangle = \hat{\alpha}|\mathrm{A}\rangle + \hat{\beta}|\mathrm{A}\rangle \text{ and} (\hat{\alpha}\hat{\beta})|A\rangle + \hat{\alpha}(\hat{\beta}|A\rangle)$$

The result of an operator $\hat{\alpha}$ acting on bra $\langle A|$ is $\langle A|\hat{\alpha}$, and is defined as

$$(\langle A|\hat{\alpha})|B\rangle = \langle A|(\hat{\alpha}|B\rangle)(\text{for any}|B\rangle)$$

The above-mentioned definitions can be used for bra too.

Obviously, $|A\rangle\langle B|$ is also a linear operator; it acts on $|P\rangle$ to get $C|P\rangle$; here $C = \langle B|P\rangle$ is a complex variable.

Till now we have established a whole set of algebraic calculation system; it covers three quantities: (1) ket vector, (2) bra vector and (3) linear operator. These quantities can multiply together using the methods mentioned earlier. The associated law of multiplication and commutative law of summation are always true for calculation but commutative law of multiplication usually cannot be used. In this calculation system, anything on the "left" and on the "right" for forming a complete bracket represents a number, and opposite order gives a linear operator. An incomplete bracket represents a vector. Among these quantities, the complex conjugate of the multiplication of any two or three of them in any sequence will be equal to the multiplication of each quantities' complex conjugates in opposite sequence.

As we mentioned before, the physical meaning of these quantities, ket and bra vector (more accurately, their directions), corresponds to a state of a mechanical system at a moment.

An adjoint of an operator $\hat{\alpha}$ is denoted as $\hat{\alpha}^+$; it is defined through the complex conjugate $\hat{\alpha}^+|A\rangle$ of $\langle A|\hat{\alpha}$. If

$$\langle P| = \langle A|\hat{\alpha}$$

then, $|P\rangle = \hat{\alpha}^+|A\rangle$

therefore, $\langle B|(\hat{\alpha}^+)^+|A\rangle = \overline{\langle A|\hat{\alpha}^+|B\rangle} = \langle B|\hat{\alpha}|A\rangle$

Because the above-menitoned equation holds for any $\langle B|$and$|A\rangle$, so

$$(\hat{\alpha}^+)^+ = \hat{\alpha}$$

This means an adjoint of the adjoint operator of an operator is primitive operator itself. Especially, if $\hat{\alpha}^+ = \hat{\alpha}$, we say $\hat{\alpha}$ is self-adjoint, or self-conjugate, or more often, called Hermite. Since

$$\overline{|A\rangle\langle B|} = |B\rangle\langle A|$$

So $|A\rangle\langle A|$ is a Hermite operator.

It is easy to prove that if $\hat{\alpha}$ and $\hat{\beta}$ are Hermite, then $\hat{\alpha}\pm\hat{\beta}$ must be Hermite, but operator $\hat{\alpha}\hat{\beta}$ is not a necessary Hermite, unless they are commutable. As $\hat{\alpha}$ and himself are commutable so $\hat{\alpha}^2$, and $\hat{\alpha}^n$ for any integer number n are Hermite. Besides, it is easy to prove that $\left(\hat{\alpha}\hat{\beta}+\hat{\beta}\hat{\alpha}\right)$ and $j\left(\hat{\alpha}\hat{\beta}-\hat{\beta}\hat{\alpha}\right)$ are Hermite. Here j is a virtual unit. This result shows that two Hermite operator's commuter $\left(\hat{\alpha}\hat{\beta}-\hat{\beta}\hat{\alpha}\right)$ will be zero (e.g., $\hat{\alpha}$ and $\hat{\beta}$ are commute) or pure virtual number; these cannot have nonzero value of real part.

3. Eigenvalue equation

Further development of linear operator theory is based on the following equation:

$$\hat{\alpha}|A\rangle = \alpha|A\rangle \tag{1.79}$$

Usually this kind of equation is in the following form: $\hat{\alpha}$ is the known operator, and number a and ket $|A\rangle$ are unknown. A is called eigenvalue of $\hat{\alpha}$, $|A\rangle$ is called eigenvector of $\hat{\alpha}$ and equation (1.79) is eigenvalue equation. In real problems, $\hat{\alpha}$ and $|A\rangle$ can be decided simultaneously when solving the eigenvalue equation.

If $\hat{\alpha}$ is Hermite, taking complex conjugate on both sides of equation (1.79), we get the following equation:

$$\alpha^*\langle A| = \langle A|\hat{\alpha}^+$$

Multiplying this equation from right by $|A\rangle$, we derive the following equation:

$$\alpha^*\langle A|A\rangle = \langle A|\hat{\alpha}^+|A\rangle = \langle A|\hat{\alpha}|A\rangle = \alpha\langle A|A\rangle$$

that leads to

$$\alpha^* = \alpha$$

This proves that the eigenvalue of a Hermite operator is a real number. Therefore, sometimes a Hermite operator is also called a real operator. As all physical quantities are real, so we are only focusing on Hermite operators and assume that all operators appearing later in this chapter are Hermites.

Assume $\alpha'$and α'' are two unequal eigenvalue of $\hat{\alpha}$; the corresponding eigenvectors are $|\alpha'\rangle$ and $|\alpha''\rangle$, so it is easy to prove that $\langle\alpha'|\alpha''\rangle = 0$.

This means eigenvectors that belong to different eigenvalues are orthogonal.

4. Observable quantities

We would like to make some assumptions based on theoretical explanation. Suppose a mechanical system is in a eigenvector state of a mechanical quantity $\hat{\alpha}$, the corresponding eigenvalue is α'. If we measure $\hat{\alpha}$, we will get a well-defined number α'; on the other hand, if the system is in such a state, measuring $\hat{\alpha}$ in such a system, we will get α', and the state of the system must be an eigenstate of $\hat{\alpha}$ and belongs to eigenvalue α'.

If system is in any arbitrary vector $|P\rangle$ state, then measuring $\hat{\alpha}$ will result in different eigenvalues with different probabilities. From this, we know that any vector $|P\rangle$ can always be expressed as linear superposition of eigenvectors of visible quantities. That is,

$$|P\rangle = \sum_n C_n|a_n\rangle$$

This means eigenvectors of an observable quantity form a complete set. Any ket vector can be expanded by using this complete set as a base. When $|a_n\rangle$ and $|P\rangle$ are normalized, measuring $\hat{\alpha}$ in $|P\rangle$ will result in an eigenvalue a_n with probability $|C_n|^2$. It is defined by the correlation between $|P\rangle$ and $|a_n\rangle$. From orthonormal of $|a_n\rangle$, it is easy to show:

$$C_n = \langle a_n|P\rangle$$

By multipole measurement of $\hat{\alpha}$ in state $|P\rangle$, the expectation values are given by $\langle P|\hat{\alpha}|P\rangle = \langle \alpha \rangle$.

It is easy to prove that commute operators have common eigenstates. If every operator in a group operators commute with all other operators, then their common eigenvector will form a complete set, and this set can be normalized. This means that in the case of two operators that are commute, if we carry two observations for two different observables represented by the two operators, the performances do not interrupt with each other, or we say it is compatible. In fact, we can treat these two observations as a single but complex observe; these observes will end up in two numbers.

If two operators are not commute, they do not have a common eigenstate. Or we would like to say if in any state we carry out an observation on the mechanical quantities represented by these two operators, we cannot get a well-defined value. This means carrying out an observation on one of the quantities will interrupt another observe later.

1.6.2 One-dimensional harmonic oscillator

Harmonic oscillator plays a very important role in the quantum theory of electromagnetic field. In this section, we will introduce some new operators, $\hat{a}$, $\hat{a}$+ and $N = \hat{a}^+\hat{a}$, then we will solve linear harmonic oscillation problems in N presentation. Finally, we will discuss the physical meaning of these two new operators.

1. Introduce $\hat{a}$, $\hat{a}$+ and $\hat{N}$ operators

Hamiltonian operator of one-dimensional harmonic oscillation with unit mass is given as follows:

$$\hat{H} = \frac{1}{2}\left(\hat{P}^2 + \omega\hat{X}^2\right) \tag{1.80}$$

where ω is angular frequency, and $\hat{X}$ and $\hat{P}$ are coordinator and momentum operators, respectively. To simplify the calculation, introduce a unitless operator $\hat{a}$ as follows:

$$\hat{a} = \frac{1}{\sqrt{2\hbar\omega}}\left(\omega\hat{X} + j\hat{P}\right) \tag{1.81}$$

And its conjugate operator $\hat{a}^+$:

$$\hat{a}^+ = \frac{1}{\sqrt{2\hbar\omega}}\left(\omega\hat{X} - j\hat{P}\right) \tag{1.82}$$

Combining equations (1.81) and (1.82) gives us the following:

$$\hat{X} = \sqrt{\frac{\hbar}{2\omega}}(\hat{a}^+ + \hat{a}\,) \tag{1.83}$$

and

$$\hat{P} = j\sqrt{\frac{\hbar\omega}{2}}(\hat{a}^+ + \hat{a}\,) \tag{1.84}$$

Multiplying equations (1.81) and (1.82) in two different orders, we find the following:

$$\hbar\omega\hat{a}\hat{a}^+ = \hat{H} + \frac{1}{2}\hbar\omega \tag{1.85}$$

and

$$\hbar\omega\hat{a}^+\hat{a} = \hat{H} - \frac{1}{2}\hbar\omega \tag{1.86}$$

Subtracting equation (1.86) from equation (1.86) gives the following:

$$\left[\hat{a}, \hat{a}^+\right] = 1 \tag{1.87}$$

Adding equations (1.85) and (1.86), we sum up with the following equation:

$$\hat{H} = \frac{1}{2}\hbar\omega(\hat{a}\hat{a}^+ + \hat{a}^+\hat{a}) = \hbar\omega\left(\hat{a}^+\hat{a} + \frac{1}{2}\right) \tag{1.88}$$

Therefore, using equations (1.83), (1.84) and (1.88), three observables $\hat{X}$, $\hat{P}$ and $\hat{H}$ can be expressed as $\hat{a}$ and $\hat{a}^+$. Using equation (1.87), we can find the following commutation relationships:

$$\left[\hat{a},\hat{X}\right]=\sqrt{\frac{\hbar}{2\omega}},\quad \left[\hat{a},\hat{P}\right]=j\sqrt{\frac{\hbar\omega}{2}},\quad \left[\hat{a},\hat{H}\right]=\hbar\omega\hat{a}$$

Furthermore, introducing operator $\hat{N}=\hat{a}^{+}\hat{a}=\hat{N}^{+}$, equation (1.88) can be rewritten as follows:

$$\hat{H}=\hbar\omega\left(\hat{N}+\frac{1}{2}\right) \tag{1.89}$$

And from equation (1.87), one can get the following:

$$\left[\hat{a},\hat{N}\right]=\hat{a},\quad \left[\hat{a}^{+},\hat{N}\right]=-\hat{a}^{+}$$

2. Energy eigenvalue of one-dimensional oscillator

According to quantum mechanics, energy of a system is measurable. So equation (1.89) indicates that $\hat{N}$ is also an observable quantity. We will discuss its physics meaning later. In addition, from equation (1.89), we know that $\hat{N}$ and $\hat{H}$ are commutes, so they should have a common eigenvectors set$|n\rangle$. Let

$$\hat{N}|n\rangle=n|n\rangle \tag{1.90}$$

We get

$$\hat{H}|n\rangle=\hbar\omega\left(n+\frac{1}{2}\right) \tag{1.91}$$

In addition, assume $|n\rangle$ is orthonormal; therefore,

$$\langle n'|n''\rangle=\delta_{n'n''}$$

where $\delta_{n'n''}$ is the Kronecker delta. The completion of ket set $|n\rangle$ is as follows:

$$\sum_{n=0}^{\infty}|n\rangle\langle n|=1 \tag{1.92}$$

From the commutation of $\hat{a}$ and $\hat{N}$, we have the following:

$$\hat{N}\{\hat{a}|n\rangle\}=\left(\hat{a}\hat{N}-\hat{a}\right)|n\rangle=(n-1)\{\hat{a}|n\rangle\}$$

This means that if $|n\rangle$ is eigenvector of $\hat{N}$ that belongs to eigenvalue n, then $\hat{a}|n\rangle$ must also be an eigenvector of $\hat{N}$, and belongs to eigenvalue $(n-1)$. However, we know that the eigenvector that belongs to eigenvalue $(n-1)$ is $|n-1\rangle$; therefore we must have $\hat{a}|n\rangle=C_n|n-1\rangle$. Hence,

$$\langle n|\hat{a}^+\hat{a}|n\rangle = n\langle n|n\rangle = |C_n|^2\langle n-1|n-1\rangle \tag{1.93}$$

So $|C_n| = \sqrt{n}$; in fact, we can choose any phase for C_n. Let us take it as zero, then

$$d|n\rangle = \sqrt{n}|n-1\rangle \tag{1.94}$$

Similarly,

$$d^+|n\rangle = \sqrt{n+1}|n+1\rangle \tag{1.95}$$

Let $(\hat{a}^+)^n$ act on $|0\rangle$ and repeat by applying equation (1.95), we then can obtain the following equation:

$$|n\rangle = \frac{(\hat{a}^+)^n}{\sqrt{n!}}|0\rangle \tag{1.96}$$

This is a very useful formula; it gives the recurrence relations of each ket in set $|n\rangle$.

From equations (1.90), (1.92), (1.94) and (1.95), we can derive the following equations:

$$\hat{N} = \hat{N}\sum_{n=0}^{\infty}|n\rangle\langle n| = \sum_{n=0}^{\infty} n|n\rangle\langle n|$$

$$\hat{a} = \sum_{n=0}^{\infty}\sqrt{n}|n-1\rangle\langle n| = \sum_{n=0}^{\infty}\sqrt{n-1}|n\rangle\langle n+1|$$

$$\hat{a}^+ = \sum_{n=0}^{\infty}\sqrt{n+1}|n+1\rangle\langle n| = \sum_{n=0}^{\infty}\sqrt{n}|n\rangle\langle n-1|$$

Orthonormal of $|n\rangle$ gives us the following equation:

$$\langle n'|\hat{N}|n''\rangle = n''\delta_{n'n''}$$

$$\langle n'|\hat{a}|n''\rangle = \sqrt{n''}\delta_{n'n''-1}$$

$$\langle n'|\hat{a}^+|n''\rangle = \sqrt{n''+1}\delta_{n'n''+1}$$

Let us examine the physics behind $\hat{N}$. First, note that both sides of $\langle n|\hat{N}|n\rangle = \langle n|\hat{a}^+\hat{a}|n\rangle = n\langle n|n\rangle$ are mode of some vector. So we have $\langle n|\hat{a}^+\hat{a}|n\rangle \geq 0$ and $\langle n|n\rangle \geq 0$; from these, $n \geq 0$. This means eigenvalues of $\hat{N}$ are nonnegative numbers. Second, repeat applying equation (1.94), we can get the following equation:

$$(\hat{a})^k|n\rangle = \sqrt{\frac{n!}{(n-k)!}}|n-k\rangle \tag{1.97}$$

where k is a nonnegative integer. Equation (1.97) indicates that $|n\rangle$, $\hat{a}|n\rangle$, $\hat{a}^2|n\rangle$, ... are all eigenvectors of $\hat{N}$; they belong to eigenvalues n, n–1, n–2 ..., respectively. In order to make n positive, n must be an integer number. This means that eigenvalues of N must be 0, 1, 2, ... , $\hat{N}$ that are called photon number operators. According to this explanation, there can be unlimited number of photons in one mechanical state, so photon is Boson.

Suppose that a mechanics system is in state $|n\rangle$, the system has n photons, and operator $\hat{a}$ acts on it, then the system will change to $|n-1\rangle$ state, which has n–1 photons. We can see operator $\hat{a}$ acts on a state that will reduce one photon of the state. Hence, $\hat{a}$ is called annihilation operator or lowing operator. On the other hand, equation (1.95) shows that operator $\hat{a}^+$ acting on $|n\rangle$ changes it to $|n+1\rangle$ state; it increases one photon, so $\hat{a}^+$ is called creation operator or raising operator.

Operators $\hat{a}$ and $\hat{a}^+$ are not Hermite $(\hat{a} \neq \hat{a}^+)$; from equation (1.97) we know, some combinations of them, for instance $(\hat{a}+\hat{a}^+)/2$ and $j(\hat{a}^+-\hat{a})/2$, are Hermite.

3. Uncertain product

Define uncertain quantities of X and P as follows:

$$\Delta X = \left(\langle X^2\rangle - \langle X\rangle^2\right)^{1/2}$$

and

$$\Delta P = \left(\langle P^2\rangle - \langle P\rangle^2\right)^{1/2}$$

It is not difficult to find their uncertain product.
First, from equation (1.83), we obtain the following equations:

$$\langle X\rangle = \sqrt{\frac{\hbar}{2\omega}}\langle n|(\hat{a}+\hat{a}^+)|n\rangle = \sqrt{\frac{\hbar}{2\omega}}[\sqrt{n}\langle n|n-1\rangle + \sqrt{n+1}\langle n|n+1\rangle] = 0$$

and

$$\langle X^2\rangle = \frac{\hbar}{2\omega}\left[\left\langle n\middle|\hat{a}^2\middle|n\right\rangle + \langle n|\hat{a}\hat{a}^+|n\rangle + \langle n|\hat{a}^+\hat{a}|n\rangle + \left\langle n\middle|(\hat{a}^+)^2\middle|n\right\rangle\right] = \frac{\hbar}{\omega}\left(n+\frac{1}{2}\right)$$

so,

$$\Delta X = \sqrt{\frac{\hbar}{\omega}}\sqrt{n+\frac{1}{2}} \tag{1.98}$$

Similarly, from equation (1.84) we obtain the following equations:

$$\langle P\rangle = j\sqrt{\frac{\hbar\omega}{2}}[\langle n|(\hat{a}^+-\hat{a})|n\rangle] = 0$$

and

$$\langle P^2\rangle = -\frac{\hbar\omega}{2}[-\langle n|\hat{a}^+\hat{a}|n\rangle - \langle n|\hat{a}\hat{a}^+|n\rangle] = \hbar\omega\left(n+\frac{1}{2}\right)$$

so

$$\Delta P = \sqrt{\hbar\omega\left(n + \frac{1}{2}\right)} \tag{1.99}$$

Multiplying equations (1.98) and (1.99), we finally obtain the following product of uncertain coordinator and momentum:

$$\Delta X\,\Delta P = \hbar\left(n + \frac{1}{2}\right) \tag{1.100}$$

It can be seen that the uncertainty increases as n increases, and when $n = 0$ (ground state), we obtain the minimum value $\hbar/2$.

1.6.3 Quantization of electromagnetic field

According to the general regulation of transfer from the classical theory to the quantum theory, electromagnetic quantization should include the following two aspects: First, using operators to express field or potential variables; second, using ket vector to express electromagnetic vector. The main aim of this section is to discuss these two questions. To do so, we need to rewrite related vectors of classical field to some more convenient forms.

1. Free classical fields

In a space where the transverse free electric current component is zero, then the electromagnetic vector potential $\boldsymbol{A}$ satisfies the following wave equation:

$$-\nabla^2\boldsymbol{A} + \frac{1}{c^2}\frac{\partial^2\boldsymbol{A}}{\partial t^2} = 0 \tag{1.101}$$

This type of field is called free field $\hat{A}$.

We now utilize quantum operator $\hat{A}$ to replace classical vector potential $\boldsymbol{A}$ to quantize the free electromagnetic field. Once the classical equation is solved and the classical field is obtained, this transformation is direct.

Consider a cubic space with side L similar to a body cavity that is used in classical electrodynamics or microwave technology. However, here the so-called cavity is just a range in space; it does not have any real boundary. Thus, by solving equations to obtain the traveling wave solution, the solution has to meet the periodical boundary conditions.

Under the above-mentioned conditions, vector potential in cavity can be expanded as Fourier series:

$$\boldsymbol{A} = \sum_{k}\{\boldsymbol{A}_k(t)exp(j\boldsymbol{k}\cdot\boldsymbol{r}) + \boldsymbol{A}_k^*(t)exp(-j\boldsymbol{k}\cdot\boldsymbol{r})\}$$

Under the Coulomb gauge, the wave vector $\boldsymbol{k}$ satisfies the following:

$$\boldsymbol{k}\cdot \boldsymbol{A}_k(t) = \boldsymbol{k}\cdot \boldsymbol{A}_k^*(\boldsymbol{t}) = \boldsymbol{0}$$

For each k, $\mathbf{A}_k(t)$ has two independent directions in the plane perpendicular to $\boldsymbol{k}$.

The components of $\boldsymbol{A}$'s Fourier transform are not related and each has to satisfy equation (1.101). So for any $\boldsymbol{K}$, we have the following equation:

$$k^2\mathbf{A}_k(t) + \frac{1}{c^2}\frac{\partial^2 \mathbf{A}_k(t)}{\partial t^2} = 0$$

or

$$\frac{\partial^2 \mathbf{A}_k(t)}{\partial t^2} + \omega_k^2 \mathbf{A}_k(t) = 0$$

Here,

$$\omega_k = ck$$

The solution of the aforementioned equation can be written as follows:

$$\mathbf{A}_k(t) = \boldsymbol{A}_k exp(j\omega_k t) \tag{1.102}$$

The total vector potential is as follows:

$$\boldsymbol{A} = \sum_k \{\boldsymbol{A}_k exp[-j(\omega_k t - \boldsymbol{k}\cdot\boldsymbol{r})] + \boldsymbol{A}_k^* exp[j(\omega_k t - \boldsymbol{k}\cdot\boldsymbol{r})]\}$$

According to the relationship between the field and potential, from equation (1.102), one can obtain components of electric field $\boldsymbol{E}$ and magnetic strength $\boldsymbol{H}$ related to the mode of $\boldsymbol{k}$:

$$\boldsymbol{E}_k = j\omega_k\{\boldsymbol{A}_k exp[-j(\omega_k t - \boldsymbol{k}\cdot\boldsymbol{r})] - \boldsymbol{A}_k^* exp[j(\omega_k t - \boldsymbol{k}\cdot\boldsymbol{r})]\}$$

$$\boldsymbol{H}_k = \left(\frac{j}{\mu_0}\right) k_x\{\boldsymbol{A}_k exp[-j(\omega_k t - \boldsymbol{k}\cdot\boldsymbol{r})] - \boldsymbol{A}_k^* exp[j(\omega_k t + \boldsymbol{k}\cdot\boldsymbol{r})]\}$$

The average energy in one period of this mode is as follows:

$$\bar{\varepsilon}_k = \frac{1}{2}\int_V (e_0 \boldsymbol{E}_k^2 + \mu_0 \boldsymbol{H}_k^2)dv = 2e_0 v\omega_k^2 \boldsymbol{A}_k \cdot \boldsymbol{A}_k^* \tag{1.103}$$

Here $V = L^3$

Consider

$$\begin{cases} \boldsymbol{A}_k = (4e_0 v\omega_k^2)^{-1/2}(\omega_k Q_k + jP_k)\boldsymbol{\varepsilon}_k \\ \boldsymbol{A}_k^* = (4e_0 v\omega_k^2)^{-1/2}(\omega_k Q_k - jP_k)\boldsymbol{\varepsilon}_k \end{cases} \tag{1.104}$$

Introducing two scalars (1) generalized position and (2) momentum, the directions of $\boldsymbol{A}_k$ and $\boldsymbol{A}_k^*$ are expressed by polarization unit vector of mode $\boldsymbol{\varepsilon}_k$.

Substituting equation (1.104) into equation (1.103), one can get the following energy in single mode:

$$\bar{\varepsilon}_k = \frac{1}{2}\left(P_k^2 + \omega_k^2 Q_k^2\right)$$

This is an expression of energy of a classical harmonic oscillator with unit mass. More equivalently, it is a general form of a Hamiltonian. Therefore, a cavity mode potential vector is equivalent to a classical harmonic oscillator.

2. Mechanical quantities' operator expression of a radiation field

We are going to connect the mode of radiation field with harmonic oscillator of quantum mechanics to realize the field quantization. First, using a direct substitute, change the generalized coordinator Q_k and momentum P_k in equation (1.104) for operators $\hat{q}_k$ and $\hat{P}_k$ of quantum mechanics:

$$\boldsymbol{A}_k = \left(4e_0 V\omega_k^2\right)^{-1/2}\left(\omega_k\hat{q}_k + j\hat{P}_k\right)\varepsilon_k$$

$$\boldsymbol{A}_k^* = \left(4e_0 V\omega_k^2\right)^{-1/2}\left(\omega_k\hat{q}_k - j\hat{P}_k\right)\varepsilon_k$$

Using equations (1.81) and (1.82), we obtain the following equations:

$$\hat{A}_k = \sqrt{\frac{\hbar}{2e_0 V\omega_k}}\hat{a}_k\varepsilon_k\,;\ \hat{A}_k^* = \sqrt{\frac{\hbar}{2e_0 V\omega_k}}\hat{a}_k^+\varepsilon_k \tag{1.105}$$

The total potential vector expressed in quantum mechanics is as follows:

$$\boldsymbol{A} = \sum_k \sqrt{\frac{\hbar}{2e_0 V\omega_k}}\varepsilon_k\left\{\hat{a}_k exp[-j(\omega_k t - \boldsymbol{k}\cdot\boldsymbol{r})] + \hat{a}_k^+ exp[j(\omega_k t - \boldsymbol{k}\cdot\boldsymbol{r})]\right\}$$

Considering a potential vector now as an operator, it has to follow the rules of calculation same as operators $\hat{a}_k$ and $\hat{a}_k^+$.

Similar substitution can be obtained form electric field and magnetic strength operators of mode $\boldsymbol{k}$:

$$\begin{cases} \widehat{\boldsymbol{E}}_k = j\sqrt{\frac{\hbar\omega_k}{2e_0 V}}\varepsilon_k\left\{\hat{a}_k exp[-j(\omega_k t - \boldsymbol{k}\cdot\boldsymbol{t})] - \hat{a}_k^+ exp[j(\omega_k t - \boldsymbol{k}\cdot\boldsymbol{t})]\right\} \\ \widehat{\boldsymbol{H}}_k = j\sqrt{\frac{\hbar c^2}{2\mu_0 V\omega_k}}k\varepsilon_k\left\{\hat{a}_k exp[-j(\omega_k t - \boldsymbol{k}\cdot\boldsymbol{t})] - \hat{a}_k^+ exp[j(\omega_k t + \boldsymbol{k}\cdot\boldsymbol{t})]\right\} \end{cases} \tag{1.106}$$

In order to express energy of mode as an operator, changing equation (1.103) to

$$\bar{\varepsilon}_k = e_0 V\omega_k^2\left(\boldsymbol{A}_k\cdot\boldsymbol{A}_k^* + \boldsymbol{A}_k^*\cdot\boldsymbol{A}_k\right)$$

and substituting equation (1.105) into the above-mentioned formula, we obtain the following formula:

$$\bar{\varepsilon}_k = \frac{1}{2}\hbar\omega_k(\hat{a}_k\hat{a}_k^+ + \hat{a}_k^+\hat{a}_k) = \hbar\omega_k\left(\hat{a}_k^+\hat{a}_k + \frac{1}{2}\right) = \hbar\omega_k\left(\hat{N}_k + \frac{1}{2}\right)$$

3. Eigenstates of quantized radiation field

We have used operators to express variables in major fields. We are going to study the second step of field quantization that is used for proper kets to express field states.

The correlation between each mode of radiation field and the harmonic oscillation in quantum mechanics to quantize the field is utilized; this means that $\hat{a}_k$ and $\hat{a}_k^+$ are operators that can make the electromagnetic mode with wave vector $\boldsymbol{k}$ decrease or increase an energy quanta $\hbar\omega_k$. These quanta are photons with momentum that equals to k. The numbers of excited photons belong to these modes inside a cavity that are defined by the eigenvalues of number operators $\hat{N}_k = \hat{a}_k^+\hat{a}_k$, and can have values such as 0, 1, 2, …. Excited states of cavity mode $\boldsymbol{k}$ are defined by eigenstates $|n_k\rangle$ of operator $\hat{N}_k$, and $\hat{a}_k|n_k\rangle = \sqrt{n_k}|n_k - 1\rangle$, $\hat{a}_k^+|n_k\rangle = \sqrt{n_k + 1}|n_k + 1\rangle$. The total field state expressed by$|n_{k_1}, n_{k_2}, n_{k_3}, \cdots\rangle$ As cavity modes are independent, the total field of each mode can be written as the product of individual state, that is,

$$|n_{k_1}, n_{k_2}, n_{k_3}, \cdots\rangle = |n_{k_1}\rangle|n_{k_2}\rangle|n_{k_3}\rangle\cdots \tag{1.107}$$

Or simplifying it we have the following equation:

$$|\{n_k\}\rangle = |n_{k_1}\rangle|n_{k_2}\rangle|n_{k_3}\rangle\cdots$$

Further, the operator associated with a specific mode k_i can only affect the photon that is in that mode. For example,

$$\hat{a}_{k_i}^+|\{n_k\}\rangle = \sqrt{n_{k_i} + 1}|n_{k_1}\rangle|n_{k_2}\rangle|n_{k_3}\rangle\cdots|n_{k_{i-1}}\rangle|n_{k_i}\rangle|n_{k_{i+1}}\rangle\cdots$$

As shown earlier, any operator in an operator set $\{\hat{a}_k, \hat{a}_k^+\}$ commutes with other operators; therefore, they have common eigenvector, that is, $|\{n_k\}\rangle$.

As any cavity mode has been assumed to be orthonormal, from equation (1.107), we can easily observe that ket corresponds to the total field and is also orthonormal, that is,

$$\langle\cdots, n_{k_3'}, n_{k_2'}, n_{k_1'}|n_{k_1}, n_{k_2}, n_{k_3}\rangle\cdots = \langle n_{k_1'}|n_{k_1}\rangle\langle n_{k_2'}|n_{k_2}\rangle\langle n_{k_3'}|n_{k_3}\rangle\cdots = \delta_{k_1'k_1}\delta_{k_2'k_2}\delta_{k_3'k_3}\cdots$$

However, if any n_k inside the set $\{n_k\}$ can be allowed to take zero and all positive integer numbers, then ket set $|\{n_k\}\rangle$ is complete. Hence, any ket vector can be expanded by using $|\{n_k\}\rangle$ as a base. According to general assumption in quantum, this means that any electromagnetic state can be expressed by $|\{n_k\}\rangle$.

Let us go back to the single mode; in the eigenstate $|n\rangle$, the operator of a single linear polarized electric field can be written as follows:

$$\hat{\boldsymbol{E}}(\boldsymbol{r},t) = j\sqrt{\frac{\hbar\omega_k}{2e_0 V}}\,\hat{a}exp\left[-j(\omega t - \boldsymbol{k}\cdot\boldsymbol{r})\right] + \boldsymbol{H}cc.$$

Here *cc* represents complex conjugate of the previous term. From evidence, the expectation value of $\hat{\boldsymbol{E}}$ is 0, that is, $\langle n|\hat{\boldsymbol{E}}|n\rangle = 0$; the expectation value of strength operator is as follows:

$$\langle n\left|\hat{\boldsymbol{E}}\right|^2 n\rangle = \left(\frac{\hbar\omega_k}{2e_0 V}\right)\langle n|\hat{a}_k\hat{a}_k^+ + \hat{a}_k^+\hat{a}_k|n\rangle \quad = \left(\frac{\hbar\omega_k}{2e_0 V}\right)[(n+1)+n] \quad = \frac{\hbar\omega_k}{e_0 V}\left(n+\frac{1}{2}\right)$$

This shows that relative to its zero ensemble average, there exists a fluctuation in the field. Note the that nonzero fluctuation is meaningful even in the vacuum state $|0\rangle$. This kind of fluctuation helps to explain many interesting phenomena in quantum optics. For instance, when we earlier discussed about stimulating radiation, we pointed out that the spontaneous emission can only be explained in pure quantum theory. Now we can think that it is the fluctuation in vacuum that promotes spontaneous radiation of an excited atom.

1.6.4 Coherent photon states

Ket $|\alpha\rangle$ defined by eigenvalue equation is as follows:

$$\hat{a}|\alpha\rangle = a|\alpha\rangle \tag{1.108}$$

This represents coherence state; solving equation (1.108), based on the completion of particle number eigen state, we can expand $|\alpha\rangle$ as follows:

$$|\alpha\rangle = \sum_{n=0}^{\infty}|n\rangle\langle n|\alpha\rangle = \sum_{n=0}^{\infty}C_n(\alpha)|n\rangle \tag{1.109}$$

where

$$C_n(\alpha) = \langle n|\alpha\rangle$$

This is the probability amplitude when measuring an oscillator and finding that it is in the state of $|\alpha\rangle$ with energy $n\hbar\omega$.

Substituting equation (1.109) into the left-hand side of equation (1.108), and considering the property of annihilate operator, we find the following:

$$\hat{a}|\alpha\rangle = \sum_{n=1}^{\infty}C_n(\alpha)\sqrt{n}|n-1\rangle$$

In the summation in the right-hand side of the above-mentioned expression, we can change $n \to n+1$, and summation limits change from 0 to ∞, then

$$\hat{a}|\alpha\rangle = \sum_{n=0}^{\infty} C_{n+1}(\alpha)\sqrt{n+1}|n\rangle \tag{1.110}$$

Substituting equation (1.109) into the right-hand side of equation (1.108), we have the following equation:

$$\hat{a}|\alpha\rangle = \sum_{n=0}^{\infty} \alpha C_n(\alpha)|n\rangle \tag{1.111}$$

Making right-hand sides of equations (1.110) and (1.111) equal to each other, we obtain the following equation:

$$\sum_{n=0}^{\infty} C_{n+1}(\alpha)\sqrt{n+1}|n\rangle = \sum_{n=0}^{\infty} \alpha C_n(\alpha)|n\rangle \tag{1.112}$$

Multiplying $\langle n'|$ to both sides of equation (1.112) and applying orthonormal of $|n\rangle$, we derive the following equation:

$$\sqrt{n+1}C_{n+1}(\alpha) = \alpha C_n(\alpha)$$

From this, we can establish a recurrence relation of $C_n(\alpha)$:

$$\begin{aligned} C_n(\alpha) &= \frac{\alpha}{\sqrt{n}}C_{n-1}(\alpha) = \frac{\alpha^2}{\sqrt{n(n-1)}}C_{n-2}(\alpha) \\ &= \frac{\alpha^3}{\sqrt{n(n-1)(n-2)}}C_{n-3}(\alpha) \end{aligned}$$

And finally we obtain the following formula:

$$C_n(\alpha) = \frac{\alpha^n}{\sqrt{n!}}C_0 \tag{1.113}$$

Substituting back to equation (1.109), we derive the following equation:

$$|\alpha\rangle = C_0 \sum_{n=0}^{\infty} \frac{\alpha^n}{\sqrt{n!}}|n\rangle$$

Here the constant C_0 can be decided by the following normalized condition:

$$\langle\alpha|\alpha\rangle = |C_0|^2 \sum_{n=0}^{\infty} \frac{\left(|\alpha|^2\right)^n}{n!} = |C_0|^2 exp\left(|\alpha|^2\right) = 1$$

That is:

$$|C_0|^2 exp\left(-\frac{1}{2}|\alpha|^2\right)$$

As the phase of C_0 is arbitrary, assume it is zero; finally we obtain the following equation:

$$|\alpha\rangle = exp\left\{-\frac{1}{2}|\alpha|^2\right\}\sum_{n=0}^{\infty}\frac{\left(|\alpha|^2\right)^n}{n!}|n\rangle \tag{1.114}$$

or from equation (1.96),

$$|\alpha\rangle = exp\left\{-\frac{1}{2}|\alpha|^2\right\}e^{\hat{\alpha}\hat{a}^+}|n\rangle \tag{1.115}$$

The above discussion assumed that coherence kets satisfy normalized condition. However, these kets are not necessary to be orthogonal; to examine this, consider the dot product of $|\alpha\rangle$ and $|\beta\rangle$. Here $|\alpha\rangle$ and $|\beta\rangle$ represent two different coherence states. From equation (1.115), we have the following equation:

$$\langle\beta|\alpha\rangle = exp\left\{-\frac{1}{2}\left(|\alpha|^2+|\beta|^2\right)\right\}\sum_{n=0}^{\infty}\frac{(\alpha\beta^*)^n}{n!} = exp\left\{-\frac{1}{2}\left(|\alpha|^2+|\beta|^2\right)+\alpha\beta^*\right\}$$

This means that coherence states are not orthogonal. As $|\langle\beta|\alpha\rangle|^2 = exp\left\{-|\alpha-\beta|^2\right\}$, when $|\alpha-\beta|$ large enough, $|\alpha\rangle$ and $|\beta\rangle$will be orthogonal.

Next, let us find the eigenvalue of annihilate operator. From the relationship similar to equations (1.83) and (1.84) between $\hat{a}$, $\hat{a}^+$, generalized coordinate $\hat{q}$ and generalized momentum $\hat{p}$, we can find the expected values of p, q in state $|\alpha\rangle$ as follows:

$$\begin{cases} \langle p\rangle = \sqrt{\frac{\hbar}{2\omega}}\left\langle\alpha\right|(\hat{a}+\hat{a}^+)\left|\alpha\right\rangle = \sqrt{\frac{\hbar}{2\omega}}(\alpha+\alpha^*) \\ \langle q\rangle = j\sqrt{\frac{\hbar}{2}}\left\langle\alpha\right|(\hat{a}^+-\hat{a})\left|\alpha\right\rangle = \sqrt{\frac{\hbar}{2}}(\alpha^*-\alpha) \end{cases} \tag{1.116}$$

From equation (1.116), we find the following:

$$\alpha = \frac{1}{2\hbar\omega}[\omega\langle q\rangle + j\langle p\rangle] \tag{1.117}$$

Similar to equation (1.116), we have the following equations:

$$\langle p^2\rangle = \frac{\hbar}{2\omega}(\alpha^2+\alpha^{*2}+2\alpha\alpha^*+1)$$

And

$$\langle q^2\rangle = \frac{\hbar\omega}{2}(\alpha^2+\alpha^{*2}-2\alpha\alpha^*+1)$$

Therefore,

$$(\Delta q)^2 = \frac{\hbar}{2\omega}, \quad (\Delta p)^2 = \frac{\hbar\omega}{2}$$

So the uncertain product is as follows:

$$\Delta p \Delta q = \hbar/2 \tag{1.118}$$

Comparing equation (1.99), we know that it is the same as the eigenvalue of a ground state of a harmonic oscillator. In addition, it is the minimum value allowed by the uncertainty principle. This means coherence state is the minimum uncertain state.

Now it is not difficult to find the statistic values of photon in coherence state. From equation (1.114), we have the following equations:

$$\hat{n}|\alpha\rangle = exp\left\{-\frac{1}{2}|\alpha|^2\right\}\sum_{n=0}^{\infty}\frac{\left(|\alpha|^2\right)^n}{n!}n|n\rangle$$

$$\hat{n}^2|\alpha\rangle = exp\left\{-\frac{1}{2}|\alpha|^2\right\}\sum_{n=0}^{\infty}\frac{\left(|\alpha|^2\right)^n}{n!}n^2|n\rangle$$

So the expectation value of photon number is as follows:

$$\langle n\rangle = exp\left\{-|\alpha|^2\right\}\sum_{n=0}^{\infty}\frac{(\alpha^*\alpha)^n}{n!}n = |\alpha|^2 = N \tag{1.119}$$

Second moment on origin is calculated as follows:

$$\begin{aligned} n^2 &= exp\left\{-|\alpha|^2\right\}\sum_{n=0}^{\infty}\frac{(\alpha^*\alpha)^n}{n!}n^2 = exp\left\{-|\alpha|^2\right\}\sum_{n=0}^{\infty}\frac{(\alpha^*\alpha)^n}{n!}[n(n-1)+n] \\ &= |\alpha|^4 + |\alpha|^2 = N^2 + N = \langle n\rangle^2 + \langle n\rangle \end{aligned} \tag{1.120}$$

The variance is given as follows:

$$\Delta n^2 = \langle n^2\rangle - \langle n\rangle^2 = |\alpha|^2 = N$$

Finally, we obtain the following:

$$P_n = |\langle n|\alpha\rangle|^2 = \frac{1}{n!}|\alpha|^{2n} exp\left\{-|\alpha|^2\right\}$$

Or express as the number of photons as follows:

$$P_n = N^n e^{-N}/n! \tag{1.121}$$

This means that the probability of n photons in coherence state follows Poisson's distribution.

We now investigate the completeness and orthogonality of eigenvector set $|\alpha\rangle$. Let $\alpha = |\alpha|\, e^{j\theta}$, from equation (1.114), we obtain the following:

$$\int |\alpha\rangle\langle\alpha| d^2\alpha = \int exp\left\{-|\alpha|^2\right\}\sum_{n=0}^{\infty}\sum_{m=0}^{\infty}\frac{(\alpha^*)^n\alpha^m}{\sqrt{n!m!}}|m\rangle\langle n| d^2\alpha$$
$$= \sum_{n,m}\frac{|m\rangle\langle n|}{\sqrt{n!m!}}\int_0^{\infty}|\alpha|^{n+m+1}e^{-|\alpha|^2}d|\alpha|\int_0^{2\pi}e^{j(m-n)\theta}d\theta \tag{1.122}$$
$$= \sum_{n,m}\frac{|m\rangle\langle n|}{\sqrt{n!m!}}\cdot\pi n!\delta_{mn} = \pi\sum_n |n\rangle\langle n|$$

From the completeness of eigenvector set of particle number, we ha ve the following equation:

$$\frac{1}{\pi}\int |\alpha\rangle\langle\alpha| d^2\alpha = 1 \tag{1.123}$$

This means that the coherence eigenvectors form a complete set.

To know if eigenvectors are orthogonal, let $|\alpha'\rangle$ and $|\alpha\rangle$ represent two eigenvectors belonging to different eigenvalues; from equation (1.114), we have the following equation:

$$\langle\alpha'|\alpha\rangle = exp\left\{-\frac{1}{2}|\alpha|^2-\frac{1}{2}|\alpha'|^2\right\}\sum_{n=0}^{\infty}\sum_{m=0}^{\infty}\frac{\left(\alpha'^*\right)^n\alpha^m}{\sqrt{n!m!}}\langle n|m\rangle$$
$$= exp\left\{-\frac{1}{2}|\alpha|^2-\frac{1}{2}|\alpha'|^2\right\}\sum_n\frac{\left(\alpha\alpha'^*\right)^n}{n!}$$
$$= exp\left\{-\frac{1}{2}|\alpha|^2-\frac{1}{2}|\alpha'|^2+\alpha'^*\alpha\right\}$$
$$= exp\left\{-\frac{1}{2}\left(|\alpha|^2-2\alpha\alpha'^*+|\alpha'|^2\right)\right\} = exp\left\{-\frac{1}{2}|\alpha-\alpha'|^2\right\} \tag{1.124}$$

Equation (1.124) shows that eigenvectors belonging to different eigenvalues are not orthogonal. However, $|\alpha-\alpha'||\alpha-\alpha'| \gg 1$, $|\alpha'\rangle$ and $|\alpha\rangle$ are orthogonal approximately.

As eigenvectors are not orthogonal to each other, any vector can be expanded by other vectors. For example,

$$|\alpha\rangle = \frac{1}{\pi}\int |\alpha'\rangle\langle\alpha'|\alpha\rangle d^2\alpha' = \frac{1}{\pi}\int d^2\alpha' exp\left\{-\frac{1}{2}|\alpha-\alpha'|^2\right\}|\alpha'\rangle \tag{1.125}$$

This means that eigenvectors belonging to different eigenvalues of a coherence state are not correlated. In order to expand any vector of a coherence state, it is not necessary to use all the base vectors inside an eigenvector set. In other words, eigenvector set is a super complete set.

1.6.5 Density operator and quantum distribution

1. Mixture state and density operator

When a mechanics system is in a specific state $|\psi\rangle$, the value of an observable quantity O of the system is not well defined; all we can be sure is only the possible values that O can have and the probability for having these values. We can also get its average value as follows:

$$\langle O\rangle_q = \langle\psi|\hat{O}|\psi\rangle \tag{1.126}$$

This represents the inherent statistics property of quantum mechanics. If we even do not know if the system is in $|\psi\rangle$ state and know only the probability P_ψ of finding that the system is in $|\psi\rangle$ state, then in this case, for finding the average value of O, we have to find the average of $\langle O\rangle_q$in an ensemble, that is,

$$\langle\langle O\rangle_q\rangle_e = \sum_\psi P_\psi\langle\psi|\hat{O}|\psi\rangle \tag{1.127}$$

Here P_ψ is assumed to be a normalized probability distribution. That is,

$$\sum_\psi P_\psi = 1$$

The state that can be applied to equation (1.126) to find the average value of a mechanical system is called statistical pure state. While the state that one has to apply to equation (1.127) to find the average value is called statistical mixed state. For statistical mixed state, equation (1.127) is a basic expression; we can predict behaviors of the radio field. However, it is easier to deal with if we change it to another form. To do so, let $|\psi\rangle$ represent a complete set of radio field state; therefore,

$$\sum_\psi |\psi\rangle\langle\psi| = 1$$

Substituting it into equation (1.127), for simplicity, use $\langle\ \rangle$to represent the calculation of finding average value.

$$\langle O\rangle = \sum_\psi P_\psi\sum_\varphi\langle\psi|\hat{O}|\varphi\rangle\langle\psi|\varphi\rangle = \sum_\varphi\left\langle\varphi\left|\left\{\sum_\psi \mathrm{P}_\psi|\psi\rangle\langle\psi|\right\}\hat{\mathrm{O}}\right|\varphi\right\rangle \tag{1.128}$$

Here,

$$\rho = \sum_\psi P_\psi|\psi\rangle\langle\psi| \tag{1.129}$$

is a density operator, so obviously it is Hermite.

Assume another orthogonal complete set of radiation field to be $|\theta\rangle$, then we have the following:

$$\langle O\rangle=\sum_{\varphi}\sum_{\theta}\sum_{\theta'}\langle\varphi|\theta\rangle\langle\theta|\hat{\rho}\hat{O}|\theta'\rangle\langle\theta'|\varphi\rangle=\sum_{\varphi}\sum_{\theta}\sum_{\theta'}\langle\theta'|\varphi\rangle\langle\varphi|\theta\rangle\langle\theta|\hat{\rho}\hat{O}|\theta'\rangle$$
$$=\sum_{\theta}\sum_{\theta'}\langle\theta'|\theta\rangle\langle\theta|\hat{\rho}\hat{O}|\theta'\rangle=\sum_{\theta}\langle\theta|\hat{\rho}\hat{O}|\theta\rangle \tag{1.130}$$

It is easy to observe that after comparing equations (1.130) and (1.128) with the help of a density operator to find the average value of a selected mechanical quantity has nothing to do with the special complete set $|\varphi\rangle$ used for calculations. This is expected. Therefore, equation (1.128) is equivalent to

$$\langle O\rangle=\boldsymbol{T}_r\left(\hat{\rho}\hat{O}\right) \tag{1.131}$$

where $\boldsymbol{T}_r$ is the trace calculation; this represents finding the summation of diagonal elements of a matrix of $\hat{\rho}\hat{O}$ in a complete set.

Trace of $\hat{\rho}$ also has nothing to do with the special complete set of state; from equation (1.129) the following can be obtained:

$$\boldsymbol{T}_r(\hat{\rho})=\sum_{\psi}P_{\psi}\sum_{\psi'}\langle\psi'|\psi\rangle\langle\psi|\psi'\rangle=\sum_{\psi}P_{\psi}=1$$

In fact, it is a special case of equation (1.128); let O be an unit operator in the equation that will lead to the above-mentioned result.

However, it should be pointed out that elements of $\hat{\rho}$'s matrix can show different properties. For instance, assume $\hat{\rho}$ is composed of eigenvectors of a state as shown in equation (1.129), then

$$\langle\psi'|\hat{\rho}|\psi''\rangle=\sum_{\psi}P_{\psi}\langle\psi'|\psi\rangle\langle\psi|\psi''\rangle=P_{\psi'}\delta_{\psi',\psi''}$$

This means only diagonal elements have nonzero values. However, for other complete set of a state, say $|\varphi\rangle$, the elements of matrix are as follows:

$$\langle\varphi|\hat{\rho}|\varphi'\rangle=\sum_{\psi}P_{\psi}\langle\varphi|\psi\rangle\langle\psi|\varphi'\rangle$$

This shows nondiagonal elements are usually not zero.

2. Quantum distribution of pure states

Pure states can be treated as a special case of statistical mixed states, one P_{ψ} equals to 1, and others are zero. Thus, from equation (1.129), we can find the density operator is equal to $\hat{\rho}=|\psi\rangle\langle\psi|$; from this, we can get $\hat{\rho}^2=\hat{\rho}$ or more general $\hat{\rho}^m=\hat{\rho}$. Here, m are positive integers. This means density operators of pure states are independent.

We have discussed in Sections 1.6.3 and 1.6.4 that the density operator of a single mode can be easily constructed. For the field of photon number state $|n\rangle$ in which n photons can appear definitely, density operator is given as follows: $\hat{\rho} = |n\rangle\langle n|$. For photon number state, only diagonal elements of $\hat{\rho}$'s matrix are not zero and $\langle n|\hat{\rho}|n\rangle = 1$. Besides, the average value of an observable quantity represented by O is as follows:

$$\langle O\rangle = \boldsymbol{T}_r\left(|n\rangle\langle n|\hat{O}\right) = \sum_{n'}\left\langle n'\right|\left(|n\rangle\langle n|\hat{O}\right)\left|n'\right\rangle = \langle n|\hat{O}|n\rangle$$

It is the same as we used to calculate average value of a mechanical quantity in $|n\rangle$state.

For coherence state$|\alpha\rangle$, density operator is as follows:

$$\hat{\rho} = |\alpha\rangle\langle\alpha| \tag{1.132}$$

From normalization of the coherence state, $\langle\alpha|\hat{\rho}|\alpha\rangle = 1$, this agrees with the photon number state. However, because eigenvector associated with different eigenvalues is not orthogonal, $\langle\alpha|\hat{\rho}|\alpha\rangle$ is not the only nonzero matrix element of $\hat{\rho}$. In fact, for any α and α', there is always $\langle\alpha|\hat{\rho}|\alpha'\rangle \neq 0$.

Now let us consider general matrix elements of density operator constructed in photon number states. From equation (1.132), we obtain the following equation:

$$\langle n|\hat{\rho}|n'\rangle = \langle n|\alpha\rangle\langle\alpha|n'\rangle$$

Substituting equation (1.114), we find the following equation:

$$\langle n|\hat{\rho}|n'\rangle = exp\left\{-|\alpha|^2\right\}\frac{\alpha^n(\alpha^*)^{n'}}{\sqrt{n!n'!}}$$

Therefore, in this case, nondiagonal elements of $\hat{\rho}$ are not zero. Especially, for example,

$$\langle n+1|\hat{\rho}|n'\rangle = exp\left\{-|\alpha|^2\right\}\frac{\alpha^{n+1}\alpha^{*n'}}{\sqrt{n!(n'+1)!}}$$

In most situations, pure state of total radiation field can be treated by using similar method. For instance, if there are defined number n_k of excited photons for a number of k cavities, the total field can be written as follows:

$$|\{n_k\}\rangle = \prod_k |n_k\rangle$$

And the density operator is

$$\hat{\rho} = |\{n_k\}\rangle\langle\{n_k\}| = \left|n_{k_1}\right\rangle\left|n_{k_2}\right\rangle\cdots\left\langle n_{k_2}\right|\left\langle n_{k_1}\right|$$

For example, every cavity mode is at a certain excited coherence state, the total field is as follows:

$$|\{a_k\}\rangle = \prod_k |a_k\rangle$$

And the density operator becomes

$$\hat{\rho} = |\{a_k\}\rangle\langle\{a_k\}|$$

Matrix elements of density operator and the average value of mechanical quantity can be calculated in a similar way as in the single mode.

3. Quantum distribution of mixed states

The most significant of density operator is that it is convenient when dealing with statistical mixed states. For example, consider a single-mode thermos-stimulated emission of a single-mode photon at temperature T. Using photon number eigenvector as the base vector set, $\hat{\rho} = \sum_n P_n|n\rangle\langle n|$; here P_n comes from Boltzmann's formula:

$$P_n = exp(-\beta n\hbar\omega)[1 - exp(-\beta\hbar\omega)] \tag{1.133}$$

Here $\beta = 1/kT$, and k is Boltzmann's constant, so

$$\hat{\rho} = [1 - exp(-\beta\hbar\omega)] \sum_n exp(-\beta n\hbar\omega)|n\rangle\langle n|$$

From equation (1.128),

$$\begin{aligned}\langle n\rangle &= \sum_n \langle n|p\hat{n}|n\rangle = [1 - exp(-\beta\hbar\omega)] \sum_{n'}\sum_n \langle n'|exp(-\beta n\hbar\omega)|n\rangle\langle n|n'\rangle n' \\ &= [1 - exp(-\beta\hbar\omega)] \sum_n \langle n|exp(-\beta n\hbar\omega)|n\rangle n \\ &= [1 - exp(-\beta\hbar\omega)] \sum_n nexp(-\beta n\hbar\omega) \\ &= [1 - exp(-\beta\hbar\omega)] \frac{exp(-\beta\hbar\omega)}{[1 - exp(-\beta\hbar\omega)]^2} \\ &= \frac{exp(-\beta\hbar\omega)}{1 - exp(-\beta\hbar\omega)} = [exp(\beta\hbar\omega) - 1]^{-1}\end{aligned}$$

Or:

$$exp(-\beta\hbar\omega) = \frac{\langle n\rangle}{1 + \langle n\rangle}$$

Substituting into equation (1.133), we derive the following:

$$P_n = \left[\frac{\langle n\rangle}{1+\langle n\rangle}\right]^n \frac{1}{1+\langle n\rangle} = \frac{\langle n\rangle^n}{(1+\langle n\rangle)^{n+1}} \tag{1.134}$$

The density operator is as follows:

$$\hat{\rho} = \sum_n \frac{\langle n\rangle^n}{(1+\langle n\rangle)^{n+1}} |n\rangle\langle n|$$

Using equation (1.134), we can easily verify normalization of P_n:

$$\sum_n P_n = \frac{1}{1+\langle n\rangle} \sum_n \left(\frac{\langle n\rangle}{1+\langle n\rangle}\right)^n = \frac{1}{1+\langle n\rangle} \frac{1}{1-\dfrac{\langle n\rangle}{1+\langle n\rangle}} = 1$$

Or equivalently

$$\boldsymbol{T}_r(\hat{\rho}) = 1$$

Consider thermos-stimulated emission of all cavity modes; the total field of the photon number operator vector $|\{n_k\}\rangle$ is a product of all eigenvectors associated with each cavity mode. As each cavity is independent, the combined density operator is related to the product of different modes. Hence, density operator in a general situation can be expressed as follows:

$$\hat{\rho} = \sum_{\{n_k\}} P_n\{n_k\} |\{n_k\}\rangle\langle\{n_k\}| \tag{1.135}$$

In a heat field, multiplying all modes' factors $\langle n_k\rangle^{n_k}/(1+\langle n_k\rangle)^{n_k+1}$ will end up with the following total probability:

$$P\{n_k\} = \prod_{n_k} \frac{\langle n_k\rangle^{n_k}}{(1+\langle n_k\rangle)^{n_k+1}} \tag{1.136}$$

where $\langle n_k\rangle$ is the average photon number at **k**th mode, and $\langle n_k\rangle$ represents summation of excited photon numbers $\langle n_{k_1}\rangle$, $\langle n_{k_2}\rangle$, $\langle n_{k_3}\rangle$, ... in each cavity mode.

Using equations (1.135) and (1.136), we can obtain the following density operator for thermos-stimulated field.

$$\hat{\rho} = \sum_{\{n_k\}} |\{n_k\}\rangle\langle\{n_k\}| \prod_{n_k} \frac{\langle n_k\rangle^{n_k}}{(1+\langle n_k\rangle)^{n_k+1}} \tag{1.137}$$

The summation of equations (1.135) and (1.137) includes all possible number sets $\{n_k\}$.

For density operator $\hat{\rho}$ defined by equation (1.137), we can also prove that $\boldsymbol{T}_r(\hat{\rho}) = 1$, and applying equation (1.137) we can get the distribution of thermal radiation; it can also apply to more general cases, as long as the statistics of light has proper random character.

1.6.6 Introduction to photon optics

In this section, we will briefly introduce the fundamental theory of different phenomena observed in optical experiments carried in the area of quanta theory of light.

There are two important advantages of using quantum mechanics to deal with optical experiments. One is it presents a deeper explanation of classical interference experiment. However, it can be seen in this kind of experiment that quantized electromagnetic field has little effect on observable phenomena. Another one is related to some new experiments based on quantized electromagnetic field that plays a very important role. We can say that only when quantum theory appeared, these experiments became possible.

1. Measurement of the photon intensity

Transition rate related to radiation field is directly proportional to

$$\sum_{\psi_f}\left|\langle\psi_f|\widehat{\boldsymbol{E}}^+(\boldsymbol{r},t)|\psi\rangle\right|^2=\sum_{\psi_f}\langle\psi|\widehat{\boldsymbol{E}}^-(\boldsymbol{r},t)|\psi_f\rangle\langle\psi_f|\widehat{\boldsymbol{E}}^+(\boldsymbol{r},t)|\psi\rangle$$

Finding the summation should consider all possible states of radiation field. So from $\sum_{\psi_f}\left|\psi_f\rangle\langle\psi_f\right|=1$, we can get the following equation:

$$\sum_{\psi_f}\left|\langle\psi_f|\widehat{\boldsymbol{E}}^+(\boldsymbol{r},t)|\psi\rangle\right|^2=\langle\psi|\widehat{\boldsymbol{E}}^-(\boldsymbol{r},t)\widehat{\boldsymbol{E}}^+(\boldsymbol{r},t)|\psi\rangle$$

This means that an atom under the influence of a radiation field in state $|\psi\rangle$ can emit a photon; the emission rate is proportional to the expectation value of operator $\hat{\boldsymbol{E}}^-(\boldsymbol{r},t)\ \hat{E}^+(\boldsymbol{r},t)$ in state $|\psi\rangle$. If the intensity of light is estimated by measuring the photoelectric current produced in a photoelectric tube, then the operator corresponding to observing the intensity will be factor times $\hat{\boldsymbol{E}}^-(\boldsymbol{r},t)\hat{E}^+(\boldsymbol{r},t)$.

For a single-mode light beam, if all photons have the same wave vector k, then the observable light intensity can be defined as follows:

$$\langle I\rangle=\left(2e_0c^2k/\omega_k\right)\langle\psi|\hat{\boldsymbol{E}}^-(\boldsymbol{r},t)\hat{\boldsymbol{E}}^+(\boldsymbol{r},t)|\psi\rangle$$

Here,

$$\hat{\boldsymbol{E}}^+(\boldsymbol{r},t)=j\left(\frac{\hbar\omega_k}{2e_0V}\right)^{1/2}\boldsymbol{A}_k exp(-j\omega_k t+j\boldsymbol{k}\cdot\boldsymbol{r})$$

$$\hat{\boldsymbol{E}}^-(\boldsymbol{r},t)=-j\left(\frac{\hbar\omega_k}{2e_0V}\right)^{1/2}a_k^+ exp(j\omega_k t-j\boldsymbol{k}\cdot\boldsymbol{r})$$

so

$$\hat{\boldsymbol{E}}^{-}(\boldsymbol{r},t)\hat{\boldsymbol{E}}^{+}(\boldsymbol{r},t)=\left(\frac{\hbar\omega_k}{2e_0V}\right)\boldsymbol{A}_k a_k^{+}=\frac{\hbar\omega_k}{2e_0V}\hat{n}_k$$

From this, we can obtain the following equation:

$$\langle\psi|\hat{\boldsymbol{E}}^{-}(\boldsymbol{r},t)\hat{\boldsymbol{E}}^{+}(\boldsymbol{r},t)|\psi\rangle=\frac{\hbar\omega_k}{2e_0V}\langle n\rangle$$

The observable light intensity is as follows:

$$\langle I\rangle=\frac{c^2\hbar k}{V}\langle n\rangle$$

Here,

$$\langle n\rangle=\langle\psi|\hat{n}_k|\psi\rangle \tag{1.138}$$

These results show that the response of photoelectric tube is directly proportional to the average density of the photon.

Although the above-mentioned calculation was carried for pure state, all the results can be expanded to statistical mixed states. Suppose that the initial state of light beam is defined by the probability P_ϕ associated with pure state $|\psi\rangle$, then from the above-mentioned discussion we can find the following:

$$\sum_{\psi}P_{\psi}\langle\psi|\widehat{\boldsymbol{E}}^{-}(\boldsymbol{r},t)\widehat{\boldsymbol{E}}^{+}(\boldsymbol{r},t)|\psi\rangle=\boldsymbol{T}_r\left[\hat{\rho}\widehat{\boldsymbol{E}}^{-}(\boldsymbol{r},t)\widehat{\boldsymbol{E}}^{+}(\boldsymbol{r},t)\right]$$

For single-mode light beam,

$$\langle I\rangle=\frac{2e_0c^2k}{\omega_k}\boldsymbol{T}_r\left[\hat{\rho}\hat{\boldsymbol{E}}^{-}(\boldsymbol{r},t)\hat{\boldsymbol{E}}^{+}(\boldsymbol{r},t)\right]$$

Substituting $\hat{E}^{-}(\boldsymbol{r},t)$, we know equation (1.138) still holds and we only need to consider $\langle n\rangle$ as the average photon number of statistical mixed state.

2. Coherence of photon

The quantum mechanic explanation of measuring light beam intensity can be expanded to any observable quantities in any experiment. For instance, the quantum theory of coherence can be expressed by a proper defined degree of coherence of quantum mechanics. We will use Young's interference experiment to prove this.

The classical treatment of Young's interference experiment has been discussed earlier. A quantum theory on this can be established by considering two radiation fields that originated from two pin holes overlaying in a photoelectric tube or other optical detectors. These optical detectors are located at an observing screen. We handled this problem very much like what we did in classical theory. For example, for a scalar field, all we need to do is to substitute an average classical light intensity of an ensemble that is proportional to intensity operator $\hat{E}^{-}\hat{E}^{+}$ by quantum detector whose response is proportional to intensity operator $\hat{E}^{-}\hat{E}^{+}$. This kind of calculation leads to first-order degree of correlation of quantum mechanics.

$$g^{(1)}(\boldsymbol{r}_1, t_1; \boldsymbol{r}_2, t_2) = \frac{\left|\langle \hat{\boldsymbol{E}}^{-}(\boldsymbol{r}_1, t_1)\hat{\boldsymbol{E}}^{+}(\boldsymbol{r}_2, t_2)\rangle\right|}{\left[\langle \hat{\boldsymbol{E}}^{-}(\boldsymbol{r}_1, t_1)\hat{\boldsymbol{E}}^{+}(\boldsymbol{r}_2, t_2)\rangle\langle \hat{\boldsymbol{E}}^{-}(\boldsymbol{r}_2, t_2)\hat{\boldsymbol{E}}^{+}(\boldsymbol{r}_1, t_1)\rangle\right]^{1/2}} \tag{1.139}$$

We used the same symbols as in the classical theory and kept their meaning: space, time, points $(\boldsymbol{r}_1, t_1)$ and$(\boldsymbol{r}_2, t_2)$. We also kept the ability of light when it undergoes superposition. Only the average has to be calculated by

$$\langle \hat{\boldsymbol{E}}^{-}(\boldsymbol{r}_i, t_i)\hat{\boldsymbol{E}}^{+}(\boldsymbol{r}_j, t_j)\rangle = \boldsymbol{T}_r\left[\hat{\rho}\hat{\boldsymbol{E}}^{-}(\boldsymbol{r}_i, t_i)\hat{\boldsymbol{E}}^{+}(\boldsymbol{r}_j, t_j)\right] \quad i, j = 1, 2$$

In the special case where photons are in pure states, the average of three ensembles involved in equation (1.139) is simplified to the expectation values of the same operators.

For example, it is easy to prove that the degree of coherence of a single-mode radiation field is 1, if it is in state $|n\rangle$. And $\langle n\rangle = \langle n|\hat{n}|n\rangle = n$, so

$$\langle\psi|\hat{E}^{-}(\boldsymbol{r}_1, t_1)\hat{E}^{+}(\boldsymbol{r}_2, t_2)|\psi\rangle = \langle\psi|\hat{E}^{-}(\boldsymbol{r}_2, t_2)\hat{E}^{+}(\boldsymbol{r}_1, t_1)|\psi\rangle = \frac{\hbar\omega}{2e_0 V}n$$

and

$$\hat{\boldsymbol{E}}^{-}(\boldsymbol{r}_1, t_1)\hat{\boldsymbol{E}}^{+}(\boldsymbol{r}_2, t_2) = \frac{\hbar\omega}{2e_0 V}n\exp\{j\omega(t_1 - t_2) - j\boldsymbol{k}\cdot(\boldsymbol{r}_1 - \boldsymbol{r}_2)\}$$

Substituting equation (1.139), we obtain $g^{(1)} = 1$, which is true for any pair of time and point in space.

As any pure state can be expressed by a complete set $|n\rangle$, so all single pure state, especially coherence between coherence states, can be found with the help of the above-mentioned result, and the answer is 1.

3. Higher order of degree of coherence

Changing $E^{*}E$ in classical expression to operator form $E^{-}E^{+}$, we can define second-order degree of coherence in quantum mechanics as follows:

$$g^{(2)}(\boldsymbol{r}_1, t_1; \boldsymbol{r}_2, t_2) = \frac{\langle \hat{\boldsymbol{E}}^{-}(\boldsymbol{r}_1, t_1)\hat{\boldsymbol{E}}^{-}(\boldsymbol{r}_2, t_2)\rangle\langle \hat{\boldsymbol{E}}^{+}(\boldsymbol{r}_2, t_2)\hat{\boldsymbol{E}}^{+}(\boldsymbol{r}_1, t_1)\rangle}{\langle \hat{\boldsymbol{E}}^{-}(\boldsymbol{r}_1, t_1)\hat{\boldsymbol{E}}^{+}(\boldsymbol{r}_1, t_1)\rangle\langle \hat{\boldsymbol{E}}^{-}(\boldsymbol{r}_2, t_2)\hat{\boldsymbol{E}}^{+}(\boldsymbol{r}_2, t_2)\rangle}$$

For different pure states, second-order degree of coherence is not difficult to calculate; the results are as follows:

1. Photon number state

$$g^{(2)} = \begin{cases} 1 - \frac{1}{n}, n \geq 2 \\ 0, n = 0, 1 \end{cases}$$

2. Coherence state $g^{(2)} = 1$; thus, $|\alpha\rangle$ is coherence in second order.

From this, we can observe that for different single-mode stimulation, there exists different values for second-order of coherence; this is different than first-order coherence where all single-mode state is 1.

In fact, it can be proved that for radiation field in a coherence state, higher order of coherence is also 1. This is why it is called the base of coherence state.

It is not difficult to notice that in the definition of quantum mechanics coherence, when finding combination of operators $\hat{\boldsymbol{E}}^-$ and $\hat{\boldsymbol{E}}^+$, it always puts positive spectrum to the right-hand side of the negative spectrum; operators having such properties are called normal form. This order is used for operators because the detector always absorbs photon from the radiation field and then gives response.

4. Photoelectron counting and photon counting

We are going to derive photoelectron counting from quantum mechanics theory.

Assume that under the action of radiation field that has a single photon $|1\rangle$, at some moment the probability of producing a photoelectron in a detector is η. Quantum efficiency η depends on the atomic character of the detector and the reaction time. If the radiation field is in $|n\rangle$ state, from simple statistical theory we know that the probability $p_m^{(n)}$ of observing m electrons will be defined by the following three factors. First, the probability of detector to absorb m photons and emitting m photoelectrons; second, the probability of n–m photons not being absorbed, $(1-\eta)^{n-m}$ and third the probability of m photons to be absorbed among n photons,

$$\binom{n}{m} = \frac{n!}{m!(n-m)!}$$

Therefore,

$$p_m^{(n)} = \binom{n}{m} \eta^m (1-\eta)^{n-m} \tag{1.140}$$

Obviously, as long as $n \geq m$, from n photons, it is possible that the detector absorbs m photons and emits m photoelectrons. Therefore, the total probability of detector emitting m photoelectrons can be calculated using equation (1.140) by adding all n that are not smaller than m, that is,

$$P_m = \sum_{n=m}^{\infty} p_m^{(n)} \rho_{nn} = \sum_{n=m}^{\infty} \binom{n}{m} \eta^m (1-\eta)^{n-m} \rho_{nn} \tag{1.141}$$

In order to get photon statistic, in equation (1.141), taking $\eta = 1$, then

$$P_m = \rho_{mm} \tag{1.142}$$

2 Optical radiation and radiation source

When a microscopic particle (atom or molecule) transitions from a high energy state to a low energy state, it radiates a photon with an energy that equals the energy difference between the two states. For atoms, the energy states are based only on electronic energy. However, in case of molecules, in addition to electronic energy, vibrational and rotational energy also need to be accounted for. Usually, the gap between the adjacent vibrational energy levels and the adjacent rotational energy levels is significantly smaller than that between the adjacent electronic energy levels. This chapter mainly focuses on atomic transition, which occurs when an atom goes from a high energy state to a low energy state, either spontaneously or on stimulation. For stimulated transition, this chapter primarily discusses the transition caused by external photons.

While light generated as a result of spontaneous emission is incoherent, stimulated emission can produce coherent light. In this chapter, Sections 2.1 and 2.2 describe spontaneous emission and incoherent light sources; Sections 2.3–2.5 explain stimulated emission and coherent light sources such as lasers. Semiconductor light emission is unique as electron mobility occurs not between energy levels but between energy bands. Examples of this type of emission include light-emitting diodes (LED) based on spontaneous emission and laser diodes (LD) based on stimulated emission. Sections 2.6–2.10 describe the semiconductor light-emitting phenomenon.

2.1 Mechanism of atomic emission

As shown in Section 1.5, radiation characteristics of matter led to the development of quantum theory. Matter is composed atoms and molecules. Therefore, the quantization of light absorption and emission from matter reflects the fact that the particles that form matter are also being quantized. This question is discussed in this section after we present a review of energy levels of atoms.

2.1.1 Scattering of alpha particles and the nuclear structure of an atom

In 1909, H. Geiger and E. Marsden started studying angular distribution of alpha particles scattered from thin gold foils. They found that, even though alpha particles are highly energetic and heavy, they still could be occasionally scattered by large angles. In 1911, E. Rutherford showed that the angular distribution observed by his collaborators could be explained by assuming that Coulomb forces were responsible for the observed scattering phenomenon, and that there were heavy scattering centers

https://doi.org/10.1515/9783110500608-002

in the atoms of the scattering material. He also posited that these centers have a positive charge proportional to the atomic number of the scattering material. Rutherford's model of a nuclear atom with positive charges concentrated around a small heavy center and negative charges distributed away from the center led to numerous predictions regarding the scattering of alpha particles. This phenomenon was finally confirmed in a series of experiments conducted by Rutherford and his coworkers between 1911 and 1925.

2.1.2 Atomic spectrum of hydrogen and Bohr's model

1. Atomic spectrum of hydrogen

Light emitted by a hot solid or liquid forms a continuous spectrum, termed a ribbon. The wavelength of the emitted light can vary within a certain range. However, if the light source is a gaseous discharge, then the spectrum comprises only a set of discrete spectral lines called an atomic spectrum. For different elements, the structure and location of the line spectra are unique, which makes them important in studying the atomic structure.

As the simplest atom with the simplest spectroscopy, hydrogen atom was among the first to be studied. In particular, the visible spectrum of atomic hydrogen was found to be consisting of a series of lines expressed by J. J. Balmer in 1885 as

$$\lambda = B\frac{n^2}{n^2 - 4} \tag{2.1}$$

Equation (2.1) is called the Balmer formula, where λ is the wavelength, $B = 364.6$ nm, and $n = 3, 4, 5, \ldots$.

The spectral lines obtained by substituting different n into equation (2.1) are referred to as Balmer series. Spectral lines in Balmer series become denser with increasing n, whereas wavelength becomes shorter. When n approaches infinity, the resulting line has the shortest wavelength, also called the limit of line series.

On substituting frequency for wavelength, equation (2.1) becomes

$$\nu = Rc\left(\frac{1}{2^2} - \frac{1}{n^2}\right) \tag{2.2}$$

where

$$R = \frac{4}{B} = 1.097 \times 10^7\ \text{m}^{-1}$$

is the Rydberg constant.

Balmer further noted that replacing 2^2 with m^2 (where m is an integer) in equation (2.2) gives the following expression for different line series of hydrogen atom

$$\nu = Rc\left(\frac{1}{m^2} - \frac{1}{n^2}\right) \tag{2.3}$$

where $m = 1, 2, 3, \ldots$; and $n = m + 1,\ m + 2,\ \ldots$. Equation (2.3) is the generalized Balmer formula. For $m = 1, 3, 4, 5, \ldots$, the lines are successively termed Lyman series, Paschen series, Brackett series, and Pfund series (Fig. 2.1), respectively, according to their discoverers.

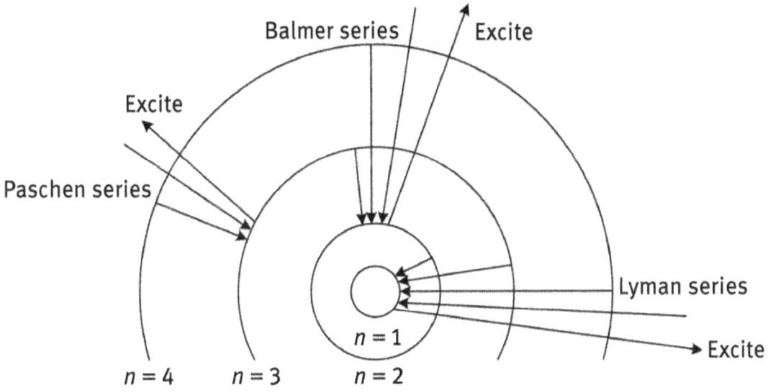

Fig. 2.1: Schematic diagram of level transition of hydrogen atom.

Generalized Balmer formula can also be used to represent the spectra of a few other elements, such as singly-ionized helium and doubly-ionized lithium; however, complex spectra cannot be expressed by Balmer formula. In 1908, W. Ritz found that the spectra of more complex elements can be decomposed into a number of line series. The frequency of each spectral line could be represented as the difference between the two terms of the spectrum, thus obtaining

$$\nu = T(m) - T(n) \tag{2.4}$$

where $T(m)$ is a constant for any one line series, whereas $T(n)$ changes with n outgoing different lines in the line series. Otherwise different $T(m)$ corresponds to different line series.

Equation (2.4) is the Ritz combination principle. It is clear from this principle that, in the presence of two lines with frequencies ν_1 and ν_2, the lines with frequencies $\nu_1 + \nu_2$ and $|\nu_1 - \nu_2|$ must also exist in the series. Further, lines with frequency $k_1\nu_1 + k_2\nu_2$ also exist in theory, where k_1 and k_2 are integer numbers.

2. Bohr's quantum theory

In 1913, Niels Bohr explained these series and described the underlying reasons of the combination principle by making three assumptions:

(1) Neglecting any motion of nucleus and considering only circular electronic orbits;
(2) Assuming that in any one of the circular electronic orbits the atom is stationary and not radiating;
(3) An electron jump from one orbit to another is followed by emission or absorption of radiation. The frequency v of this emitted radiation and the energy change $E_n - E_m$ in the atom were assumed to be related in the same manner that Planck and Einstein used to explain black-body spectrum and photoelectric emission, namely

$$v = (E_n - E_m)/h \tag{2.5}$$

On comparing equations (2.5) and (2.3), it is clear that, based on Bohr's assumption, the energy for the nth stationary state of a hydrogen atom is

$$E_n = -Rch/n^2; \quad n = (1, 2, 3, \ldots) \tag{2.6}$$

In addition, all the spectral lines in the line series for hydrogen atom can be understood as the corresponding radiation as a result of transition from one energy level E_n to another E_m. Bohr's quantum theory successfully explained the stationary structures of atoms and their line spectra. He also gave an expression for the energy E_n of stationary states as

$$E_n = -\frac{1}{\varepsilon_0^2}\frac{m_e e^4}{8n^2h^2} \tag{2.7}$$

and the radii of the electronic orbits in these states as

$$r = \varepsilon_0 \frac{n^2h^2}{\pi m_e e^2} = n^2 r_0 \tag{2.8}$$

Among them, $\varepsilon_0 = 8.85 \times 10^{-12}\,\mathrm{C^2N^{-1}m^{-2}}$; $m_e = 9.11 \times 10^{-31}\,\mathrm{kg}$; $e = 1.60 \times 10^{-19}\,\mathrm{C}$

where C and N represent electric charge in Coulomb and force in Newton, respectively. Substituting these values and $n = 1$ into equation (2.8) results in

$$r_0 = 5.3 \times 10^{-11}\,\mathrm{m}$$

which is the radius of Bohr's first orbit.

Substituting these values into equation (2.7) gives

$$E_n = -\frac{2.18 \times 10^{-18}}{n^2}\,\mathrm{J} = -\frac{13.6}{n^2}\,\mathrm{eV} \tag{2.9}$$

When an electron rotates around the nucleus in the first orbit (n=1), the atom has minimum energy. This minimum energy state is called the ground state, whereas the states at $n > 1$ are referred to as the excited states. The orbital radius of an electron in the excited state increases with n^2, and the absolute value of atomic energy decreases with n^2. Atom can transform from a high energy state to a low energy state accompanied by light emission. When an electron jumps from an excited state of $n \geq 2$ to the ground state, it results in Lyman lines, and when the electron jumps from an excited state of $n \geq 3$ to the ground state, it results in Balmer series (Fig. 2.1).

The next breakthrough in the theory atomic structure occurred 10 years after Bohr first proposed his atomic structure model. In 1923, de Broglie suggested that on theoretical grounds, the wave–particle duality may be common. He said that if classical physics had overemphasized the wave nature of light while ignoring its particle nature, and conversely placed too much emphasis on its particle character while ignoring its wave nature, the kind of material made is a mistake. Based on this, W. Heisenberg, E. Schrodinger, P. Dirac, among others developed nonrelativistic solutions for atomic structure problems and obtained the same energy levels for atoms similar to hydrogen. This is referred to as quantum mechanics which will be discussed in the following sections.

2.1.3 Quantum mechanics and atomic emission

As discussed in the previous section, Bohr's quantum theory can be used to explain the line spectrum of atomic hydrogen. However, his theory can only be used to calculate the frequency of atomic hydrogen emission, and not the related intensity of spectral lines. For atoms of other elements, even radiation frequency cannot be defined using his theory; hence, these issues can only be resolved using quantum mechanics.

1. Wave function and Schrödinger equation

As mentioned in the previous section, de Broglie suggested that the wave–particle duality may be common for particles with energy E and momentum p. Frequency and wavelength of such a particle is given by the so-called de Broglie relationship

$$\nu = E/h, \ \lambda = h/p \tag{2.10}$$

The momentum and energy of a free particle remain constant; therefore according to equation (2.10), the wavelength and frequency of its associated wave will also be constant, i.e., a plane wave. Assuming it travels in x direction, the wave can be expressed as

$$u(x,t) = a \exp\left[-j2\pi\left(\nu t - \frac{x}{\lambda}\right)\right] \tag{2.11}$$

Substituting (2.10) into (2.11) results in

$$u(x,t) = a \exp\left[-j\frac{2\pi}{h}(Et - px)\right] \tag{2.12}$$

Differentiating (2.12) with respect to t gives

$$\frac{\partial u}{\partial t} = -j\frac{2\pi}{h}Eu(x,t)$$

Then differentiating (2.12) with respect to x we get

$$\frac{\partial^2 u}{\partial^2 x^2} = \left(j\frac{2\pi}{h}\right)^2 p^2 u(x,t)$$

Using the relationship $E = p^2/2m$, we finally obtain

$$-j\frac{\partial u}{\partial t} = \frac{\hbar}{2m}\frac{\partial^2 u}{\partial^2 x^2} \tag{2.13}$$

where $\hbar = h/2\pi$. Equation (2.13) is the simplest form of the Schrödinger equation, and the unknown function $u(x,t)$ is the simplest wave function. If the particle moves in a three-dimensional space under the effect of a potential field $U(r,t)$, then its energy can be calculated by

$$E = \frac{p^2}{2m} + U(r,t)$$

Equation (2.13) can then be written as

$$j\hbar\frac{\partial}{\partial t}\psi(r,t) = \left[-\frac{\hbar^2}{2m}\nabla^2 + U(r,t)\right]\psi(r,t) \tag{2.14}$$

Equation (2.14) is the general form of the Schrödinger equation, where function u has been replaced by ψ, which is most commonly used in quantum mechanics, and $\nabla$is the Laplace operator in a three-dimensional space. Here, the meaning of ψ is essentially different from the traditional wave function. According to the statistical interpretation by Born in 1926, ψ represents a probability wave. It implies that the strength of the wave function at a point is proportional to the probability that a particle appears at that point (coordinate representation, (x,y,z)). Suppose a wave function of a particle at point (x,y,z) and time t is $\psi(x,y,z)$, then the probability of finding a particle at that moment within an infinitesimal region $x{\sim}x+dx$, $y{\sim}y+dy$, and $z{\sim}z+dz$, i.e., in a small volume $v{\sim}v+dv$ is

$$\mathrm{d}w(x,y,z;t) = |\psi(x,y,z;t)|^2\ \mathrm{d}x\mathrm{d}y\mathrm{d}z = |\psi(x,y,z;t)|^2\ \mathrm{d}v$$

Also, the probability per unit volume

$$w(x,y;t) = |\psi(x,y,z;t)|^2 \tag{2.15}$$

is the probability density at that point at time t. Because the total probability of finding a particle in the entire universe is 1, the probability density satisfies the normalization condition, that is,

$$\int w(x, y, z; t) dv = 1 \tag{2.16}$$

One of the basic characteristics of wave function is that it obeys the principle of superposition. Therefore, if a system is in a series of states represented by $\psi_i(1, 2, \ldots)$, then the system must be in the state represented by

$$\psi = \Sigma_i c_i \psi_i \tag{2.17}$$

Here c_i is a constant.

2. Observable and its operator representation

As mentioned earlier, if a particle is in a state whose wave function is $\Psi(x, y, z)$, then quantum mechanics cannot determine the exact position of the particle but can provide a probability of that particle appearing at a point (x, y, z). The average value of an arbitrary function of coordinate $F(x, y, z)$ for this particle can then be expressed as

$$\overline{F(x, y, z)} = \int \psi^* F \psi dv \tag{2.18}$$

However, if we need to find the average value of a function $G(p_x, p_y, p_z)$ of momentum p, then there is no simple expression like equation (2.18). This is due to the uncertainty principle, which states that a quantum state does not exist when both position x and momentum p_x have well-defined values. Thus, quantum mechanics required new mathematical tools that are different from those in the classical theory, namely, operators. An operator is a symbol for a calculation and has the ability to change a function ψ to Φ. Consider the following relation of an operator $\hat{L}$:

$$\phi = \hat{L}\psi \tag{2.19}$$

If we replace observable quantities with proper operators, equation (2.18) will remain the same. For example, let

$$\widehat{p}_x = -j\hbar\frac{\partial}{\partial x}; \quad \widehat{p}_y = -j\hbar\frac{\partial}{\partial y}; \quad \widehat{p}_z = -j\hbar\frac{\partial}{\partial z} \tag{2.20}$$

represent three components of linear momentum in cartesian coordinates. The average value of a function $\hat{G}$ of the momentum can be calculated as

$$\overline{G(p_x, p_y, p_z)} = \int \psi^*(x, y, z)\hat{G}(p_x, p_y, p_z)\psi(x, y, z) dv \tag{2.21}$$

In general, function ϕ may be entirely different from ψ in equation (2.19). However, in some cases, the difference between ϕ and ψ can only be a constant, for example,

$$\hat{L}\psi = L\psi \tag{2.22}$$

In this case L is an eigenvalue of operator $\hat{L}$, and ψ is an eigenfunction of $\hat{L}$ that belongs to the eigenvalue L. Equation (2.22) is the eigenvalue equation of the operator $\hat{L}$.

Quantum mechanics equates the set of eigenvalues of the operator with the set of all possible values of observable quantities, providing a universal method for determining the possible values of observable quantity, namely, convert it into a problem of solving the equation in the form of equation (2.22).

Operators in quantum mechanics must have the following two features:

(1) Linearity

An operator acts as a sum of set of functions, which is equal to the same operator acting on individual function, and then add them up, that is

$$\hat{L}\sum_i c_i u_i = \sum_i c_i \hat{L} u_i$$

where c_i is a constant.

(2) Hermiticity, namely

$$\int \psi_1^*(x)\hat{L}\psi_2(x)dx = \int \psi_2^*(x)\left[\hat{L}\psi_1^*(x)\right]^* dx$$

Any meaningful observable needs to be real, and it is known that only the eigenvalues of a Hermitian operator are real. By combining these two properties, it is clear that operators used in quantum mechanics must be linear self-conjugated operators. The eigenfunction of a linear self-conjugate operator has the following important properties:

(i) The eigenfunctions ψ_n and ψ_m belong to two different eigenvalues L_n and L_m that are orthogonal to each other and satisfy the normalization condition giving equation (2.23)

$$\int \psi_m^* \psi_n dx = \delta_{mn} \tag{2.23}$$

where δ_{mn} is the Kronecker symbol.

(ii) Eigenfunctions constitute a complete assembly, that is, if any function $\psi(x)$ in the same definition area meets the same boundary conditions with the eigenfunction $\psi_n(x)$, then $\psi(x)$ can be expressed as

$$\psi(x) = \sum_i c_n \psi_n \tag{2.24}$$

where

$$c_n = \int \psi_n^*(x)\psi(x)dx$$

Equation (2.24) shows that an arbitrary state characterized by the wave function $\psi(x)$ can always be expressed as a superposition of a series of stationary states where the observable quantity L has certain values L_n. In addition to the coordinate represented by coordinate operator $(r = r)$ and momentum operator, given by equation (2.20), another important operator is the energy operator or the Hamiltonian operator

$$\hat{H} = -\frac{\hbar^2}{2m}\nabla^2 U(\boldsymbol{r}) \tag{2.25}$$

3. The time-independent Schrödinger equation

In general, the potential field $U(\boldsymbol{r}, t)$ in equation (2.14) is time-dependent, and its solution $\psi(\boldsymbol{r}, t)$ cannot be expressed as the product of two parts: one is a function of $\boldsymbol{r}$ only and the other is a function of t only. However, in an important case wherein the potential field is time-independent, we can separate equation (2.14) using a separation constant E and obtain

$$j\hbar\frac{df}{dt} = Ef(t) \tag{2.26}$$

and

$$-\frac{\hbar^2}{2m}\nabla^2\psi + U(\boldsymbol{r})\psi = E\psi \tag{2.27}$$

The solution of equation (2.26) is

$$f(t) = Ce^{-j\frac{E}{\hbar}t}$$

Here C is a constant and the wave function

$$\Psi(\boldsymbol{r}, t) = \psi(\boldsymbol{r})e^{-j\frac{E}{\hbar}t} \tag{2.28}$$

represents a particular solution of equation (2.14).

Equation (2.27) is the time-independent Schrödinger wave equation, whose solutions determined by equation (2.28) are time-independent energy wave functions. The set of allowed values of E are the eigenvalues of energy.

If we write the wave function representing the nth stationary state of the system as

$$\Psi_n(\boldsymbol{r}, t) = \psi_n(\boldsymbol{r})e^{-j\frac{E_n}{\hbar}t}$$

then the general solution of the Schrödinger equation (2.14) is the linear superposition of the time-independent wave functions, namely,

$$\Psi(\mathbf{r}, t) = \sum_n c_n \psi_n(\mathbf{r}) e^{-j\frac{E_n}{\hbar}t}$$

where c_n remain constant. According to the superposition principle, $|c_n|^2$ represents the probability of that system in the time-independent state Ψ_n.

4. Time-dependent perturbation theory and quantum transition

(1) Time-dependent perturbation theory

It can be recalled that, in the matter of atom, subatomic levels could, in principle, be addressed in the context of quantum mechanics. However, the subjects/fields that can be really solved by quantum mechanics are strictly numbered. A majority of the problems have to be solved by various approximation methods. Light emission of concern in this chapter is one example. The following section will first introduce a common approximation in quantum mechanics—the perturbation theory. Next, the absorption and emission of light will be described.

This section aims to solve the Schrödinger equation

$$j\hbar\frac{\partial \Psi}{\partial t} = \hat{H}(t)\Psi \tag{2.29}$$

and obtain the probability of a system transiting from one quantum state to another. This differs from the time-independent Schrödinger wave equation. In equation (2.29), the Hamiltonian operator $\hat{H}(t)$ is time-dependent. Supposing that $\hat{H}(t)$ can be divided into two parts as follows:

$$\hat{H}(t) = \hat{H}^{(0)} + \hat{H}'(t) \tag{2.30}$$

where $\hat{H}'(t)$ is the perturbed Hamiltonian operator, while $\hat{H}^{(0)}$ is the nonperturbed Hamiltonian obeying the following condition:

$$\hat{H}^{(0)}\psi_n^{(0)} = E_n^{(0)}\psi_n^{(0)}; \quad (n = 1, 2, \ldots)$$

Substituting (2.30) in (2.29), we obtain

$$j\hbar\frac{\partial \Psi}{\partial t} = \left[\hat{H}^{(0)} + \hat{H}'(t)\right]\Psi \tag{2.31}$$

Further assuming that the system before the perturbation effect for the Hamiltonian $\hat{H}^{(0)}$ is in the time-independent state shown by the wave function

$$\Psi_k = \psi_k^{(0)} e^{-\frac{j}{\hbar}E_k^{(0)}t} \tag{2.32}$$

Under the perturbation effect, the system transits to the state Ψ, which does not belong to the set of

$$\left\{\Psi_n^{(0)} e^{-\frac{j}{\hbar}E_n^{(0)}t}\right\}$$

but can be expanded in terms of a complete orthonormal set as

$$\Psi = \sum_n c_n(t)\psi_n^{(0)} e^{-\frac{j}{\hbar}E_k^{(0)}t} \tag{2.33}$$

Using the coefficients of expansion $c_n(t)$, we can obtain the probability $|c_n|^2$ of the perturbed system. Therefore, the next step is to solve the Schrödinger equation (2.31).

Substituting (2.33) into (2.31), we obtain

$$j\hbar\sum_n \Psi_n \dot{c}_n(t) = \sum_n c_n(t)\hat{H}'(t)\Psi_n \tag{2.34}$$

Multiplying both sides of equation (2.34) from the left-hand side by Ψ_m^* and integrating, we obtain

$$j\hbar\dot{c}_m(t) = \sum_n c_n(t)\hat{H}_{mn}e^{-j\omega_{mnt}}(t) \tag{2.35}$$

where

$$H_{mn}{}' = \int \Psi_m^{(0)*}\hat{H}'\Psi_n^{(0)}dv \tag{2.36}$$

and

$$\omega_{mn} = \frac{E_m^{(0)} - E_n^{(0)}}{\hbar} \tag{2.37}$$

At this point the analysis is straightforward and solving equation (2.35) is equivalent to obtaining a solution of Schrödinger equation. Assuming that the perturbation is at t=0, we obtain

$$c_n(0) = \delta_{nk} \tag{2.38}$$

Taking it as the 0th-step alternate solution of $c_n(t)$ and substituting it into equation (2.35), we obtain the 1st-step alternate solution of $c_n(t)$ as

$$c_n^{(t)}(t) = -\frac{j}{\hbar}\int_0^t H'_{nk}\, e^{j\omega_{nk}t'}dt' \tag{2.39}$$

The first-rank approximation of the probability of transition from the initial state to the final state under perturbation effect is given by

$$W_{kn} = \left|c_n^{(1)}\right|^2 = \frac{1}{\hbar^2}\left|\int_0^t H'_{nk}e^{j\omega_{nk}t'}dt'\right|^2 \tag{2.40}$$

(2) Transition caused by perturbation of monochromatic plane wave

As a special case, we consider the perturbation of a monochromatic plane wave with a wavelength far larger than the scale of an atom. The wave can be considered uniform within the scope of the atom and is defined by

$$\varepsilon(t) = \frac{\varepsilon_0}{2}\left(e^{j\omega t} + e^{-j\omega t}\right) \tag{2.41}$$

Here ε is an independent spatial coordinate. In the dipole approximation, the perturbation can be written as

$$\hat{H}' = \varepsilon \cdot \boldsymbol{D} \tag{2.42}$$

where $\boldsymbol{D}$ ($\boldsymbol{D} = -e\boldsymbol{r}$) is the electric dipole moment of the atom and is independent of the time coordinates. Therefore, integration of variables v and t can be done. Substituting (2.42) in (2.36) and integrating with respect to spatial coordinates gives

$$H' = -e\varepsilon \int \Psi_n^{(0)*} \hat{r} \Psi_k^{(0)} dv = -e\varepsilon \cdot \boldsymbol{r}_{nk} \tag{2.43}$$

Substituting (2.43) in (2.39) and on time-domain integration, we obtain

$$c_n^{(1)}(t) = \frac{j}{2h} e\varepsilon_0 \cdot r_{nk} \int_0^t \left(e^{j\omega t'} + e^{-j\omega t'}\right) e^{j\omega_{nk} t} dt' = -\frac{e}{2\hbar}\varepsilon_0 \cdot \boldsymbol{r}_{nk} \left[\frac{e^{j(\omega_{nk}+\omega)t} - 1}{\omega_{nk}+\omega} + \frac{e^{j(\omega_{nk}-\omega)t} - 1}{\omega_{nk}-\omega}\right] \tag{2.44}$$

Thus, the transition probability from state k to n is

$$W_{nk} = -\frac{e^2}{4\hbar^2} |\varepsilon_0 \cdot \boldsymbol{r}_{nk}|^2 \left|\frac{e^{j(\omega_{nk}+\omega)t} - 1}{\omega_{nk}+\omega} + \frac{e^{j(\omega_{nk}-\omega)t} - 1}{\omega_{nk}-\omega}\right|^2 \tag{2.45}$$

5. Light absorption and emission

Based on the above-mentioned solutions, the problems of light absorption and emission will be discussed in this section. First, if $\omega \neq \pm\omega_{nk}$, the quantity order of the denominator on the right side of equation (2.45) is 10^{15}. The numerator, on the other hand, has a quantity order of 10^0. Thus, we have $W_{nk} \approx 0$. This implies that, when the light wave frequency deviates significantly from atomic resonance, light neither causes atomic energy-level transition nor atomic absorption or emission of light.

Otherwise, for $\omega \approx \omega_{nk} = (E_n - E_k)/\hbar$ or $E_n - E_k = \hbar\omega$, light is absorbed. As per the above-mentioned reason, the first term of equation (2.45) can be neglected, and the probability of absorption can be given by the second term as

$$W_{nk} = \frac{e^2}{4\hbar^2} |\varepsilon_0 \cdot \boldsymbol{r}_{nk}|^2 \frac{\left|e^{j(\omega_{nk}-\omega)t} - 1\right|}{(\omega_{nk}-\omega)^2} = \frac{e^2}{\hbar^2} |\varepsilon_0 \cdot \boldsymbol{r}_{nk}|^2 \frac{\sin^2 \frac{1}{2}(\omega_{nk}-\omega)t}{(\omega_{nk}-\omega)^2} \tag{2.46}$$

Using the relation

$$\delta(x) = \lim_{t\to\infty} \frac{t}{\pi}\left(\frac{\sin xt}{xt}\right)^2$$

equation (2.46) can be rewritten as

$$\begin{aligned} W_{nk} &= \frac{e^2}{4\hbar^2}|\varepsilon_0 \cdot \boldsymbol{r}_{nk}|^2 t\delta\left(\frac{\omega_{nk}-\omega}{2}\right) = \frac{\pi e^2}{2\hbar^2}|\varepsilon_0 \cdot \boldsymbol{r}_{nk}|^2 t\delta(\omega_{nk}-\omega) \\ &= \frac{\pi e^2}{2\hbar^2}|\varepsilon_0 \cdot \boldsymbol{r}_{nk}|^2 t\delta(E_n - E_k - \hbar\omega) \end{aligned} \tag{2.47}$$

If $\omega \approx -\omega_{nk}$ or $E_k - E_n = \hbar\omega$, then the atom transition from state Ψ_k to Ψ_n is accompanied by a light emitting frequency of $\omega = (E_k - E_n)/\hbar$ with the probability

$$W_{nk} = \frac{\pi e^2}{2\hbar^2}|\varepsilon_0 \cdot \boldsymbol{r}_{nk}|^2 t\delta(E_n - E_k + \hbar\omega) \tag{2.48}$$

It can be observed from equations (2.47) and (2.48) that the probabilities of atomic absorption-transition and emission-transition are the same owing to the monochromatic radiation field.

To describe the transition probabilities between two energy levels, three parameters A_{nk}, B_{nk}, and B_{kn} were introduced by Einstein in 1917. These parameters are called the spontaneous emission coefficient, stimulated emission coefficient, and stimulated absorption coefficient, respectively, and their relationship can be expressed as

$$A_{nk} = B_{nk}, A_{nk} = \frac{\hbar\omega_{nk}^3}{\pi^2 c^3} B_{nk}$$

Suppose the light energy density in frequency interval $\omega \infty \omega + d\omega$ is $I(\omega)\mathrm{d}\omega$, which is in accordance with Einstein's interpretation that $B_{nk}I(\omega_{nk})$ represents the probability that an atom absorbs photonic energy and transits from the state Ψ_k to Ψ_n in per unit time. At the same time, $B_{kn}I(\omega_{kn})$ represents the probability that an atom transits from the state Ψ_n to Ψ_k and emits a photon with energy $\hbar\omega_{kn}$ per unit time. This leads to

$$B_{nk}I(\omega_{nk}) = \frac{\pi e^2}{2\hbar^2}|\varepsilon_0|^2 |r_{nk}|^2 \cos\theta \tag{2.49}$$

where θ is the angle between ε_0 and r.

Because ε_0 has a determinate direction and because electric dipole moment of an atom may be randomly oriented, an average over $\cos^2\theta$ results in

$$\overline{\cos^2\theta} = \frac{1}{4\pi}\iint \cos^2\theta\, \mathrm{d}\Omega = \frac{1}{4\pi}\int_0^{\pi} \cos^2\theta \sin\theta\, \mathrm{d}\theta \int_0^{2\pi} \mathrm{d}\phi = \frac{1}{3}$$

and

$$I(\omega_{nk}) = \frac{|\varepsilon|^2}{4\pi} = \frac{|\varepsilon_0|^2}{8\pi}$$

Substituting them into equation (2.49) gives

$$B_{nk} = \frac{4\pi^2 e^2}{3\hbar^2} |\boldsymbol{r}_{nk}|^2 \tag{2.50}$$

This shows that the intensity of the spectral line is proportional to $|\boldsymbol{r}_{nk}|^2$ under the dipole approximation. In particular, if $|\boldsymbol{r}_{nk}|^2 = 0$, then the corresponding transition probability is zero, the so-called forbidden transition. The rule that determines which transition belongs to the forbidden transition is called the selection rule.

Because

$$\boldsymbol{r}_{nk} = \int \psi_n^{(0)*} \boldsymbol{r} \psi_k^{(0)} dv$$

for further calculation we need to know the specific form of the wave function. The following is a simple example.

Example 2.1 Determine the emission intensity of hydrogen atom corresponding to the transition from the first excited state ($n = 2$, $l = 1$, $m = 1$) to the ground state ($n = 1$, $l = 0$, $m = 0$).

Solution: Using quantum mechanics, the wave function of the first excited state and the ground state of the hydrogen atom is in the order of

$$\psi_{2,1,1} = \frac{1}{8\pi^{1/2} a_H^{5/2}} r e^{-r/2a_H} \sin\theta e^{j\phi}$$

and

$$\psi_{1,0,0} = \frac{1}{\pi^{1/2} a_H^{3/2}} r e^{-r/a_H}$$

where $a_H = 5.3 \times 10^{-11} m$ is the first orbital radius of the hydrogen atom.

Using polar coordinates $x = r \sin\theta \cos\phi$ and $dv = r^2 \sin\theta dr d\theta d\phi$, we have

$$\int \psi_{2,1,1}^* a \psi_{1,0,0} dv = \frac{1}{8\pi a_H^4} \int\int\int r^4 e^{-3r/2a_H} dr \sin^3\theta d\theta \cos\phi e^{-j\phi} d\phi = 4\left(\frac{2}{3}\right)^5 a_H$$

and

$$\int \psi_{2,1,1}^* y \psi_{1,0,0} dv = -4j\left(\frac{2}{3}\right)^5 a_H$$

$$\int \psi_{2,1,1}^* z \psi_{1,0,0} dv = 0$$

such that

$$|r_{nk}|^2 = \left|\int \psi_{2,1,1}^* x \psi_{1,0,0} dv\right|^2 + \left|\int \psi_{2,1,1}^* y \psi_{1,0,0} dv\right|^2 + \left|\int \psi_{2,1,1}^* z \psi_{1,0,0} dv\right|^2 = 0.555 a_H^2$$

2.1.4 Spectral line broadening

1. The line broadening concept

According to quantum mechanics, when an atom transits from an energy level E_k to an energy level E_n, it is accompanied by a photon that is either absorbed ($E_k < E_n$) or emitted ($E_k > E_n$) with a frequency of

$$\nu = \frac{E_n - E_k}{\hbar} \tag{2.51}$$

Till now we have assumed that the values of E_k, E_n, and ν are precise. The transitions corresponding to the absorption or emission of photon are shown in Fig. 2.2.

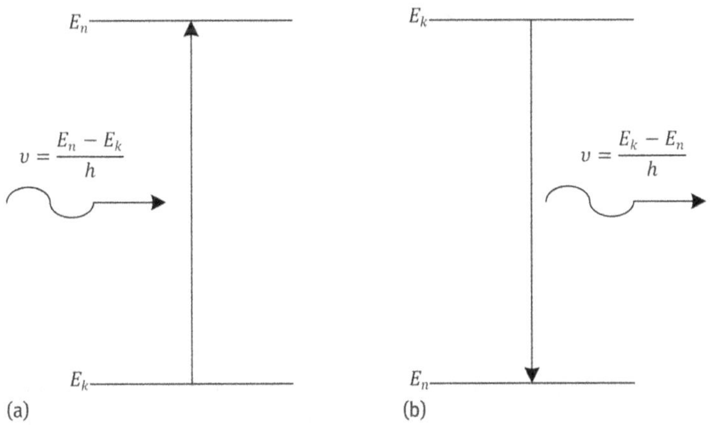

Fig. 2.2: Light absorption and emission excluding energy level broadening: (a) absorption; (b) emission.

In fact, light frequency ν cannot strictly have an accurate value, and there is always a dispersion range $\Delta\nu$. This is because the energy levels E_n and E_k both have certain width ΔE_k and ΔE_n and it turns to the spectral line broadening, as shown in Fig. 2.3. The frequencies of light emission corresponding to the maximum energy gap and the minimum energy gap are ν_+ and ν_-, respectively, and therefore, emission line broadening can be expressed as

$$\Delta\nu = \nu_+ - \nu_- = \frac{\Delta E_k + \Delta E_n}{\hbar} \tag{2.52}$$

2. Line broadening mechanism

Two types of line broadening have been described, namely, homogeneous broadening and inhomogeneous broadening. The essential feature of homogeneous

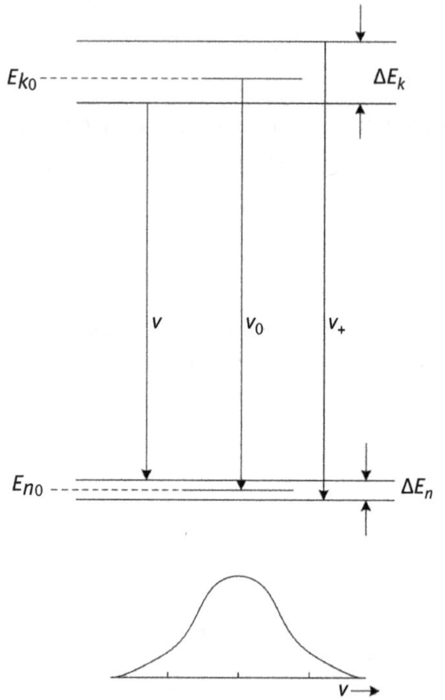

Fig. 2.3: Energy levels and spectral line broadening.

broadening is that every atom has the same atomic line shape and frequency response such that a signal applied to the transition has exactly the same impact on all atoms. This implies that, within the line width of the energy level, each atom has the same probability function for the transition. Mechanisms of inhomogeneous broadening tend to displace the center frequencies of individual atoms. Different atoms have different resonance frequencies during the same transition. We will now consider lifetime broadening and collision broadening as examples of homogeneous broadening and Doppler broadening as an example of inhomogeneous broadening. Before that it is necessary to introduce the concept of atomic line shape function (ν, ν_0).

(1) Concept of atomic line shape function

Suppose radiation is emitted during the atomic transition from E_2 to $E_1(<E_2)$. If we neglect the energy level width, then the radiation is monochromatic, and its full power is concentrated at a single frequency $\nu_0 = (E_2 - E_1)/\hbar$. Whereas, in line broadening mechanism, the radiated power is no longer concentrated at a single frequency but is distributed in accordance with frequency. On representing the radiation power at the frequency ν as $P(\nu)$, the ratio of $P(\nu)$ to the total power P is defined as a line shape function and expressed as $g(\nu, \nu_0)$, namely,

$$g(\nu,\nu_0)=P(\nu)/P \tag{2.53}$$

Here ν_0 is the center frequency of the spectrum. It can be easily observed from equation (2.53) that $g(\nu,\nu_0)$ must be normalized to unity:

$$\int_{\infty}^{\infty} g(\nu,\nu_0)d\nu=1$$

From the above discussion it follows that $g(\nu,\nu_0)$ can be defined as the probability of emission or absorption per unit of frequency interval. It has the maximum value of $g(\nu,\nu_0)$ at $\nu=\nu_0$ and reduces to $g(\nu,\nu_0)/2$ at $\nu=\nu_0\pm\Delta\nu/2$. Then the value $\Delta\nu$ is called the line width.

After the introduction of line function, radiation power distributed in $\nu \sim \nu + d\nu$ can be written as

$$P(\nu)d\nu=g(\nu,\nu_0)Pd\nu$$

(2) Natural broadening

Natural broadening is caused by the decay mechanisms of the atomic system. Spontaneous radiation or fluorescence has a radiative lifetime τ_N and the natural broadening ν_N is related by the equation

$$\Delta\nu_N=1/2\pi\tau_N \tag{2.54}$$

The line shape function is referred to as the Lorentz shape

$$g_N(\nu,\nu_0)=\frac{\Delta\nu_N/2\pi}{(\nu-\nu_0)^2+(\Delta\nu_N/2)^2} \tag{2.55}$$

Equation (2.55) is also the spectrum function type of any homogeneous broadening. If $g_H(\nu,\nu_0)$ and $\Delta\nu_H$ represent spectral line functions and line width of homogeneous broadening, respectively, we obtain

$$g_H(\nu,\nu_0)=\frac{\Delta\nu_H/2\pi}{(\nu-\nu_0)^2+(\Delta\nu_H/2)^2} \tag{2.56}$$

(3) Collision broadening

At higher pressures, atomic collision results in transition to other levels, which may be spontaneous radiative transition or nonradiative transition.

Collision of radiating atoms and the consequent random interruption of the radiative process leads to broadening. Because atomic collision interrupts either the emission or the absorption of radiation, the long wave train becomes truncated. After the collision, the atom resumes its motion with a random initial phase.

In gas-emitting media, collision broadening $\Delta\nu_L$ is proportional to the pressure p:

$$\Delta\nu_L = \alpha p$$

where the proportionality coefficient α is related to collision section and temperature among other factors, and can be measured experimentally.

(4) Doppler broadening

The random motions of atoms result in Doppler broadening. Because atoms move in different directions, the centers of the natural broadening lines vary. When monochromatic light passes through a gas, only those atoms can interact for which the frequency of the light lies within the natural line width as shown in Fig. 2.4.

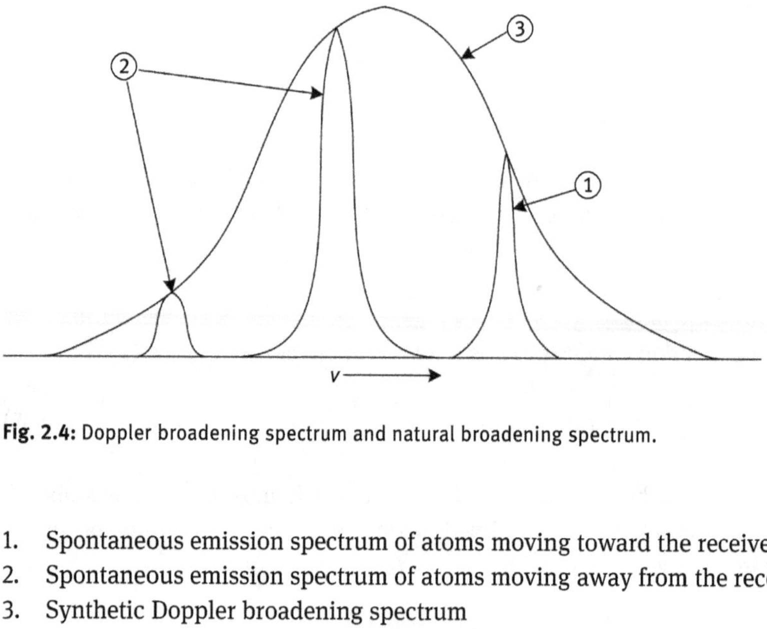

Fig. 2.4: Doppler broadening spectrum and natural broadening spectrum.

1. Spontaneous emission spectrum of atoms moving toward the receiver
2. Spontaneous emission spectrum of atoms moving away from the receiver
3. Synthetic Doppler broadening spectrum

An atom moving with a velocity v relative to an observer will radiate at a frequency measured by the observer as

$$\nu_{\pm} \approx \nu_0(1 \pm v/c) \tag{2.57}$$

If the Doppler broadening is larger than the natural broadening, then the resulting line is Gaussian:

$$g_D(\nu, \nu_0) = \frac{2}{\Delta\nu_D}\sqrt{\frac{\ln 2}{\pi}}\exp\left[\frac{-4(\ln 2)(\nu - \nu_D)^2}{\Delta\nu_D{}^2}\right] \tag{2.58}$$

With the full half-width

$$\Delta\nu_D = 2\nu_0 \left[\frac{2(\ln 2)KT}{Mc^2}\right]^{1/2}$$

Here M is the atomic mass.

2.2 Spontaneous radiation and its sources

The theory of radiation describes the processes wherein electromagnetic rays are either emitted or absorbed. Although emission or absorption can involve any elementary particle, we confine our discussion to the electron as it the theory of main concern.

An excited state E_u of an atom may decay to another state E_l of lower energy by the emission of electromagnetic radiation. In the absence of external photons, such processes are referred to as spontaneous radiation transitions, which are accompanied by emission of photons with frequency (as shown in Fig. 2.5).

$$\nu = \frac{E_u - E_l}{h} \tag{2.59}$$

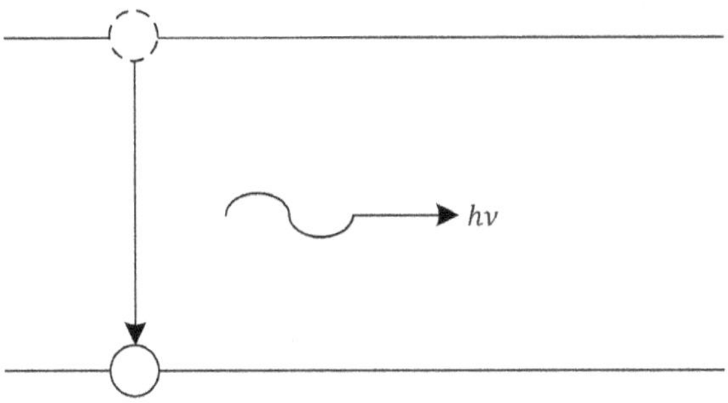

Fig. 2.5: Spontaneous radiation transitions.

Atomic spontaneous emission is random, i.e., the atoms involved, the time of emission, and the energy levels involved are all occasional and random. Therefore, if different light-emitting atoms have sufficient distance between them such that the interaction between them can be ignored, the frequency, phase, polarization state, and direction of propagation of the radiations emitted by them have no determinate relationship. Even for a single atom, the frequency, phase, polarization state, and direction of propagation of the light emitted by double transition are different. Therefore, in general, the light emitted from spontaneous atomic emission is incoherent.

The light generated by a single atom in a single shot should be coherent in nature. However, the order of magnitude of ordinary light-emitting duration τ_0 is only 10^{-8} s. In many cases, the receiver response time $\tau \gg \tau_0$ and the field received is an average of the interference field. Consequently, the cross-term effect disappears and only noncoherent superposition is displayed by the receiver. For quick response of the receiver of $\tau < \tau_0$ to time average does without and interference pattern will be recorded.

In short, spontaneous emission from a common source results in noncoherent light. Before the invention of laser, light sources, both natural and man-made, were basically incoherent. Light emitted from the sun and other stars are among the natural self-emitting light sources; other light emitters, including the moon, are solar reflectors in a majority of cases. Some examples of artificial light sources are tungsten lamps, mercury lamps, sodium lamps, and arc lamps.

Invention of lasers provided mankind with a light source having an unprecedented high coherence and high brightness. It is no exaggeration to say that various fields have undergone a fundamental shift because of its applications. The following few sections will briefly introduce the laser mechanism, its main characteristics, main devices, and its applications in fields such as optoelectronics.

2.3 Laser mechanism

As a major invention in the history of humans, laser has been widely used in various fields owing to its characteristics that cannot be compared with ordinary sources. The following sections describe the characteristics of the laser along with the conditions for producing a laser. To facilitate this discussion, the concept of laser resonator and mode will be introduced first.

2.3.1 Concept of laser resonator and modes

A typical laser consists of three components, namely, laser materials, pumping systems, and optical resonator. Laser material may be a gas, solid, or semiconductor; the corresponding lasers are referred to as gas lasers, solid state lasers, semiconductor lasers. A pumping system provides energy such that the two levels (referred to as higher and lower level) of laser material distribute contrary to equilibrium distribution. The simplest optical resonator consists of two terminal mirrors separated by a distance L. The resonator in which there is no active element is a passive resonator. Introducing an active element, such as a laser material, into the cavity produces an active resonator (Fig. 2.6).

The symbols TEM_{mnq} are used to describe the variation of the electromagnetic field inside the optical resonator. Here, TEM is "transverse electromagnetic waves" and the first two indices, m and n, refer to a particular transverse mode, while q describes a longitudinal mode. The typical laser resonators are considerably longer than the laser wavelength. Therefore, the index q is generally very large.

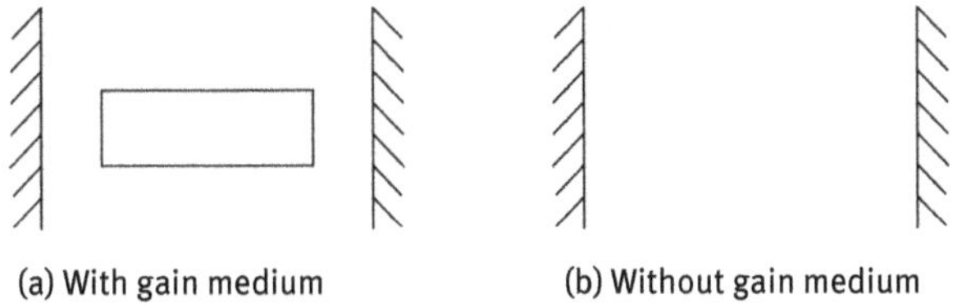

Figure 2.6: Simplest optical resonator. (a) Active resonator; (b) passive resonator.

Transverse modes are defined by TEM_{mn} for cartesian coordinates. The integers m and n represent, respectively, the number of nodes or zeros of intensity transverse to the beam axis in vertical and horizontal directions. The larger the values of m and n, the higher the mode order. The lowest-order mode is the TEM_{00} mode, which has a Gaussian intensity profile with its maximum on the beam axis.

2.3.2 Necessary conditions for producing a laser

A laser is essentially produced as a result of the interaction between radiation and matter. In this section, we will introduce the three kinds of transitions between atomic energy levels and their interaction with radiation. Based on this, we will describe a condition for light amplification, which is a necessary condition for producing a laser.

The actual structure of the atomic energy levels is often very complex; however, there are only two energy levels directly related with laser production. For the sake of simplicity, this chapter will discuss the two-level system and as the case of homogeneous, unless stated otherwise. The higher energy level and lower energy level are represented, respectively, as u and l. N_u and N_l represent the number of atoms per unit volume in the upper and lower levels of the transition, respectively.

1. Kinds of transitions of the two-level system

Following are the three possible radiative interactions between the two levels u and l, as presented in Fig. 2.7, where E_u and E_l are the corresponding energies per unit volume of the medium. Radiative transition is accompanied by a photon emission with frequency

$$\nu = \frac{E_u - E_l}{h} \tag{2.59}$$

(1) Spontaneous transition

Figure 2.7(a) shows spontaneous downward emission from level u to level l at a spontaneous emission rate A_{ul}. Thus, the number of downward transitions per unit volume per unit time occurring due to spontaneous emission from level u to level l is

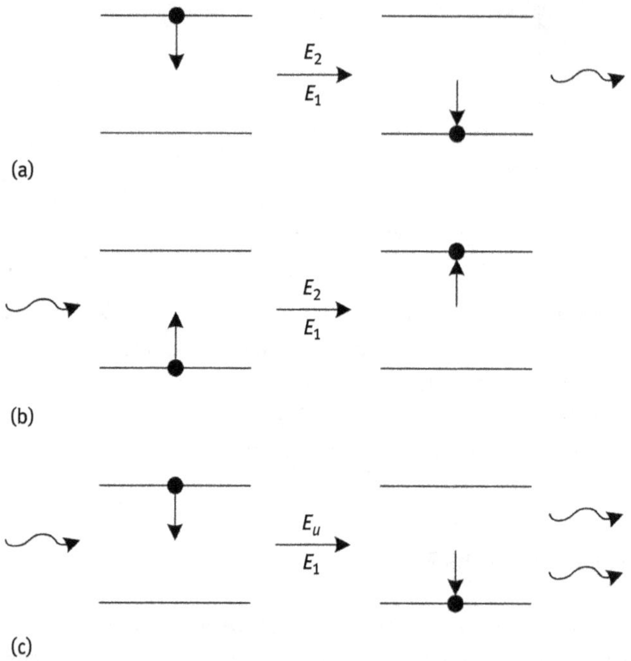

Fig. 2.7: Transitions between the two levels of a two-level system: (a) spontaneous emission; (b) stimulated absorption; (c) stimulated emission.

$$\left(\frac{dN_u}{dt}\right)_{\text{sp}} = -A_{ul}N_u \tag{2.60}$$

Here, the subscript "sp" indicates spontaneous emission. The solution to equation (2.60) is given by

$$N_u(t) = N_{u0}e^{-A_{ul}t} = N_{u0}e^{-\frac{t}{\tau_N}} \tag{2.61}$$

where N_{u0} is the value of N_u at time $t = 0$, and

$$\tau_N = \frac{1}{A_{ul}} \tag{2.62}$$

is the spontaneous transition lifetime of level u.

If the transition at lower levels between level u and i occurs more than once and the spontaneous emission rate from level u to level i is A_{ui}, equation (2.62) should be replaced by the more general relationship:

$$\tau_N = \frac{1}{\sum_i A_{ui}} \tag{2.63}$$

In addition to spontaneous emission, lifetime is related to a number of other processes such as atomic collisions. In combination with various factors, upper level lifetime of some common lasers is listed in Table 2.1.

Table 2.1: Upper level lifetime of some common lasers.

Laser type	He-Ne	Ar^+	CO_2	N_2	Nd:YAG	Nd:Glass	Ti: Sapphire	Er:fiber
λ / nm	632.8	488/514	10600	337	1064	1054–1062	660–1180	1530–1560
τ_u / s	1.7×10^{-7}	1.0×10^{-8}	4×10^{-4}	4×10^{-8}	2.3×10^{-4}	3×10^{-4}	3.8×10^{-6}	1.1×10^{-2}

(2) Stimulated transitions

The other two processes shown in Fig. 2.7 are stimulated, and are proportional to the energy density $\rho$of the beam at frequency ν.

(i) Stimulated absorption

An atom at level l absorbs a photon at frequency ν, and may transit to level u (Fig. 2.7b) with the probability

$$W_{lu} = B_{lu}\rho \tag{2.64}$$

where B_{lu} is the Einstein stimulated absorption coefficient.

Stimulated absorption results in the reduction of the number of incident photons, n, and the number of atoms at the level u increases with the rate of

$$\left(\frac{dN_u}{dt}\right)_{ab} = -\left(\frac{dn}{dt}\right)_{ab} = W_{lu}N_l \tag{2.65}$$

Here, the subscript “ab” indicates absorption process.

(ii) Stimulated emission

Owing to the action of incident photons, downward transitions occur between the two levels per unit volume per unit time as

$$W_{ul} = B_{ul}\rho \tag{2.66}$$

and photons just like incident photons (Fig. 2.7c) are emitted, where B_{ul} is the Einstein stimulated emission coefficient. This process is referred to as stimulated emission which was first proposed by Einstein in 1917.

Stimulated emission results in reduction in the number of atoms N_u at the level u and increase in the number of photons n at the rate of t

$$\left(\frac{dN_u}{dt}\right)_{st} = -\left(\frac{dn}{dt}\right)_{st} = -W_{ul}N_u \tag{2.67}$$

Here, the subscript “st” indicates stimulated emission process.

2.3.3 Relationship between radiation coefficients

The three proportionality coefficients concerning the interaction of radiation with atoms A_{ul}, B_{lu}, and B_{ul} are called Einstein coefficients. They depend only on the nature of the atoms, and regardless of the radiation field, there is a certain link among the three.

When the three processes occur concurrently, the total change N_u at the rate of t is given by

$$\frac{dN_u}{dt} = \left(\frac{dN_u}{dt}\right)_{sp} + \left(\frac{dN_u}{dt}\right)_{ab} + \left(\frac{dN_u}{dt}\right)_{et}$$

Under the thermal equilibrium conditions, we have $dN_u/dt = 0$. By substituting equations (2.60), (2.65), and (2.67) into the above equation, we obtain

$$N_u A_{ul} + N_u B_{ul} \rho = N_l B_{lu} \rho \tag{2.68}$$

On setting the energy levels u and l with the degeneracy g_u and g_l, respectively, the Boltzmann distribution leads to

$$N_l = \frac{g_l}{g_u} N_u e^{\frac{hv}{\kappa T}} \tag{2.69}$$

Here $\kappa = 1.38 \times 10^{-23}\,\mathrm{J\,K^{-1}}$ is the Boltzmann constant. The value of g_l/g_u is within 0.5~2 for most transitions. Therefore, $g_l/g_u = 1$ is used.

In the laser process, the first optical signal is from the heat radiation of the working substance. Radiation energy–density in the frequency component v is given by the Planck equation as

$$\rho = \frac{8\pi h v^3}{c^3}\left(e^{\frac{hv}{\kappa T}} - 1\right)^{-1} \tag{2.70}$$

Substituting equations (2.69) and (2.70) into (2.68) leads to

$$\frac{B_{ul}}{A_{ul}}\left(\frac{B_{lu} g_l}{A_{lu} g_u} e^{\frac{hv}{\kappa T}} - 1\right)\frac{8\pi h v^3}{c^3} = \left(e^{\frac{hv}{\kappa T}} - 1\right) \tag{2.71}$$

Taking the limit $T \to \infty$ on both sides, finally we obtain

$$B_{lu} g_l = B_{ul} g_u \tag{2.72}$$

or

$$B_{lu} = \frac{g_u}{g_l} B_{ul} \tag{2.72a}$$

Substituting equation (2.72) into (2.71) we get

$$A_{ul} = \frac{8\pi h v^3}{c^3} B_{ul} \tag{2.73}$$

Equations (2.72) and (2.73) show the relationship of the three Einstein coefficients.

2.3.4 Necessary conditions for laser production

Consider a light beam passing through a volume of a medium having a length dz and a cross-sectional area dA (Fig. 2.8). We will estimate the energy that can be added to or subtracted from the beam. The contribution of spontaneous emission is neglected for this calculation because spontaneous emission in all directions into a solid angle is 4π and its contribution is very small in the direction of the incident beam.

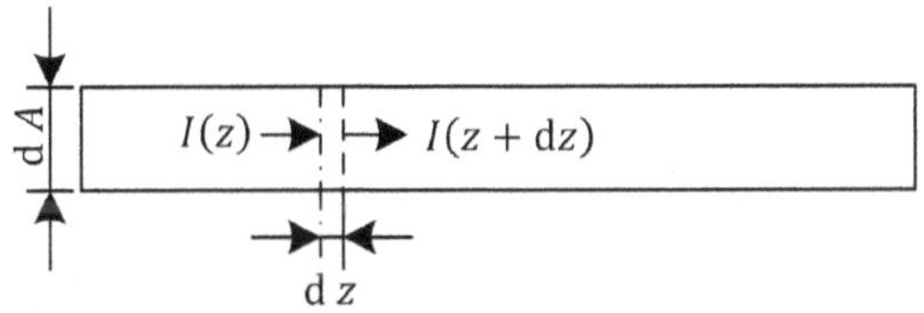

Fig. 2.8: Light propagates through the working substance.

Suppose that the intensity before and after the light beam passes through the medium are $I(z)$ and $I(z+dz)$, respectively. The amount of energy added to the beam per unit time can be expressed as the difference between the number of upward and downward transitions per unit time multiplied by the energy of the photon:

$$[I(z+dz)-I(z)]dA=(N_uB_{ul}+N_lB_{lu})\rho hvdAdz$$

where $\rho=I/c$.

Dividing throughout by $dAdz$ gives

$$\frac{dI}{Idz}=(N_uB_{ul}-N_lB_{lu})\frac{hv}{c} \tag{2.74}$$

The solution to equation (2.74) can be expressed as

$$I=I_0e^{Gz} \tag{2.75}$$

where I_0 represents the initial beam intensity, and

$$G=\left(N_u-\frac{g_u}{g_l}N_l\right)B_{ul}\frac{hv}{c} \tag{2.76}$$

is referred to as the gain coefficient and has dimensions of m^{-1} in the MKS system of units. The gain coefficient decreases with the increase in intensity and reaches its maximum G^0 when I is extremely small. G^0 is referred to the small-signal gain coefficient, which is listed in Table 2.2 for some laser materials.

Let us define the first factor on the right-hand side of equation (2.76) as population difference ΔN_{ul}.

Table 2.2: Small signal gain for some laser materials.

Laser type	He-Ne	Ar^+	CO_2	N_2	Nd:YAG	Nd: Glass	Ti: Sapphire	Er:filer	DL
G^0 / m^{-1}	0.15	0.5	0.9	10	10	3	20	1.35	10^4–10^5

$$\Delta N_{ul} = N_u - \frac{g_u}{g_l} N_l \tag{2.77}$$

It can be observed that, if ΔN_{ul} is positive

$$\Delta N_{ul} > 0 \tag{2.78}$$

then the beam intensity will increase and provide amplification. Conversely, if ΔN_{ul} is negative, then the beam intensity will decrease. When the upper level is more populated than the lower level (taking into account statistical weights), the scenario is referred to as population inversion as the state is not normal under the condition of thermal equilibrium. A population inversion is a necessary condition for producing a laser but not a sufficient condition, which we will introduce briefly in the following section.

2.3.5 Sufficient condition for producing a laser

As discussed in the previous section, only when population inversion exists in the medium, it is possible to generate stimulated emission amplification, which may lead to laser production. Here "it is possible" always means that it is a necessary condition for lasing; then, what is the sufficient condition for lasing? This is the condition to be studied in this section. Before doing so, we must consider the saturation effect of the laser beam within a gain medium.

1. Concept of saturation intensity

According to equation (2.78), as long as $G > 0$, or $\Delta N_{ul} > 0$, the beam intensity will exponentially increase. Thus, if the media is sufficiently long, it will inevitably generate a laser radiation. It can be observed, however, that the condition to obtain equation (2.75) from equation (2.74) is G; therefore, ΔN_{ul} is kept constant. In fact, to increase the beam intensity we have to just decrease ΔN_{ul}. When the beam intensity is increased to a certain value, the equation (2.75) does not hold, which means intensity I no longer grows exponentially. The beam intensity at this moment is called saturation intensity, and is expressed as I_s, whereas the length of the media at which that saturation effect occurs can be expressed as the saturation length L_s.

2. A simple calculation to estimate

Assume that a population inversion between levels u and l exists as a result from the pump flow W_u and level u is in a steady-state equilibrium, so we can write a rate equation taking into account all of the population changes affecting level u and equate it to zero, since "steady state" implies that there no net changes in N_u. This equation can be expressed as

$$\frac{dN_u}{dt} = W_u - N_u\left(\frac{1}{\tau_u} + \frac{B_{ul}I}{c}\right) = 0 \tag{2.79}$$

Equation (2.79) suggests that the steady-state solution for N_u is

$$N_u = \frac{W_u}{\frac{1}{\tau_u} + \frac{B_{ul}I}{c}} \tag{2.80}$$

In case of "small signal," N_u is given by

$$N_u \approx W_u \tau_u$$

It can be observed that N_u would decrease as I increases. We will arbitrarily define I_s as the intensity at which the downward stimulated rate equals the normal decay rate:

$$\frac{B_{ul}I_s}{c} = \frac{1}{\tau_u}$$

Using the Einstein relationship, we find that

$$I_s = \frac{8\pi\nu^2}{c^2 A_{ul}}\frac{h\nu}{\tau_u} = \frac{h\nu}{\sigma_{ul}\tau_u} \tag{2.81}$$

where

$$\sigma_{ul} = \frac{c^2}{8\pi\nu^2}A_{ul} \tag{2.82}$$

is the stimulated emission cross-section. Thus, equation (2.76) can be rewritten as

$$G = \sigma_{ul}\Delta N_{ul} \tag{2.83}$$

3. Sufficient condition for producing a laser

If the beam intensity can reach I_s from a small signal within the effective length of the gain medium, it is thought to be sufficient for generating a laser. There are two points worth noting: First, the definition of the saturation intensity given here has a certain arbitrariness; second, after the intensity reaches I_s, it does not mean that it is no longer growing, but means that the growth rate has slowed considerably. We cannot derive an exact value of the gain required to reach the saturation intensity I_s, but can derive an approximate range of values that are dependent on the length and diameter of the gain medium.

Consider a cylindrical gain medium, as shown in Fig. 2.9, that has a length L, cross-sectional diameter d_a, and area A. It is assumed that the population inversion is large enough that N_l can be neglected, or in other words, we have the relationship $\Delta N_{ul} \approx N_u$.

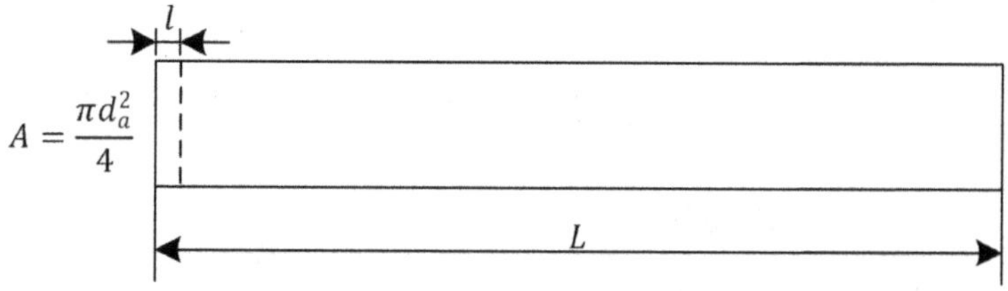

Fig. 2.9: Beam increase in the gain medium.

Consider that the beam starts from one end of the medium in a region of length l. The total energy radiated per unit time into a 4π solid angle is $(A \cdot l)N_u A_{ul} h\nu$. The fractional portion of this energy that would reach the opposite end of the medium can be expressed as $\frac{\Omega}{\pi} = \frac{A}{4\pi L^2}$. Divide that energy by the area A to obtain intensity and then equate it to the saturation intensity.

$$\frac{1}{A}(A \cdot l)N_u A_{ul} h\nu \frac{A}{4\pi L^2} e^{GL} = I_s \tag{2.84}$$

For the simple case,

$$l = 1/G$$

Further simplification leads to

$$e^{GL} = \left(\frac{4L}{d_a}\right)^2 \tag{2.85}$$

Example 2.2 Let us estimate the diameter d_a of gain medium which meets the sufficient conditions for generating a laser using the following data:

$$G = 100\,\mathrm{m}^{-1}$$
$$L = 0.08\,\mathrm{m}$$

Substituting the above conditions into equation (2.85) gives

$$16\left(\frac{L}{d_a}\right)^2 = e^8 = 2981$$

$$\frac{L}{d_a} = \frac{\sqrt{2981}}{4}$$

$$d_a = 0.0059(\text{m}) = 5.9\ \text{mm}$$

Example 2.3 A Nd:YAG laser gain medium with G =10 m^{-1}, L =0.5 m, d_a =0.1 m, and would have a small-signal beam reaching saturation intensity as it emerges from the medium after having made a single pass through the medium.

Using these data in (2.85) gives

$$\left(\frac{4 \times 0.5}{0.1}\right)^2 = 400$$

and

$$e^5 = 148 < 400$$

Thus, a small-signal beam cannot reach the saturation intensity as it emerges from the medium after making a single pass through the medium.

For the above-mentioned reasons, in most practical lasers, at one or both the ends of the laser medium a coating film or reflective mirror is generally added to increase its effective length.

2.4 Physical properties of the lasers

The physical characteristics include monochromatic, directionality, coherence, and high brightness. In fact, monochromatic and directionality are determined by the first-order coherence. Moreover, the coherence should also include higher-order coherence.

2.4.1 Monochromatic and temporal coherence

The magnitude of coherent time is inverse of the bandwidth $\Delta\nu$ of the beam:

$$\tau_c = \frac{1}{2\pi\Delta\nu} \tag{2.86}$$

It implies that the narrower the line width of the beam, the longer the coherent time. In other words, the better the monochromatic, the better the temporal coherence. Under the conditions of stable oscillation, bandwidth $\Delta\nu$ of a laser beam of a single longitudinal-mode laser output can theoretically reach

$$\Delta v = 2\pi h v_0 \frac{(\Delta v_c)^2}{P_0}\left(\frac{n_2}{n_2 - n_1}\right) \tag{2.87}$$

where v_0 is the center frequency of the laser output; P_0 output power; n_1, n_2 are population density of laser at upper and lower levels, respectively, and hs is the Planck constant and

$$\Delta v_c = \frac{c\delta}{2\pi L}$$

is the bandwidth mode of the corresponding passive cavity, with cavity length L and cavity loss δ.

As an example, consider a He–Ne laser having milliwatt output power level and $\delta = 0.01$, $L = 0.5\,\text{m}$. From equation (2.87) we obtain Δv with the order of magnitude 10^{-4} Hz. Substituting it into equation (2.86) gives coherent time as $10^3{\sim}10^4$ s.

In a practice laser, various uncertainties lead to fluctuations in the resonance frequency, Δv is considerably larger than the theoretical value. For most stringent frequency stabilization, it has been observed in the He–Ne laser to 2 Hz bandwidth, while Δv of a typical frequency-stabilization single-mode gas laser can reach $10^6{\sim}10^3$ Hz.

The bandwidth of the beam emitted by natural sources has the same order of magnitude as the center frequency which is approximately 10^{11} Hz. Before the invention of the laser, a mercury lamp line spectrum was considered to be the best source of coherent light. The bandwidth of a mercury lamp line spectrum is still more than 10^8 Hz, which is five orders of magnitude higher than that of frequency-stabilization single-mode gas laser. Correspondingly, the coherent time of the former is only approximately 10^{-5} of the latter. This shows that highly monochromatic or temporal coherence is a characteristic of the laser beam.

2.4.2 Directivity and spatial coherence

The conditions of apparent coherence for the light wave emitted from a source with a width d and propagating inside the angle θ can be represented as

$$d \le \frac{a\lambda}{\theta} \tag{2.88}$$

where λ is the wavelength of the light and a is a constant, the value of which is in the range of approximately 1. Equation (2.88) can also be written as

$$d^2 \le \left(\frac{\lambda}{\theta}\right)^2 \tag{2.89}$$

Alternatively, the conditions of apparent coherence for the light wave propagating inside the angle θ is that the area of the light source not greater than

$$A_c = \left(\frac{\lambda}{\theta}\right)^2 \quad (2.90)$$

A_c is referred to as coherence area of the light source. Equation (2.90) shows that the coherence area of the light source is inverse to the square of the divergence angle θ. Thus, the smaller the divergence angle, the larger the coherence area, and the better the spatial coherence of the source.

A natural source has a divergence angle of 4π; therefore, the spatial coherence is considerably poorer than that of lasers. Thus, having a small divergence angle or high spatial coherence is a characteristic of lasers.

2.4.3 Higher-order coherence

For natural sources, it is possible to improve spatial coherence by reducing size or adding a stop. We can also increase coherence time by using an optical filter to reduce Δv. However, the intensity will severely reduce; therefore, it is not desirable in practice. Therefore, it can be said that theoretically, lasers not only have temporal coherence but also spatial coherence. However, it should be noted that the temporal coherence and spatial coherence of light waves belong to the first-order coherence. It can be shown that only laser beams have the higher-order coherence. So, we can say that higher-order coherence is an essential characteristic of laser beams.

2.4.4 High brightness

The brightness of a monochromatic source with a cross-sectional area A can be expressed as

$$B_v = \frac{P}{A\Delta v\pi\theta^2} \quad (2.91)$$

where P is the power of the output beam within the solid angle $\pi\theta^2$ and frequency interval of $v{\sim}(v+\Delta v)$.

From the discussion in the previous section, it is clear that the frequency interval Δv and the divergence θ of laser irradiation is much smaller than that of a natural source. In other words, the optical power emitted by lasers in a very small frequency interval to a very small solid angle is much higher than the normal. Therefore, the monochrome brightness of the laser is significantly higher than ordinary light. Therefore, we can say that high brightness is a fundamental characteristic of laser.

2.5 Introduction to the operating characteristics of laser

Wide-reaching applications of laser are due to its unique physical properties as well as its work characteristics. This section briefly describes briefly the ultrashort pulse and frequency stability characteristics of lasers.

2.5.1 Ultrashort pulse characteristics

In the laser pulsed manner, the interval between two half-peak energy points is called pulse width and expressed as τ_p (shown in Fig. 2.10), whereas the number of pulses per second is referred to as pulse repetition rate N.

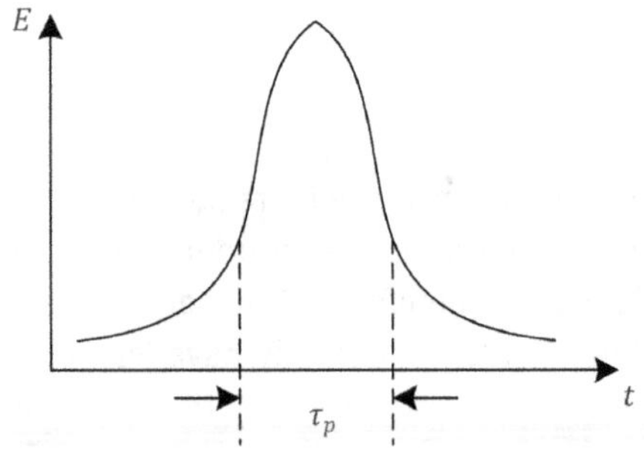

Fig. 2.10: Schematic diagram of pulse width.

With respect to applications of pulse laser, its output power level is of particular. Assuming that the number of laser pulses output per unit time is N, the energy per pulse is E, then define

$$\bar{P} = NE \tag{2.92}$$

as the average power of pulsed laser and

$$P_p = \frac{E}{\tau_p} \tag{2.93}$$

is referred to as peak power.

It can be observed from equation (2.93) that to increase the peak power of the laser pulse, the pulse energy should have raised or compressed pulse width. Because improved laser pulse energy is limited by many factors and because the short pulse is significant in many laser applications, the compressed pulse width of laser has been emphasized more. In fact, just over a year after invention of laser, the technique of Q-switching based on the loss control of cavity was invented wherein laser pulse width was compressed by three orders of magnitude or more to reach the $ns(10^{-9}\,s)$ level. A few years later, the mode-locking technology was successfully introduced and continues to improve, with the current mode-locking technology generating a pulse of $fs(10^{-15}\,s)$. Working toward narrower pulse efforts, the peak power of some solid-state lasers can reach or even surpass the levels of 10^{14} W.

2.5.2 Frequency stability characteristics

As stated in the previous section, the laser has a high monochromatic, i.e., $\Delta v/v_0$ is very small or frequency stability is very good. For example, assume $\Delta v = 10^{16}$ Hz, in the visible spectrum, with frequency stability up to 10^{-8}. However, some applications require a special light source that has higher frequency stability. Hence, frequency stabilization is also an important research area in laser technology and various frequency stabilization techniques have been discovered to meet different requirements. For example, based on the Lamb-hollow technique, frequency stability better than 10^{-9}has been achieved, and based on Zeeman effect, we have been able to achieve a stability of 10^{-10}~10^{-11}; and by the use of saturated absorption $\Delta v/v_0 = 10^{-11}$~10^{-12} can be realized.

2.6 Band structure and electronic states of semiconductors

The semiconductor light source, whether it is an LED based on spontaneous emission or an LD based on stimulated emission, has a very important application. The following few sections of this chapter will discuss the applications, but in the meantime, we first introduce and discuss semiconductors.

2.6.1 Introduction to the band concept

Quantum mechanics clarifies that electron can take only certain discrete energy values in isolated atoms. For some simple atoms, energy levels and electron wave functions can be accurately determined.

Consider a system composed of N identical particles. If these atoms are far apart from each other, such as that the interaction between them can be ignored, then each atom can be regarded as isolated and have exactly the same energy level structure, i.e., each electron energy level is N-fold degenerate. When these atoms are gradually close and form condensed matter, then each level corresponding to isolated atoms (mostly the level corresponding outer electrons) split into N strips. Because the value of N is typically large (e.g., approximately 10^{23} cm^{-3}), the resulting levels are very intensive and form a quasi-continuous energy band called allowed energy bands. The band between the allowed energy bands that arise from different energy levels of atoms is referred to as the forbidden band.

As described above, the N atoms form a solid, and each band comprises N different energy levels. According to Pauli exclusion principle (each electron energy level can hold up to two electrons of opposite spin), the band can hold $2N$ electrons. However, because of "hybrid orbital" (wave function combinations), the reality is not so simple. Take two important semiconductor materials Si and Ge as examples. Each atom has four valence electrons, and in atomic state, there are two s states and two p states. In the crystal state, atoms may wish to produce two bands, one corresponding to s state, that contains N states, and another corresponding to triplet p state containing $3N$ states. In fact, each of the two bands contain $2N$ states due to the hybrid orbital. The lower band holds $4N$ valence electrons, called the valence band, and the upper one is empty and is called the conduction band. At higher temperatures, some electrons are stimulated from the valence band to the conduction band and the materials exhibit electrical conductivity.

2.6.2 Electronic state in semiconductor

We can determine electron energy and motion state of an isolated atom in quantum mechanics by solving the Schrödinger equation. However, because solids have more number of atoms, it is simply impossible to solve Schrödinger equation for each electron. So the only way is through some approximation methods. The band theory based on a single electron played an important role in the development of semiconductor physics. This theory solves Schrödinger equation for a single electron given by

$$\nabla^2\Psi(r) + \frac{2m}{\hbar^2}[E - V(r)]\Psi(r) = 0 \tag{2.94}$$

with regard to the interaction of all other electrons with some electrons equivalent to mean field superimposed on the real periodic potential field of atoms, and are indicated by $V(r)$. For simplicity, consider the one-dimensional case, then equation (2.94) becomes

$$\frac{d^2\Psi(x)}{dx^2} + \frac{8\pi^2 m}{h^2}[E - V(x)]\Psi(x) = 0 \tag{2.95}$$

Let the period of potential field be the lattice constant a, that is,

$$V(x) = V(x + na) \tag{2.96}$$

where n is an integer. For the wave function $\Psi(x)$ satisfying equation (2.95), $\Psi(x + a)$ must also be a wave function belonging to the same energy eigenvalue of $\Psi(x)$, and that the two functions can only differ by a phase-factor mode equal to one. Writing $\Psi(x)$ as

$$\Psi(x) = u_k(x)e^{ikx} \tag{2.97}$$

here k is the wave number, then it is easy to prove that $u_k(x)$ is also a function with the cycle of a, that is,

$$u_k(x+a) = u_k(x) \tag{2.98}$$

The right-hand side of equation (2.97) is often referred to as the Bloch function, where e^{ikx} represents a plane wave, and its corresponding energy eigenvalue is

$$E = \frac{\hbar^2 k^2}{2m_e} + V$$

where

$$E = \frac{\hbar^2 k^2}{2m_e} \tag{2.99}$$

is the kinetic energy, and m_e is the electron mass.

For issues discussed in this section, and in a small enough range of k, E_k can be expanded using Maclaurin series, and only the first two terms are retained to obtain

$$E(k) = E(0) + \frac{\hbar^2 k^2}{2m_{eff}} \tag{2.100}$$

where m_{eff} is called the effective mass of an electron. It is different from m_e; m_{eff} is either a positive or a negative value.

Equation (2.100) shows that, in the vicinity of $k = 0$, $E(k)$ with k varies based on the parabola, in which the opening direction is determined by m_{eff} symbols. When $m_{\text{eff}}>0$, the parabola opens upward, and the corresponding band is called the conduction band. The equation (2.100) then becomes

$$E(k) = E_c(0) + \frac{\hbar^2 k^2}{2m_{\text{eff}}^c}, \quad m_{\text{eff}}^c > 0 \tag{2.100a}$$

In the case of $m_{\text{eff}}<0$, the parabola opens downward, and the corresponding band is called the valence band, then the E–k relationship is given by

$$E_v(k) = E_v(0) + \frac{\hbar^2 k^2}{2m_{\text{eff}}^v}, \quad m_{\text{eff}}^v < 0 \tag{2.100b}$$

where the subscripts c and v refer to the conduction and valence bands, respectively.

Comparison of equations (2.100a) and (2.100b) shows that the minimum value of conduction band and the maximum value of valence band occur at the same point in k space. These materials are called direct bandgap semiconductor, whose E –k relationship is shown in Fig. 2.11(a). The difference between the minimum conduction band and the maximum valence band refers to the forbidden band

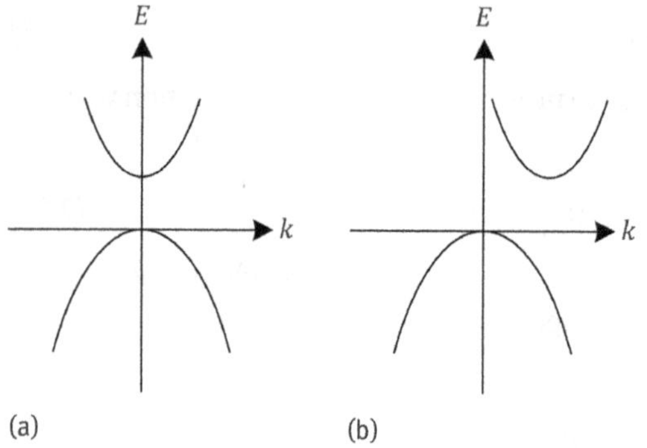

Fig. 2.11: Direct bandgap (a) and indirect bandgap (b) of the semiconductor energy band diagram.

pitch. There is another kind of semiconductor material called an indirect bandgap material whose E–k relation is shown in Fig. 2.11(b). The semiconductor LED usually involves only the direct bandgap material. Thus, unless otherwise noted, this chapter will not discuss indirect bandgap semiconductor material.

It should be noted that the above result is greatly simplified. In fact, because effective mass is not constant and because of the influence of spin-orbit coupling along with other factors, the actual band structure is much more complex. However, problems associated with the semiconductor LED mostly involves only the state at the top and bottom of bands, and it can be shown that the equation (2.100) is established under appropriate conditions.

2.7 Excitation and recombination radiation

2.7.1 Direct transition and the semiconductor light-emitting material

Electronic transitions between different states in semiconductors can result in light absorption or emission. In intrinsic semiconductors, the free carriers and impurity atoms are small in number, hence the corresponding absorption process is extremely weak. The main absorption caused by the transition from the valence band to the conduction band is called the ground state absorption or intrinsic absorption, which will be discussed later, and it commonly represents the intrinsic absorption coefficient α_0. Obviously, the photon energy that can cause intrinsic absorption must be greater than a threshold, which is substantially equal to the bandgap E_g. Absorption spectrum in the vicinity of intrinsic absorption threshold is called the absorption edge.

The process corresponding to absorption is emission. During emission, electrons can decay from the conduction band to the valence band by recombining within holes, which results in emission of photons, and is called recombination radiation.

The electron transition process must satisfy the conservation of momentum; hence, in light radiation process, it is

$$\hbar \boldsymbol{k}' - \hbar \boldsymbol{k} = \hbar \boldsymbol{k}_{pt}$$

where $\boldsymbol{k}$ and $\boldsymbol{k}'$ are the wave vectors of electrons in the initial and final state, respectively, and $\boldsymbol{k}_{pt}$ is the photon wave vector. In general, $\boldsymbol{k}_{pt}$ is approximately four orders of magnitudes smaller than $\boldsymbol{k}$, and can be considered as a pure optical transition to satisfy the selection rule

$$\hbar \boldsymbol{k}' = \hbar \boldsymbol{k}$$

Electronic transition occurs at the same point in k space, and is called vertical jump or direct transition (Fig. 2.12a). In direct transition, the radiation photons meet

$$E_g = h\nu \text{ or } \lambda = hc/E_g$$

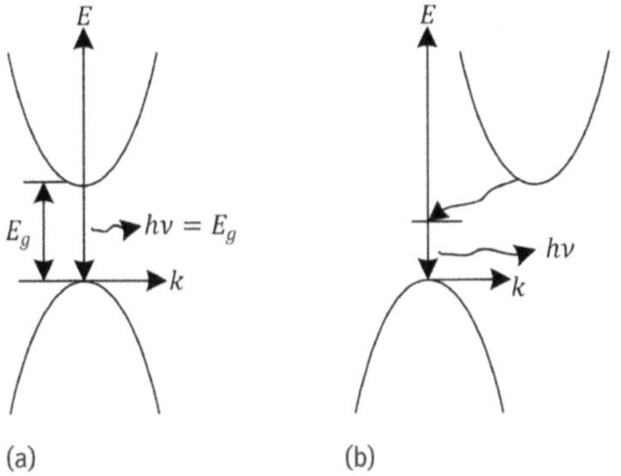

Fig. 2.12: Examples of direct (a) and indirect (b) transition between the energy bands of semiconductor materials.

If the transition occurs between the two states with the wave vector of the larger difference, it indicates that there are phonons participating in the process (Fig. 2.12b). The conservation of momentum can then be expressed as

$$\hbar \boldsymbol{k}' - \hbar \boldsymbol{k} \approx \pm \hbar \boldsymbol{k}_{pn}$$

where $\boldsymbol{k}_{pn}$ is the phonon wave vector, which is usually an order of magnitude smaller than $\boldsymbol{k}$. It is called a nonvertical electronic transition or indirect transition wherein the initial and final states correspond to different points in $\boldsymbol{k}$ space.

2.7.2 Density of states and electronic excitation

The probability that an electronic state at energy E is occupied by an electron is given by the Fermi–Dirac law

$$\rho_e(E) = \frac{1}{1 + e^{\frac{E - E_F}{KT}}} \tag{2.101a}$$

Thus, the probability that it is occupied by a hole is given by

$$\rho_h(E) = 1 - \rho_e(E) = \frac{1}{1 + e^{\frac{E_F - E}{KT}}} \tag{2.101b}$$

where E_F is the Fermi energy defined as the energy at which there is a probability of 0.5, such that a specific level is occupied. Above that level, the probability decreases, and below that level, it rapidly approaches the maximum. In Fig. 2.13, the Fermi energy for an intrinsic semiconductor at absolute zero occurs midway between the valence and conduction bands.

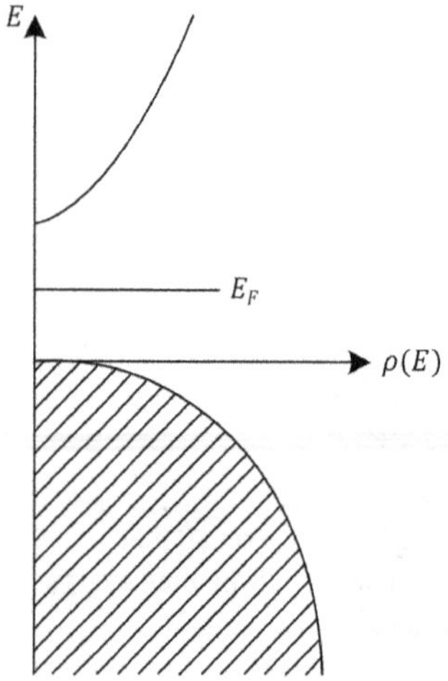

Fig. 2.13: Distribution function of electrons in an intrinsic semiconductor at absolute zero temperature.

The probability of the conduction band energy level being occupied by electrons can be expressed as

$$\rho_{ce}(E) = \frac{1}{1 + e^{\frac{E_C - E_F}{KT}}}$$

and when it is occupied by holes is expressed as

$$\rho_{ch}(E) = \frac{1}{1 + e^{-\frac{E_C - E_F}{KT}}}$$

The probability of the valence band being occupied by electrons can be expressed as

$$\rho_{Ve}(E) = \frac{1}{1 + e^{-\frac{E_F - E_V}{KT}}}$$

and when it is occupied by holes is given by

$$\rho_{Vh}(E) = \frac{1}{1 + e^{\frac{E_F - E_V}{KT}}}$$

At $T = 0$ K, we obtain from (2.101a)

$$\rho_{ce}(E) = 0; \quad \rho_{ch}(E) = 1$$

From (2.101b)

$$\rho_{Ve}(E) = 1; \quad \rho_{Vh}(E) = 0$$

Therefore, at a temperature of 0 K, the intrinsic semiconductor electrons are located in the valence band, and in the conduction band, there are almost no electrons. The electron distribution function is shown in Fig. 2.13.

When T>0 K, as k is of the order of 10^{-4} eV K, even the temperature reaches 300 K, KT is only of the order of 10^{-2} eV, which is generally assumed that

$$E_C - E_F \gg KT, \quad E_F - E_V \gg KT$$

Thus, equation (2.101a) becomes

$$\rho_{ch}(E) = 1$$

$$\rho_{ce}(E) = e^{-\frac{E_C - E_F}{kT}} \ll 1$$

This implies that only a small amount of electrons in the conduction band obey Boltzmann distribution law, and are concentrated relatively close to the bottom of the conduction band of E_F. On the contrary, equation (2.101b) gives

$$\rho_{Ve}(E) = 1$$

$$\rho_{Vh}(E) = e^{-\frac{E_F - E_V}{kT}} \ll 1$$

This implies that the valence band is substantially filled with electrons and there are only small holes located in the valence band. According to the Boltzmann distribution law, electrons are concentrated relatively near the top of the valence band E_F. When the material is excited, a portion of electrons transit from the valence to the conduction band (Fig. 2.14), which leads to electron–hole pairs as a hole is created in the valence band for every electron that transits to the conduction band. Electrons decaying back to the valence band may lead to recombination radiation by recombining with holes.

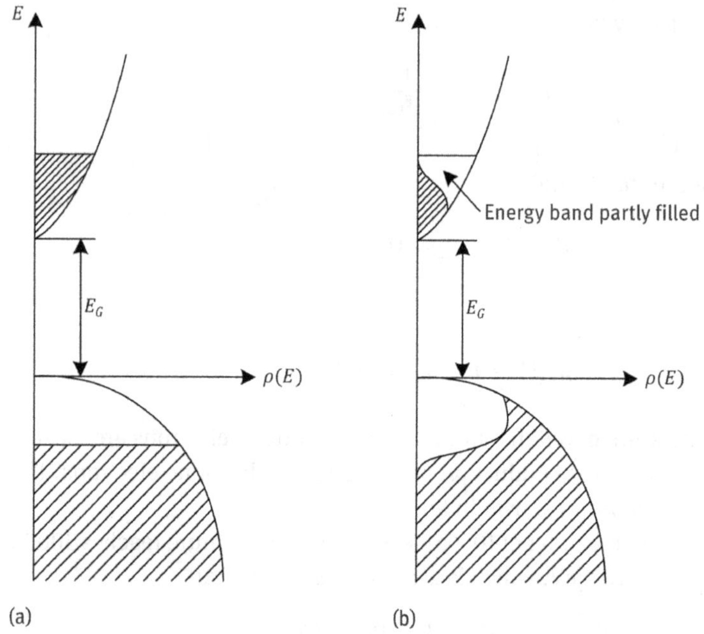

Fig. 2.14: Semiconductor excitation at (a) $T = 0$ K and (b) $T > 0$ K.

2.7.3 p–n Junction in extrinsic semiconductor materials

Electric current in semiconductor light-emitting materials is essential to provide excitation energy to excitation levels. However, as mentioned above, in pure semiconductors (often referred to as intrinsic semiconductors), the conduction band is empty while the valence band is filled with electrons, which do not have electrical conductivity. Doping small amounts of impurity atoms (i.e., formation of extrinsic semiconductors) provides additional free electrons (n-type) and holes (p-type) that increase the conductivity and enable current formation. Such doping produces additional energy levels within the energy band structure, usually within the bandgap and close to the band edges. For n-type materials, these levels are near the conduction band, and the corresponding Fermi level shifts upward (or into) to the conduction band. For p-type materials, these levels are near the valence band, and the corresponding Fermi level shifts downward to the valence band (Fig. 2.15a). Thus, the net effect of doping is the formation of additional free carriers in the conduction and valence band. In n-type materials, electrons are **the majority carriers**, or simply multi carriers, and the holes are minority carriers; in p-type material, the holes are majority carriers and electrons are minority carriers. By establishing contact between n-type material and p-type material, a p–n junction is formed. If the two materials are of the same matrix doped with donor or acceptor impurities, respectively, the p–n junction called a homojunction is formed; if the two materials are different, the resulting junction is called a heterojunction.

When the two materials are in contact, excess electrons flow into the p-type material resulting in an increase in space charge. The consequent voltage V_0 that results from this space charge stops the flow of electrons, as shown in Fig. 2.15(b). However, by applying an electric field (voltage V_f) across the junction, with the positive voltage attached to the p-type material, the energy barrier is reduced significantly, as shown in Fig. 2.15(c), and the current is maintained.

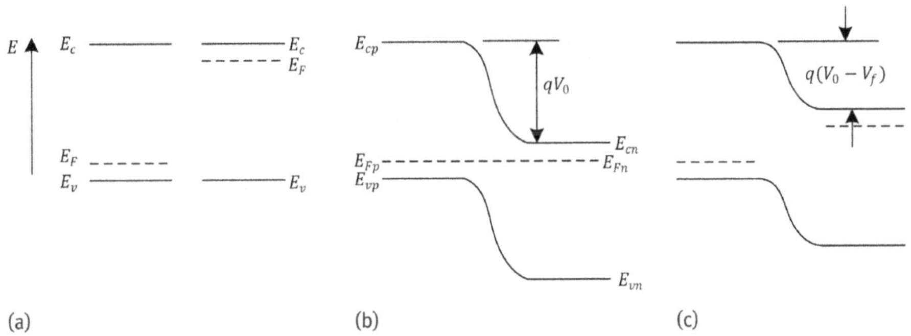

Fig. 2.15: Relevant energy levels for valence band, conduction band, and Fermi level of a p–n junction for (a) separated materials, (b) p and n materials in contact, and (c) p and n materials in contact with an applied forward bias.

Large quantities of electrons flow through the junction, and recombination radiation occurs in proportion to the current flow. This radiation leads to the generation of light-emitting semiconductors. Thus, the region becomes an active region, and the thickness has the same magnitude with the diffusion length of the carriers; for early homojunction semiconductor light-emitting materials, this is generally in the range of 2.4 μm.

2.8 Working mechanism of LEDs

LED is a light-emitting optoelectronic device generated as a result of electro-optical conversion. The most common type of LED is the p–n junction LED. Materials for manufacturing semiconductor LEDs are usually heavily doped. Under thermal equilibrium, in the n-region mobility of electrons is high, whereas the p-region has lower mobility of holes. On the role of the p–n junction as barrier layer, these excess carriers cannot lead to normally radiative recombination. However, by applying an electric field across the junction, with the positive voltage attached to the p-type material, energy barrier is reduced significantly; the n-region with excess electrons are injected into the p-region where they recombine with holes in the junction slightly close to the p-region. The excess energy is released in the form of luminescence. For direct bandgap material, recombination radiation energy is the material bandgap E_g.

The light emitted by an LED does not have a single wavelength, and typically has a spectral half-width of several tens of nanometer. However, a relative emission intensity peak exists whose corresponding wavelength λ_p is called the peak wavelength and is given by

$$\lambda_p = 1239/E_g \tag{2.102}$$

where λ_p is in units of nm and E_g is in units of eV.

Equation (2.102) shows that LED light emission peak wavelength is determined by the bandgap of the material used, or conversely, for manufacturing an LED, suitable bandgap of the material is determined by the desired wavelength. For example, if a red light wavelength of 0.7–0.6 μm is required, the bandgap E_g of the diode material should be in the range of 1.77–2.03 eV. It should be noted that the discussion is for pure semiconductor materials. For impure materials, λ_p will be longer than that obtained from equation (2.102).

At present, the material used for the visible light range of the LED is III–IV valence compounds, such as gallium phosphide (GaP), gallium arsenide (GaAs), and a mixture of gallium, arsenide, as phosphide ($GaAs_{1-x}P_x$), where x refers to the fraction of phosphide replacing arsenide in the mixture. The bandgap material of GaP at room temperature is 2.26 eV, theoretical peak wavelength is approximately 550 nm, for the reasons described earlier, the actual emission peak wavelength is approximately 700 nm, and spectral half-width is approximately 100 nm. The peak wavelengths of $GaAs_{1-x}P_x$ vary with the value of x in the 620–680 nm region and half-width of approximately 20–30 nm.

In general, in addition to useful radiative recombination, there are also various complex methods. LED is concerned with that of fraction of radiative recombination in all recombinations, which is commonly described in the quantum efficiency, defined as the ratio of the number of photons generated per unit time to the number of electrons flowing through the diode.

2.9 Semiconductor diode laser

2.9.1 Semiconductor optical gain

As mentioned earlier, the conduction and valence band of the semiconductor material correspond to the upper and lower levels, respectively, in an ordinary laser. Thus, when a light with suitable wavelength interacts with materials, the stimulated absorption and emission processes occur simultaneously. In the stimulated absorption process, the electron transits from the valence to conduction band by the absorption of a photon and leaving a hole in the valence band. In the stimulated emission process, electrons transits from the conduction to valence band and recombine with the hole, thus radiating a photon identical to incident photons. Obviously, the gain G_1 of the stimulated emission process to the incident light is positive and proportional to the probability of the conduction band being occupied by electrons and valence band being occupied by holes, that is,

$$G_1 = a_0(hv)P_{Ce}P_{Vh}$$

On the contrary, the gain G_2 of the stimulated absorption process to the incident light is negative (attenuation) and proportional to the probability of the conduction band being occupied by holes and valence band being occupied by electrons, that is,

$$G_2 = -a_0(hv)P_{Ch}P_{Ve}$$

where a_0 is the gain with all the conduction band levels occupied by electrons (P_{Ce} =1), while the valence band levels are occupied by holes (P_{Vh} =1). a_0 can also be referred to as the intrinsic absorption with all the conduction band levels occupied by holes (P_{Ch} =1), while the valence band levels are occupied by electrons (P_{Ve} =1).

Notice that

$$P_{Ce} + P_{Ch} = 1, \quad P_{Ve} + P_{Vh} = 1$$

It can be observed that the total effect of stimulated emission and stimulated absorption to light can be expressed as

$$G = \alpha_0(hv)(P_{Ce} - P_{Ve})$$

and the condition for the net gain is

$$P_{Ce} - P_{Ve} > 0 \tag{2.103}$$

Thus, the probability of conduction band level being occupied by electrons is greater than that of valence band level being occupied by electrons. We know, however, that under electronic equilibrium, the electrons in an intrinsic semiconductor are almost entirely occupied in the valence band, whereas in the conduction band is basically completely empty. Thus, the distribution satisfying equation (2.103) belongs to the reverse distribution, corresponds to the reverse distribution as

$$\left(N_u - \frac{g_u}{g_l}N_l\right)$$

for a two-level atomic system.

Inverted distribution state is a nonequilibrium state. In this state, the conduction and valence band have respective Fermi level referred to as the conduction band quasi-Fermi levels and the valence band quasi-Fermi levels, and are expressed as E_F^C and E_F^V.

It is worth noting that, in the stimulated absorption process, the valence band electrons with the energy of $E_V = E - hv$ can only absorb 1 hv energy photon and transit to an empty level in the conduction band with the energy of $E_c = E$. In the stimulated emission process, otherwise, the conduction band electrons of $E_c = E$ can only transit to the valence band with the energy of $E_V = E - hv$ and has not been occupied by electrons, while radiating a photon of hv energy.

Thus, equation (2.103) can be written as

$$P_{Ce}(E) > P_{Ve}(E - hv) \tag{2.103a}$$

Equation (2.101a) and (2.101b) is substituted to obtain

$$\frac{1}{1 + e^{\frac{E_C - E_F^C}{kT}}} > \frac{1}{1 + e^{\frac{E_V - hv - E_F^V}{kT}}}$$

The population inversion condition can then be expressed as

$$E_F^C - E_F^V > hv \approx E_g \tag{2.104}$$

Visible in the nonequilibrium state, the quasi-Fermi level is no longer in the forbidden band, but is visible separately in the conduction band and the valence band. In addition, to make the p–n junction to achieve population inversion distribution, the following two conditions must be satisfied:

(1) Doping concentration is high enough, so that quasi-Fermi level, respectively, get into the conduction and valence bands
(2) The forward bias voltage V is high enough, so that $eV > E_g$, thereby

$$E_F^C - E_F^V = eV > hv$$

2.9.2 Loss and oscillation threshold condition

Similar to other lasers, semiconductor diode laser also includes an optical resonator and active medium. When light propagates into the cavity, in addition to the gain described earlier, it will also experience various losses. Only when the gain is greater than the loss to be overcomed, the light can be enlarged or sustains oscillations. Losses α_i (cm^{-1}) including diffraction, nonintrinsic absorption are caused by free carriers as well as by facet reflectivity R_1 and R_2 and the length L of the cavity. The light with initial intensity I_0 after a round-trip in the cavity becomes

$$I = I_0 R_1 R_2 e^{(G - \alpha_i)2L}$$

where the threshold gain is obtained as

$$G_{\text{th}} = \alpha_i + \frac{1}{2L} \ln \frac{1}{R_1 R_2} \tag{2.105}$$

where α_i is caused mainly by the free carrier absorption and its size is proportional to the carrier concentration n. At room temperature the empirical formula for GaAs material is

$$\alpha_i \approx 0.5 \times 10^{-17} n \,[\text{cm}^{-1}]$$

Here n is the background carrier concentration n_0 and the injection carrier density at threshold condition n_{th} has a dimension of cm^{-3}. For typical values of $n \sim 2\times10^{18}\ \text{cm}^{-3}$, $\alpha_i \sim 10\ \text{cm}^{-1}$.

In practice, the light field cannot be completely bound in the active region. However, only the light propagating in the active region can obtain a gain. The light propagating from the active region cannot obtain a gain experiences propagation loss caused by absorption and reflection, which must be compensated by corresponding gain in the active region under oscillating conditions. If the introduction of the optical confinement factor Γ represents the ratio of the energy in active region to the total energy both in the active region and the passive region, and set the absorption coefficient and facet reflectivity in passive zones as α'_i and R'_1, R'_2, respectively, equation (2.105) will be replaced by a more universal relationship

$$\Gamma G_{th} = \Gamma\left(\alpha_i + \frac{1}{2L} ln \frac{1}{R_1 R_2}\right) + (1-\Gamma)\left(\alpha'_i + \frac{1}{2L} ln \frac{1}{R'_1 R'_2}\right) \tag{2.106}$$

It also can be expressed as

$$G_{th} = \delta + \delta' \tag{2.107}$$

where $\delta = \left(\alpha_i + \frac{1}{2L}\ln\frac{1}{R_1R_2}\right)$ and $\delta' = \frac{1-\Gamma}{\Gamma}\left(\alpha'_i + \frac{1}{2L}\ln\frac{1}{R_1R_2}\right)$ is the total loss factor inside and outside the active region, respectively. When the light confinement effect is very strong, as a result $\Gamma \approx 1$, $\delta' \approx 0$, then the formula (2.107) becomes (2.106), with the optical confinement reducing, Γ reducing, the role of δ' becomes increasingly evident. In general, $R'_1 = R_1, R'_2 = R_2$ can be set, and further assuming $\alpha'_i = \alpha_i$, then equation (2.106) becomes

$$G_{th} \approx \frac{1}{\Gamma}\left(\alpha_i + \frac{1}{2L} ln \frac{1}{R_1 R_2}\right) \tag{2.108}$$

For GaAs, $R_1 = R_2 = 0.319$, let $L = 500\ \mu\text{m}$, $\alpha_i = 10\ \text{cm}^{-1}$, then when $\Gamma \approx 1$, we have $G_{th} \approx 32.8\ \text{cm}^{-1}$, and when $\Gamma = 0.5$ we obtain $\delta' = \delta$, $G_{\text{th}} = 2\delta = 65.6\ \text{cm}^{-1}$.

In the semiconductor laser, the small-signal gain is generally between 5000 and 10,000 m^{-1}, therefore, a single pass is sufficient to overcome the inherent distribution losses in gain medium.

The threshold current density oscillation condition can also be expressed as

$$j_{th}\left(\frac{G_{th}}{\beta}+j_0\right)\frac{d}{\eta}$$

where d is the active region thickness at the current direction, η is the ratio of the radiative recombination rate to the total recombination rate, called the internal quantum efficiency. These two parameters are temperature-dependent. Table 2.3 gives their values at different temperatures in the intrinsic GaAs.

Table 2.3: Values of β and j_0 at different temperatures in intrinsic GaAs.

T/K	80	160	250	300	350	400
β/cm A^{-1}	0.160	0.080	0.057	0.044	0.039	0.036
j_0/A $cm^{-2}\cdot\mu m^{-1}$	600	1600	3200	4100	5200	6200

2.10 Heterojunction semiconductor lasers

The p–n junction described above is incorporated into donor impurity and an acceptor impurity in different regions of the same substrate to form n-type region and the p-type region. Because two regions have the same kind of substrate, it is called a homojunction, the lasers based on which can operate only at very low temperatures. To make room-temperature lasers, it was necessary to develop a hetero-junction. This section will discuss some characteristics of the hetero-junction and then introduce the hetero-junction lasers.

2.10.1 Hetero-junction semiconductor

As the name suggests, the hetero-junction comprises different materials. The two materials that formed junctions have the similar structural along the interface, however, different bandgap and electron affinity energy.

There is a good match between the GaAs and $Al_xGa_{1-x}As$. Figure 2.16 shows a diagram of a simple hetero-junction, consisting of an upper p-type layer of $Al_xGa_{1-x}As$ followed by a p-type layer of GaAs and then a substrate of n-type GaAs. The p-type GaAs layer has an active region that is only 0.1–0.2 μm thick (or smaller in quantum-well devices). This is the only region where the excitation and recombination radiation can occur and the current can flow. Typically, the threshold current densities of single hetero-junction laser is of the order of $1\times10^4\cdot$A cm^{-2}.

Double hetero-junction structure provides even more control over the size of the active region. Figure 2.17 shows a double hetero-junction composed of various

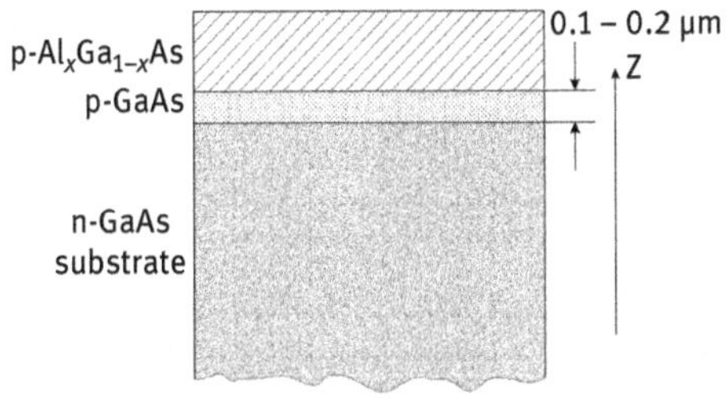

Fig. 2.16: A simple hetero-junction semiconductor material.

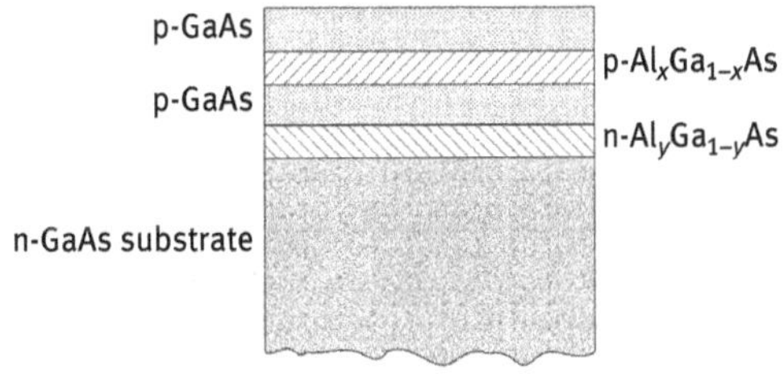

Fig. 2.17: A double hetero-junction semiconductor material.

doping combinations of GaAs and $Al_xGa_{1-x}As$. In addition to the control of layer thickness, the side walls of the active region are narrowed, confining the laser to a narrower region, thus reducing the threshold current densities further to the order of $1{\times}10^3{\cdot}A{\cdot}cm^{-2}$.

If the thickness of the active layer of the hetero-junction is reduced even further to dimensions of 5–10 nm, then the energy levels exhibit quantum behavior, which is referred to as a quantum-well structure. Owing to the quantum effect that occurs only in the direction perpendicular to the plane of the junction, it can be obtained by considering the electron as a particle in a one-dimensional potential well. The solution of quantum theory for such a system suggests a series of discrete energy levels described by

$$E_n = \frac{(\pi\hbar)^2 n^2}{2m_c d^2}$$

where d is the layer thickness and n (=1, 2, . . .) is a quantum number.

Under the condition that both carriers and optical field have been strongly constrained, quantum-well lasers can achieve very high gain and as low as 0.5 mA cm^{-2} of the threshold current density.

2.10.2 Laser structure

To avoid thermal damage and loss of energy inside the semiconductor laser, we should try to limit the electron–hole recombination zone. There are two types of structures used. In one case, the region over which gain is produced is limited and this is referred to as gain guiding. Alternatively, an index-of-refraction change is fabricated into the laser such that the beam is confined by reflection at that interface; this is referred to as index guiding.

In a gain waveguide structures such as shown in Fig. 2.18, confinement of the current to a narrow strip will limit the amount of current flowing in the laser and thus prevent thermal damage to the semiconductor laser. A disadvantage of which is that effective width of the gain region broadens as the current applied is increased.

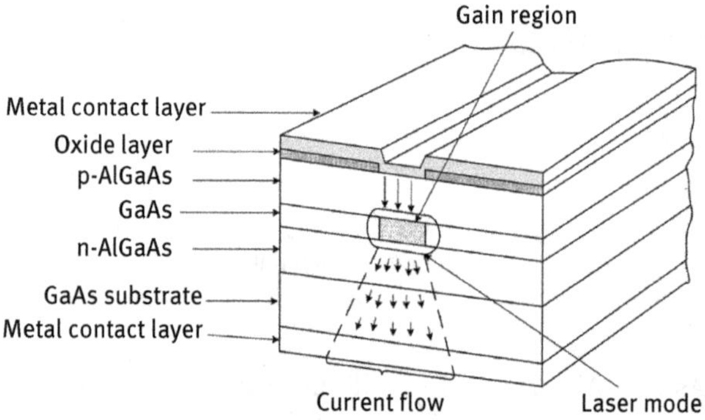

Fig. 2.18: Gain-guided laser structure.

Limiting the laser mode can also be achieved by fabricating stripes of material with low index adjacent to the gain region of high index material as shown in the ridge waveguide structure of Fig. 2.19. The laser mode extending above the active region is confined by this index barrier

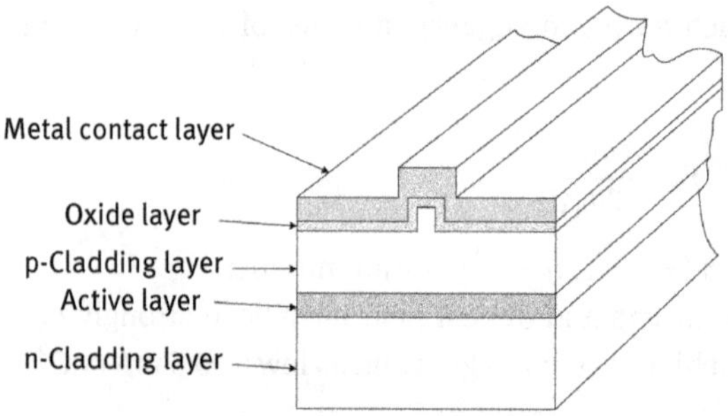

Fig. 2.19: Index-guided laser structure.

In addition to that shown in Fig. 2.19, the specific form of the index-guided structure also include the buried hetero-junction structure, the channeled-substrate planar structure, the buried-crescent laser, and the dual-channel planar-buried hetero-junction laser.

3 Bulk solid-state lasers

3.1 Overview

Solid-state lasers use solid-state laser medium as the working substance. Usually, solid work materials are incorporated in the matrix with a small amount of activated ions. Laser transition occurs between different energy levels during ion activation. Matrix materials are mainly crystal, glass, and ceramics, and activated ions can be categorized as trivalent rare earth metal ions, divalent rare earth metal ions, transition metal ions, and actinide metal ions.

Solid-state lasers are mostly optically pumped, and the light sources for pumping mainly include flash and semiconductor LD or other lasers. The main problem of the flash pump is its low conversion efficiency and the fact that a large amount of pump energy is wasted, which not only causes considerable energy wastage but the thermal stress and thermal lens effect severely hampers beam quality. Thus, it was not long after the introduction of the first Gallium–Arsenide (GaAs) LD in 1962 that the output line of a LD was narrowed (usually a few nanometers), thus enabling a spectral comparison with the solid-state laser working substance. LD has several advantages, such as highly efficient electro-optical conversion, long life, small volume, and light weight, and is a promising pump source for solid-state lasers. However, because the early LD could not work at room temperature and the output power and reliability did not the optimum requirements, they could not be used effectively. With the rapid development of LD manufacturing technology, at room temperature, line-array LD with an output power of tens of watts have emerged. LD-pumped solid-state laser technology has also developed rapidly. In the last ten years, when pump source spectrum and laser medium spectrum are better matched, the output power of LD-pumped solid-state lasers has exceeded 1012 W, and in some cases even 1015 W. LD-pumped solid-state lasers have become the mainstay in the development of solid-state lasers, which will be described briefly in Section 3 of this chapter.

Although LD-pumped solid-state lasers have much higher conversion efficiency than flash-pumped, considerable waste heat generation occurs as pump power increases. The traditional laser rod with a small surface area to volume ratio is not sufficient for heat dissipation. To increase this ratio of the working substance, research has been conducted in two opposing directions. One is to use a thin-sheet laser material with a large radius to length ratio, while the other is to use a fiber material with a large length to radius ratio. Section 3.3 of this chapter describes sheet lasers. Another promising laser type, the slab-like lasers, is presented in Section 3.4, and fiber lasers are covered in Chapter 4.

Using work material of different shapes significantly improves the laser heat dissipation performance; however, when the output power is increased

https://doi.org/10.1515/9783110500608-003

further, the heat problem still continues to affect the beam quality. As an efficacious thermal management technology of solid-state laser, the heat capacity laser technology developed at the end of the 20th century was an important accomplishment. Section 3.6 discusses heat-capacity lasers with a thin-sheet as the working substance. The thermal stress behavior of solids is introduced in Section 3.5.

3.2 LD-pumped solid-state lasers

3.2.1 Comparison with the flash lamp pump

For flash-pumped solid-state lasers, the poor match between the flash light emission spectrum and the absorption spectrum of the laser species results in the low pumping efficiency of the device. For the xenon-lamp-pumped Nd^{3+}:YAG laser, the emission spectrum of the xenon lamp is shown in Fig. 3.1 (top part). The bottom part of Fig. 3.1 shows the absorption spectrum of Nd^{3+}. It is clear from the figure that xenon lamp emission spectrum is a continuous spectrum, whereas the absorption spectrum of Nd^{3+} comprises few discrete lines with a strong peak. Thus, when a Nd^{3+}:YAG laser is pumped with a xenon lamp, only a small part of the radiant energy is absorbed by the lower laser energy level of Nd^{3+} resulting in transit to the upper laser energy level, emitting a laser when during the transit back to the lower laser energy level. However, in this process, a majority of energy is absorbed by other energy levels leading to the generation of waste heat.

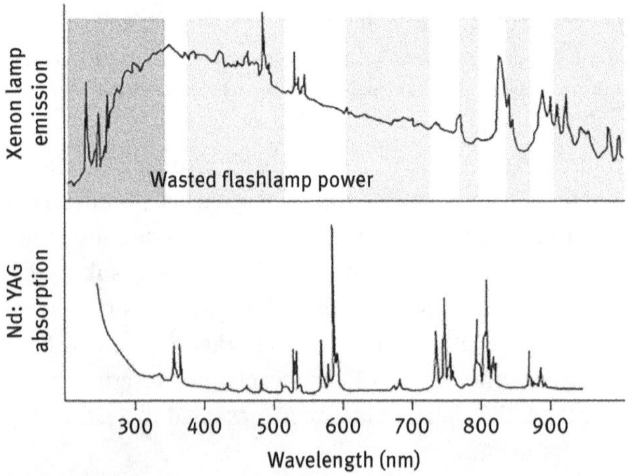

Fig. 3.1: Emission spectrum of xenon lamp (top) and the absorption spectrum of Nd:YAG (bottom).

In contrast, the radiation-peak wavelength of LD can be locked in a very small range near the absorption peak of the working medium, which significantly reduces the generation of unwanted waste heat. In addition, in comparison to flash pump, the LD is smaller, more robust, reliable, and durable; and therefore, most modern solid-state lasers use LD pumps and are referred to as diode-pumped solid-state lasers (DPSSL) The unwanted waste heat in DPSSL mainly arises from the "quantum defects" in the working medium. A quasi three-level system can reduce the unwanted waste heat generated as a result of quantum defects. At present, the most suitable laser material is Yb^{3+}:YAG. Compared with Nd^{3+}-doped materials, laser upper level of Yb^{3+}-doped materials have a longer lifetime (1 to 2 ms), smaller quantum defects, and similar emission wavelengths. In addition, the absorption band of Yb^{3+} ion is in the 915~980 nm wavelength range; with the use of an LD as a pump source, the total electro-optical conversion efficiency can be up to 70%.

Yb^{3+}:YAG is one of the most important solid-state laser materials. Recently, ytterbium-doped alkaline earth fluorides, such as Yb^{3+}:CaF^2, Yb^{3+}:SrF^2, and Yb^{3+}:BaF^2, have shown good performance in the field of LD-pumped solid-state lasers and amplifiers. These crystalline materials have been shown to compete with oxide crystals and glasses in terms of thermal conductivity. Because large, single crystals and ceramic materials can be prepared, they are suitable for high-energy and high-power lasers. In addition, Yb^{3+}:CaF^2 can be applied for both 940 nm and 980 nm wavelength pump, and does not require complex wavelength stabilization for pump diode.

As the absorption band of Yb^{3+}:YAG is at 940 nm and the laser transition is at 1030 nm, its energy level structure is considerably simpler than that of Nd^{3+}:YAG (Fig. 3.2). At room temperature, Boltzmann population number in laser lower level is only 5%. At the same input power, unwanted waste heat generation in Yb^{3+}:YAG is only one-fourth of that in Nd^{3+}:YAG.

3.2.2 Threshold power and above threshold operation

In this section, we introduce the threshold power of the diode laser-pumped solid-state laser and the operation above the threshold in four-level systems.

1. Threshold absorption power

It can be seen from the rate equation that the change in the population number in upper energy level per unit time is

$$\frac{dN_u(r,z,t)}{dt} = Rf_uI_p(r,z) - \frac{N_u(r,z,t)}{\tau} \tag{3.1}$$

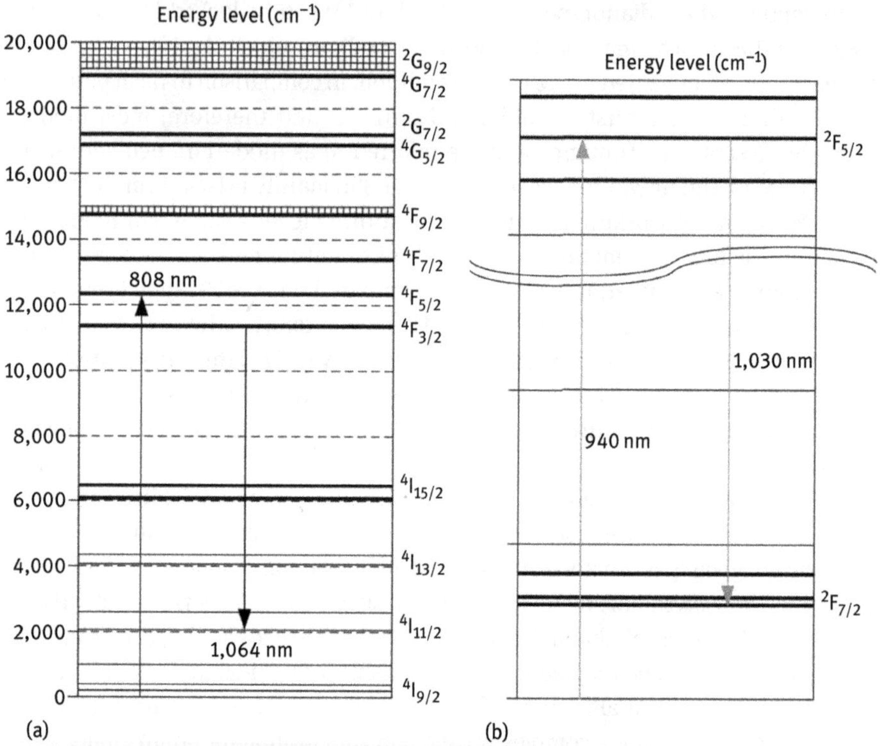

Fig. 3.2: Comparison of energy levels of Nd^{3+}:YAG (a) and Yb^{3+}:YAG (b).

where f_u is the ratio of the population number in the upper energy level to the total population number; τ is the lifetime of the upper energy level; and $I_p(r,z)$ is the pump intensity distribution in the gain medium, which satisfies the normalization condition

$$\iiint I_p(r,z)dV = 1 \tag{3.2}$$

R is the total pumping rate and is expressed as

$$R = \eta_p P_a / h\nu_p \tag{3.3}$$

where P_a is the absorbed pump power, $h\nu_p$ is the pump photon energy, and η_p is the pumping quantum efficiency.

With $N_l(r,z)$ expressing the population number of the lower laser level, the inversed population is expressed as $\Delta N(r,z) = N_u(r,z) - N_l(r,z)$, and the round-trip gain can be expressed by $2\sigma L\Delta N(r,z)$, where σ is the stimulated emission cross-section and L is the gain medium length. Note the steady state

$$\frac{dN_u(r,z,t)}{dt}=0$$

Equation (3.1) leads to

$$N_u(r,z)=\eta_p P_a f_u I_p \tau / h\nu_p \tag{3.4}$$

Considering the spatial distribution of pump and cavity modes, the total gain G can be expressed using the integral in the entire space for the inversed population as

$$G=2\sigma L\iiint I_0(r,z)[N_u(r,z)-N_l(r,z)]dV$$

Substituting equation (3.4), we get the threshold condition

$$2\sigma L\iiint I_0(r,z)\left[\eta_p P_{a,th} f_u I_p \tau / h\nu_p - N_l\right]dV=\delta \tag{3.5}$$

Here $P_{a,th}$ is P_a at the threshold; δ is the round-trip total loss, which arises owing to doping absorption and internal scattering, which is proportional to the length of the medium expressed as $2\alpha_i L$; output coupling T, and the external loss resulting from interface scattering and Fresnel reflection. The latter two are used instead of δ_e as N_l does not vary with the position at the threshold condition; therefore, for equation (3.5) to be

$$P_{a,th}\iiint I_0(r,z)I_p(r,z)dV=\frac{h\nu_p}{\eta_p\sigma_e\tau}\left(\frac{\delta_e}{2L}+\alpha_i L+\alpha_l\right) \tag{3.6}$$

Among them

$$\alpha_l=\sigma N_l$$

indicates the absorption coefficient of the lower-level populations, and

$$\sigma_e=\sigma f_u$$

is the effective stimulated emission cross-section.

Further calculations need to know the function form of intensity $I_0(r,z)$ and $I_p(r,z)$. At the end of pumping, the laser is typically operated in a mode TEM_{00}, so

$$I_0(r,z)=\frac{2}{\pi\omega_0^2 L}exp\left(\frac{-2r^2}{w_0^2}\right) \tag{3.7}$$

and satisfies the normalization condition

$$\iiint I_0(r,z)dV=1$$

In addition, as threshold does not strongly depend on the distribution of pump power, it can be assumed to follow a Gaussian model:

$$I_p(r,z) = \frac{2\alpha}{\pi\omega_p^2[1-exp(-\alpha L)]} exp(-\alpha z) exp\left(\frac{-2r^2}{\omega_p^2}\right) \tag{3.8}$$

and satisfies the normalization condition

$$\iiint I_0(r,z)dV = 1$$

Substituting equations (3.7) and (3.8) into (3.6), the absorption pump power at the threshold can be solved as

$$P_{a,th} = \frac{\pi h\nu_p}{2\sigma_e\eta_p\tau}\left(\omega_0^2+\omega_p^2\right)\left(\frac{\delta_e}{2}+\alpha_i L+\alpha_l L\right) \tag{3.9}$$

Equation (3.9) shows that the smaller the loss and spot size, the greater is the effective emission cross-section and pump quantum efficiency, the longer the lifetime of the upper energy level, and the lower the threshold pump power.

2. Laser oscillation above the threshold

Above the threshold, the rate equation can be expressed as

$$\frac{d\Delta N(r,z)}{dt} = (f_u+f_l)RI_p(r,z) - \frac{\Delta N(r,z)-\Delta N^{(0)}}{\tau} - \frac{(f_u+f_l)c\sigma\Delta N(r,z)}{n} SI_0(r,z) \tag{3.10}$$

$$\frac{dS}{dt} = \frac{c\sigma}{n}\iiint \Delta N(r,z)SI_0(r,z)dV - \frac{c\delta}{2nL}S \tag{3.11}$$

where $\Delta N^{(0)}$ is the population difference in case of equilibrium; c is the velocity of light; n is the refractive index of the medium; and S is the total number of photons in the cavity whose change is represented by equation (3.11), where the first term on the right-hand side denotes increase in photon number as a result of stimulated emission and the second term denotes reduction in photons as a result of loss. To derive equations (3.10) and (3.11), it is assumed that there is only a TEM_{00} model within the cavity. In case of end pumps, single-mode operation is easy as long as the pump region is within the model volume of TEM_{00}.

Under steady-state conditions,

$$\frac{d\Delta N(r,z)}{dt} = 0 \tag{3.10$'$}$$

$$\frac{dS}{dt} = 0 \tag{3.11$'$}$$

Equation (3.10)' leads to

$$\Delta N = \frac{(f_u + f_l)R\tau I_p(r,z) + \Delta N^{(0)}}{1 + \frac{(f_u + f_l)c\sigma\tau S I_0(r,z)}{n}}$$

substituting equation (3.10)' into equation (3.11)' and using $\Delta N^{(0)} = -\frac{\alpha_l}{\sigma}$ gives

$$\frac{1}{2\sigma L}\left[\delta + 2\alpha_l L \iiint \frac{I_0(r,z)dV}{1 + \frac{(f_u + f_l)c\sigma\tau S I_0(r,z)}{n}}\right] = (f_u + f_l)R\tau \iiint \frac{I_0(r,z)dV}{1 + \frac{(f_u + f_l)c\sigma\tau S I_0(r,z)}{n}} \tag{3.12}$$

The second term on the left-hand side of the equation is the absorption loss of the laser lower level to radiation wavelength, which is negligible for a four-level system, which is expressed as

$$(f_u + f_l)R\tau \iiint \frac{I_0(r,z)dV}{1 + \frac{(f_u + f_l)c\sigma\tau S I_0(r,z)}{n}} = \frac{\delta}{2\sigma L} \tag{3.13}$$

the time spent for a round-trip pass is $\frac{2L}{c/n}$, and the photon number outside the cavity in per unit time can be expressed as

$$S_0 = \frac{Sc}{2nL} \tag{3.14}$$

The output power can then be obtained as

$$P_0 = h\nu \frac{TSc}{2nL} \tag{3.15}$$

For a quasi three-level system, the absorption of radiation by the laser lower level is called the saturation loss:

$$\delta_{sat} = 2\alpha_L L \iiint \frac{I_0(r,z)dv}{1 + \frac{(f_u + f_l)c\sigma\tau}{n} S I_0(r,z)} \tag{3.16}$$

Assuming that the laser cavity mode is TEM_{OO}, on substituting I_O (r, z) from equation (3.7) into equation (3.16), the right-hand side of which is reduced to a single integral, it is not difficult to obtain

$$\delta_{sat} = \alpha_L L \frac{\pi w_0{}^2 nL}{S(f_u + f_l)c\sigma\tau} ln\left[1 + \frac{2S(f_u + f_l)c\sigma\tau}{\pi w_0{}^2 nL}\right] \tag{3.17}$$

With respect to light intensity, equation (3.17) can be written as

$$\delta_{sat} = \frac{\alpha_l L I_{sat}}{I} ln\left(1 + \frac{2I}{I_{sat}}\right) \tag{3.18}$$

where

$$I = \frac{Schv}{\pi w_0{}^2 nL} \tag{3.19}$$

is the average light intensity of the Gaussian beam and

$$I_{sat} = \frac{hv}{(f_u + f_l)\sigma\tau} \tag{3.20}$$

is the saturation light intensity.

3.2.3 Structure of LD-pumped solid-state laser

The working shape of DPSSL mediums include rod, sheet, and slab. This section describes the basic structure of rod-like working laser materials. The pump coupling mode is divided into direct end-pumped, fiber-coupled end-pumped, and side-pumped.

1. Direct end pump

Direct end-pumped pumping is common for low-power DPSSL. A typical direct end-pumped DPSSL structure is shown in Fig. 3.3.

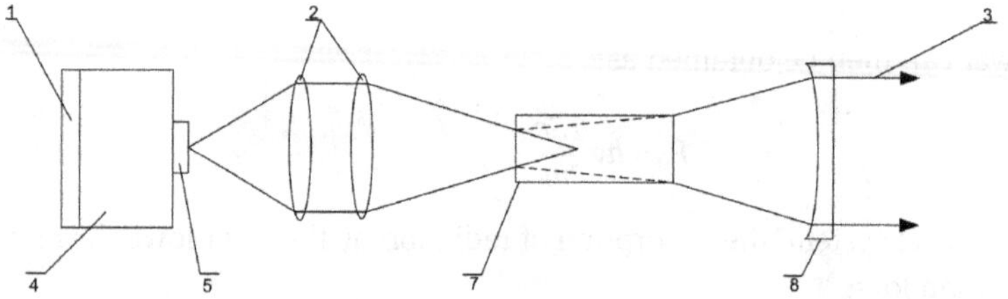

1.TE refrigerator 2.Focuser 3.Output beam 4.Radiator 5.LD arrayal 6.Pump
7.Tail reflector (HR@1.06 μm,AR@0.81 μm) 8.Output reflector

Fig. 3.3: Structure of a direct end-pumped DPSSL.

Direct end-pumped DPSSL mainly comprises an LD pump source, a coupling optical system, and a solid laser resonator. A coupling optical system couples pump light efficiently into a solid working substance. In general, a resonator is a half-external-cavity or half-inner-cavity structure Coating film on the pump-coupled end of the solid-state laser material has complete reflectivity for the laser wavelength and anti-reflex for the pump wavelength, such that the end face of the solid-state laser rod is referred to as the complete reflectivity end of the resonator. The output mirror has an appropriate transmittance for the laser wavelength.

In comparison with the other two pumping methods, direct end-pump has the most compact structure and the highest overall efficiency. The high efficiency of the direct end-pump is mainly because of the fact that the coupled optical system efficiently couples the pumping light with the solid working substance resulting in only a little loss when the pump laser mode is optimum. On the other hand, when the solid-state laser oscillation mode is closely related to the pump light pattern, the resulting matching effect is good and the utilization ratio of the solid working substance to the pump light is high.

One of the drawbacks of direct end-face pumping is that the size of the end face limits the power of the pump light, thus limiting the output power of the solid-state laser. The reason is that the emission aperture of the LD array is directly proportional to its emission power, and therefore, it is difficult to effectively couple the pump light having a large emission aperture with the smaller end face.

2. Fiber-coupled, end-pumped

Here, the pumping method aims to couple the pumping light to the solid working substance via an optical fiber or an optical fiber bundle. The typical structure of a fiber-coupled, end-faced pump is shown in Fig. 3.4.

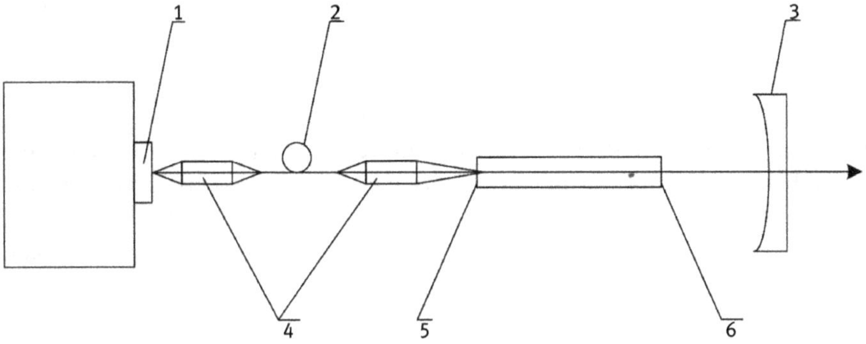

1.LD arrayal 2.Coupling fiber 3.Output reflector
4.Coupling system 5.Optical film (HR@1.06 μm,AR@0.81 μm) 6.Optical Film (AR@0.81 μm)

Fig. 3.4: Configuration of a fiber-coupled, end-faced pump.

A fiber-coupled, end-faced pump comprises a coupling optical system and a coupling fiber compared with the direct end-pump. The beam quality of the pump is improved when it is transmitted through an optical fiber. As the pump well matches with the laser, the resulting pump efficiency is high.

Coupled fiber can isolate heat transfer between the LD and the solid-state laser. As a result, the mutual influence of the thermal effect is reduced and the laser output is improved. In addition, the coupling of the LD pump light with the fiber is easier than direct coupling with the working substance. Therefore, device adjustment requirements can be reduced.

3. Side-pump

In the side-pumped solid-state laser, the laser head is usually surrounded by three diode pumps (Fig. 3.5).

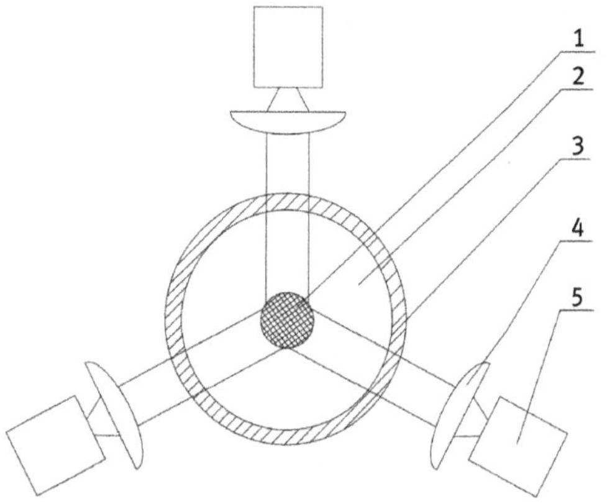

1.Laser medium; 2.Cooler; 3.Glass pipe; 4.Prismoid; 5.LD.

Fig. 3.5: LD side-pumped solid-state laser.

The three beams pairs emitted by the diode laser pump source are converged into the three areas of the antireflection film. The converged beam then passes through the tube wall and is absorbed by the crystal. Because most of the tube surface s plated with a high reflection film, the pump light entering the pump chamber reflects back and forth until it is sufficiently absorbed by the crystal and results in uniform gain distribution in the cross-section of the crystal.

3.3 Thin-disc laser

A thin-disc laser can be scaled to very high output power and energy along with very high efficiency and good beam quality.

Since the late 1980s, many groups have worked on DPSSL to increase the efficiency of solid-state lasers and improve beam quality. However, almost all groups worked on the traditional rod material as the working substance. In January 1992, researchers at the University of Stuttgart in Germany designed a flake-like activation medium for the first time. The primary thin-disc laser design was developed in March 1992, and in the late spring of 1993, the first demonstration was realized with an output power of 2 W. Today, thin-disc lasers have become a new class of solid-state lasers with high power, high efficiency, and high beam quality.

3.3.1 Thin media and pumping

The core of the thin-disc laser is a thin, disc-shaped active medium with a diameter of several millimeters (depending on the required output power or energy) and a typical thickness of 100–200 μm (depending on the material properties, doping concentration, and pump design). Because the thickness of the disc is much smaller than its diameter, heat flow is very short when the waste heat is led out from one surface of the sheet medium. Hence, a large temperature gradient is not generated in the disk even with a large pumping energy. The resulting power is the output from the other surface of the sheet, as shown in Fig. 3.6. Thus, the heat flow is nearly uniform and axial one-dimensional parallel to the laser. This considerably reduces the thermal lens effect and lateral temperature gradient, minimizing the phase distortion on the beam. Therefore, the slice laser can effectively remove thermal deposition of the gain medium and achieve high-power laser output while maintaining high efficiency and high beam quality.

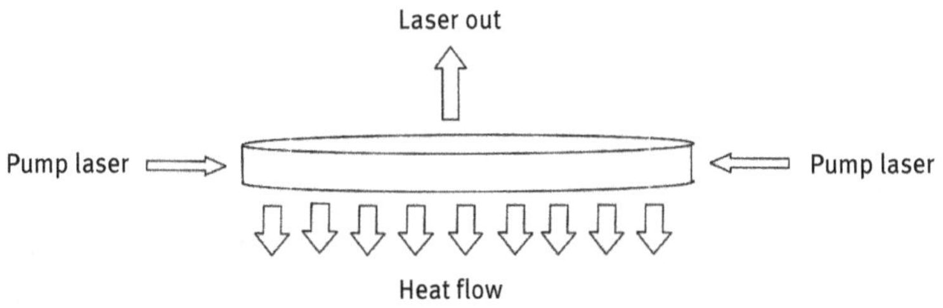

Fig. 3.6: Schematic of conduction heat of thin-disc medium.

To achieve a compact, thin-slice laser medium, an LD side pump mode is usually employed, as shown in Fig. 3.6. Side-pumping can reduce the complexity of the coupling system and because of its long absorption path reduces the required doping concentration of the gain medium. The disc material suitable for the end-pumped Nd^{3+}:YAG and Yb^{3+}:YAG is Yb^{3+}:gadolinium gallium garnet (GGG) at a wavelength of 885 nm. Here, it is important to control the influence of parameters such as the number of diode lasers, pump spacing and laser gain medium absorption system on pump uniformity.

3.3.2 Principle of thin-disc laser

Figure 3.7 shows the basic structure of a thin-disc laser. The front face of the disc is an anti-reflective (AR) coating for both the laser and pump wavelengths, whereas the

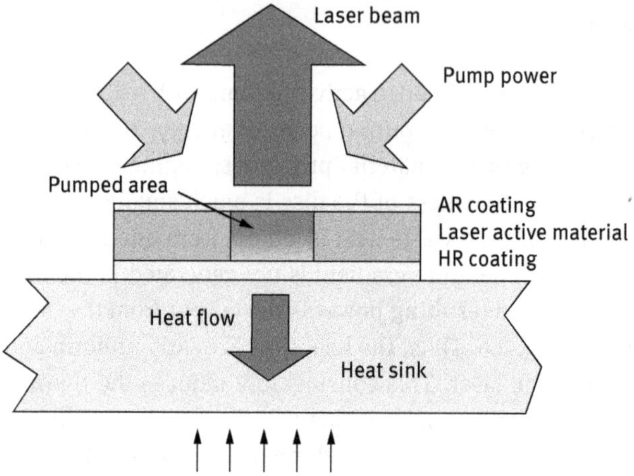

Fig. 3.7: Principle of the thin-disc laser (from High-power laser handbook).

back face is high-reflective (HR) coating for both wavelengths. This disc is mounted with its back on a water-cooled heat sink using indium tin or gold tin solder, which allows for a very stiff fixation of the disc to the heat sink without causing disc deformation.

It can be seen from Fig. 3.7 that the direction of the heat flux and the output laser are collinear and perpendicular to the working medium surface, and the heat sink surface is large, with a negligible boundary effect. Therefore, the temperature gradient is in a direction perpendicular to the beam, and the beam wave front distortion is extremely small. In the beam propagation direction, even though the temperature gradient is relatively large, the material thickness is very small, and therefore, the total temperature difference is not large. Consequently, a high beam quality can be maintained at high output power.

The small disc thickness and the resulting small temperature difference causes the sheet laser to produce a high quality beam. On the other hand, as the working medium has a small thickness and a short optical path length, the pump light absorption and the laser gain are small. Therefore, it is necessary to increase the number of times the pump light and laser light pass through the sheet. To achieve this, replace the planar mirror with a parabolic reflector and a roof prism, as shown in Fig. 3.8. Using this configuration, the pumping light is repeatedly reflected back to the laser gain medium resulting in an absorption rate of more than 90%.

The above-mentioned characteristics of the disc laser medium determine its working characteristics, i.e., the energy of the single pulse is small but can operate at high repetition frequency. For example, according to "High Power Laser Science and Engineering," "high average power pulsed laser" (HiLASE) project team received

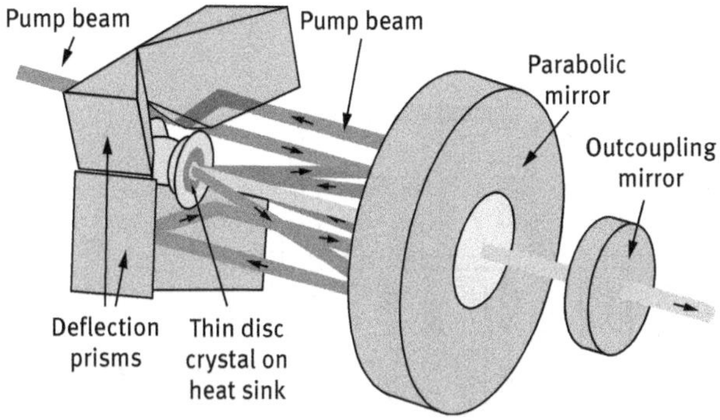

Fig. 3.8: Improved thin-disc laser (from High-power laser science and engineering).

a single-pulse (some mJ) using a pulse repetition frequency of kHz output. The next goal is to achieve a single pulse of 0.5 J and a pulse repetition frequency of 100 kHz, which will improve the average power to tens of kW.

3.3.3 "Liquid" lasers

"Liquid" lasers employ a liquid cooling technique. In early 2003, General Atomic and Aeronautical Systems (GA-ASI) in the United States began to develop high-performance "liquid" lasers using a submerged liquid-cooled flake design. The core concept of the flake laser design approach was to use a thin, disc-shaped activation medium that is cooled through a plane of a sheet. This method ensures a large ratio of surface to volume, providing efficient thermal management. The basic structure of a liquid laser is shown in Fig. 3.9.

The module shown in Fig. 3.10 represents the combat light source of the entire laser weapon system. No external power supply is needed in this module. The module carries enough lithium ion battery and can meet a certain number of shots power supply.

3.4 Slab lasers

3.4.1 Introduction

To generate high-energy pulses with low or moderate repetition rates, it is necessary to adopt medium geometry and an effective cooling mechanism. One of the solutions is to use an active medium in slab geometry along with active cooling of the slab faces. One of the materials of choice for next-generation, high-energy solid-state lasers is Yb^{3+}:YAG ceramic. However, Yb^{3+}:YAG is a quasi three-level system that requires a high pump threshold, thus increasing the number and cost of pumping

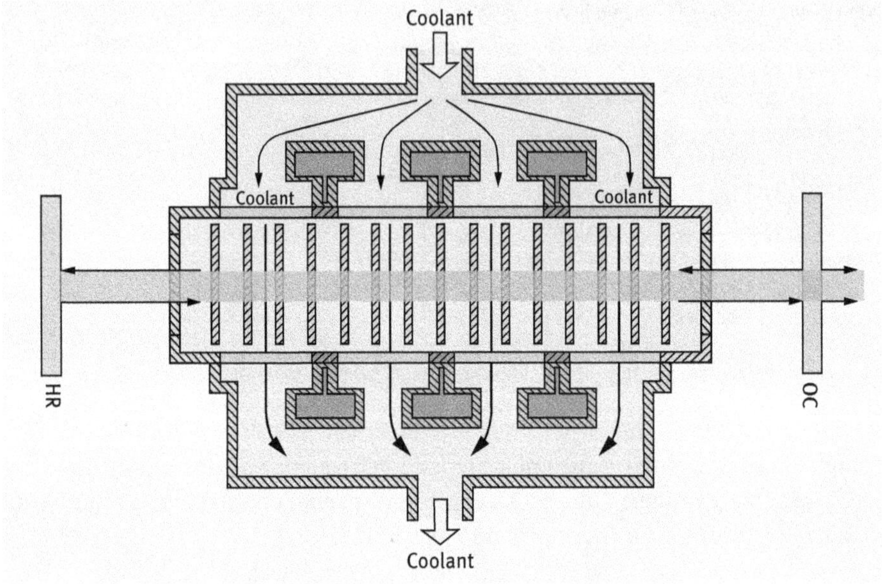

Fig. 3.9: "Liquid" lasers resonator.

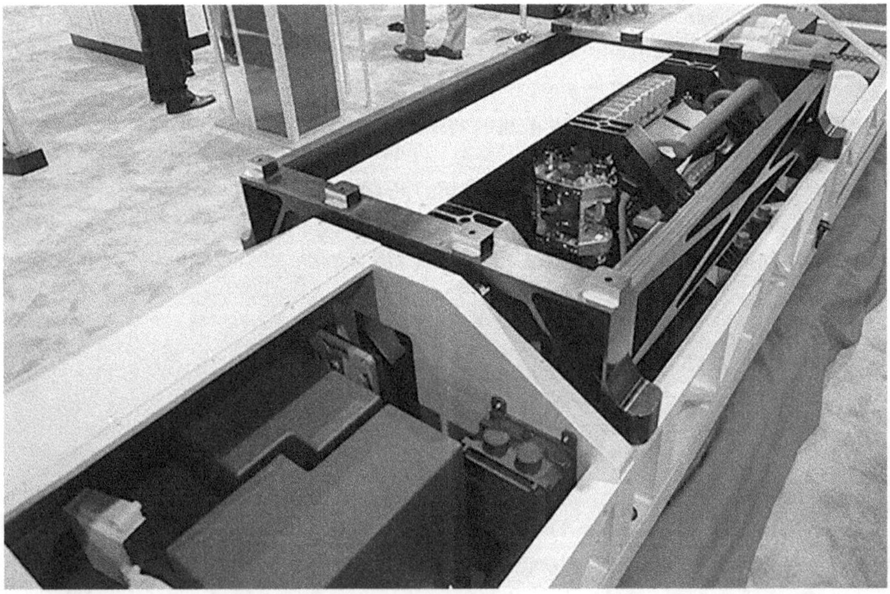

Fig. 3.10: Third-Generation HELLADS Laser Module.

diodes. By cooling the slab to a low temperature, the energy scheme changes to four levels, resulting in an decrease in the pump threshold by several orders of magnitude.

There are basically two different geometric configurations of the pump in a typical slab laser, as shown in Fig. 3.11.

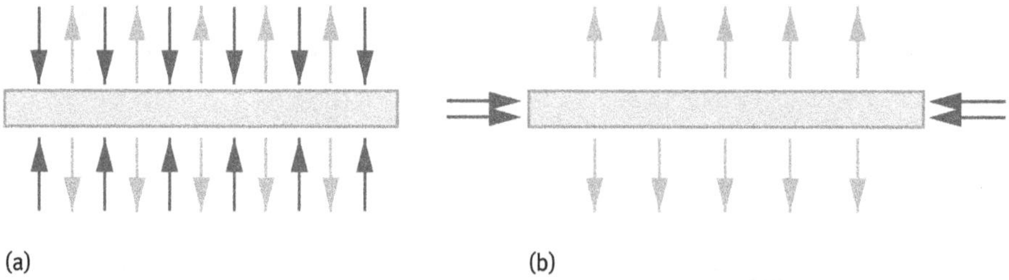

Fig. 3.11: Two types of slab laser pumping: (a) surface-pumped structure; (b) edge-pumped structure.

The surface-pumped structure is typically used for lamp-pumped lasers, where the pump (red arrow) and the heat-derived (orange arrow) pass through the large surface. The edge-pumped structure is more suitable for a diode pump, where the pumping (red arrow) and heat export (orange arrows) are in a mutually perpendicular plane. In both the geometries, the laser beam propagates along a zigzag path in a direction perpendicular to the surface.

For the diode pump of interest, the edge-pumped structure is more suitable, as mentioned above.

The high spatial coherence of a high-power LD(such as a diode bar or a diode bunch) makes it possible to inject all the pump light through a relatively narrow edge.

The pump diode can be placed directly on the side of the slab without any optical elements between the two. Because large surfaces need not be transparent, different cooling mechanisms are available, including conduction of metal heat sinks with internal water channels. Water channels inside the heat sink can keep the cooling water containing contaminants away from the optical surface.

In addition to the practical advantages of cooling and pumping at different locations, this geometry provides a long absorption path along the width of the sheet, thereby allowing the use of a very thin slab without compromising the pump uptake efficiency. Moreover, it is also possible to reduce the doping concentration of the material and increase pump strength, making it easier to use a quasi three-level gain medium. The reflection film is coated on the pump side, and pump light is injected only through the slit, further improving the absorption efficiency and pump uniformity.

There are two possible geometries for edge pumping. The first geometry is in the direction perpendicular to the laser beam pumping, which does not require any separation components to achieve sufficient separation of the pump and signal light. However, the disadvantage here is that the distribution of pump intensity is not ideal, and the intensivist pump light appears at the periphery of the signal light, resulting in the light beam to have a straight path instead of a zigzag path.

The other geometry is zigzag which allows the pump light and signal path to follow the same direction and to overlap well throughout. Figure 3.12 shows a typical slab laser where the temperature gradient occurs in the thin dimension of the slab (the two sides along the width of the slab are thermally insulated).

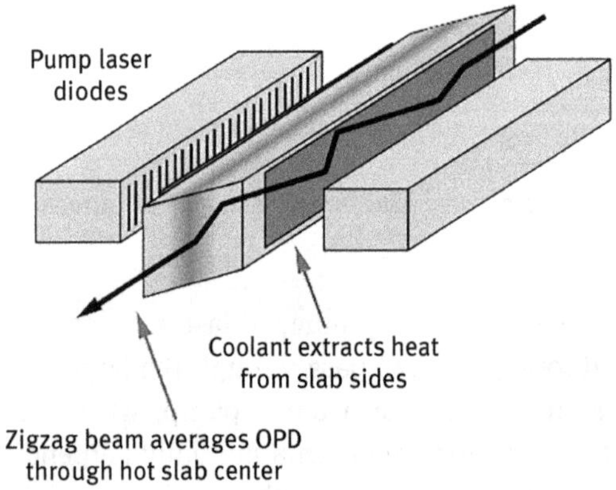

Fig. 3.12: Schematic of a traditional side-pumped zigzag slab. OPD: optical path difference.

In general, the slab is cut to have angled input faces and polished sides, and is cooled from the latter. The laser beam is injected into the slab such that it allows the beam to make multiple internal reflections from the polished sides. The main goal of the zigzag geometry is to average over the temperature gradients in the thin dimension of the slab.

3.5 Solid heat capacity

This section discusses the heat capacity of solids based on both the classic theory and the quantum theory intends to prepare for the discussion of thin-disc and slab lasers operating at the thermal mode in the next section.

3.5.1 The classic theory of solid heat capacity

In an ideal solid model, all make small amplitude, harmonic vibrations near their equilibrium position. There are 3N degrees of freedom for a solid containing N atoms. The energy of each vibration freedom is expressed as

$$\varepsilon = \frac{1}{2m}p^2 + \frac{1}{2}m\omega^2q^2 + \varepsilon_0 \tag{3.21}$$

where m is the atomic mass; q is the coordinate of the atom relative to the equilibrium position; p is the momentum conjugating to q; ω is the circular frequency of the vibration; and ε_0 is the potential energy of each freedom degree when the atom is located in its equilibrium position. According to the energy-sharing theorem, the average of each square term in the energy equals 0.5 kT, where k is the Boltzmann constant. Equation (3.21) contains two square terms whose mean equal 1 kT. The entire solid has 3N degrees of freedom, hence, the total energy is expressed as

$$\bar{E} = 3NkT + E_0$$

where $E_0 = 3N\varepsilon_0$ is independent of the temperature. The equivoluminal heat capacity of a solid can be obtained as

$$C_u = \left(\frac{\partial \bar{E}}{\partial T}\right)_u = 3Nk \tag{3.22}$$

It is consistent with the laws published by Dulong and Petit in 1818 and coincide with experimental results at both room temperature and high temperature. However, at low temperatures, the test value is small and rapidly declines with temperature, even approaching 0 when temperature approaches 0 K. This phenomenon is not consistent with the classic theory, and can only be explained by the quantum theory described below.

3.5.2 Quantum theory of solid heat capacity

1. The Einstein model of lattice heat capacity

To solve the contradictions pointed out in the previous section, Einstein first applied Planck's quantum theory to the vibrations of atoms in solids in 1907. Assuming that all atoms vibrate at the same frequency ν, the quantized energy eigenvalues of harmonic vibrations of the jth atom are

$$\varepsilon_j = \left(n_j + \frac{1}{2}\right)h\nu$$

While the statistical average of energy is

$$E_j(T) = \frac{1}{2}h\nu + \frac{\sum_{n_j} n_j h\nu e^{-n_j\beta h\nu}}{\sum_{n_j} e^{-n_j\beta h\nu}} \tag{3.23}$$

where n_j is constant

and

$$\beta = \frac{1}{kT}$$

Thus, equation (3.23) can be rewritten as

$$\overline{E_j}(T) = \frac{1}{2}h\nu - \frac{\partial}{\partial\beta} ln \sum_{n_j} e^{-n_j\beta h\nu} \tag{3.24}$$

Where the summed term is a geometric progression with a common ratio less than unity, so

$$\sum_{n_j} e^{-n_j\beta h\nu} = \frac{1}{1 - e^{-n_j\beta h\nu}}$$

$$\ln \sum_{n_j} e^{-n_j\beta h\nu} = -\ln\left(1 - e^{-\beta h\nu}\right)$$

$$\frac{\partial}{\partial\beta}\ln \sum_{n_j} e^{-n_j\beta h\nu} = -\frac{h\nu e^{-\beta h\nu}}{1 - e^{-\beta h\nu}} = -\frac{h\nu}{e^{\beta h\nu} - 1}$$

Substituting these equations into equation (3.24), we obtain

$$\overline{E_j}(T) = \frac{1}{2}h\nu + \frac{h\nu}{e^{\beta h\nu} - 1} \tag{3.25}$$

Where the first term on the right-hand side is zero-point energy and the second term is the average thermal energy.

Einstein assumed that the vibrations of the atoms in the lattice can independent of each other. Considering that each atom can vibrate in three directions, the system having N atoms has a total of 3N vibrations of frequency ν with the total energy of

$$\bar{E}(T) = 3N\overline{E_J}(T)$$

While the heat capacity of the lattice is

$$C_V = \frac{d\bar{E}(T)}{dT} 3Nh\nu \frac{d}{dT}\left[\left(e^{\beta h\nu} - 1\right)^{-1}\right] = 3Nk\frac{\left(\frac{h\nu}{kT}\right)^2 e^{\frac{h\nu}{kT}}}{\left(e^{\frac{h\nu}{kT}} - 1\right)^2} \tag{3.26}$$

Below we discuss the asymptotic behavior of equation (3.26) at high and low temperatures.

At high temperatures $kT \gg h\nu$, the exponential in equation (3.26) is expanded in the neighborhood of $\frac{h\nu}{kT} = 0$ as progression

$$C_V = 3Nk\frac{\left(\frac{h\nu}{kT}\right)^2\left(1 + \frac{h\nu}{kT} + \ldots\right)}{\left[\frac{h\nu}{kT} + \frac{1}{2}\left(\frac{h\nu}{kT}\right)^2 + \ldots\right]^2} \approx 3Nk \tag{3.27}$$

It is consistent with equation (3.22). This is because the quantization effect is negligible when the thermal energy of the vibrator is significantly larger than the energy-quantum $h\nu$. Thus, the results of the quantum theory should naturally return to the classic theory, which is experimentally correct.

When the temperature is very low $h\nu \gg kT$ equation (3.26) gives

$$C_V = 3Nk\left(\frac{h\nu}{kT}\right)^2 e^{-\frac{h\nu}{kT}} \xrightarrow{(T \to 0k)} 0 \tag{3.28}$$

This is because, when the vibrational energy level is quantized and the absolute temperature approaches zero, the vibration becomes "frozen" in the ground state and it is very difficult to be excited by heat; hence, the contribution of heat capacity also approaches zero.

It can be seen that the results obtained from the Einstein model are in good agreement with the test at high temperatures. At low temperatures, it is also consistent with the change in C_V with temperature. However, the predicted rate of decline is considerably larger than the experimental results. The reason for this contradiction is that the model assumes that there is a strong interaction between the atoms in the solid and that there is a certain distribution of vibrational frequencies. The solution to this problem leads to the Debye model discussed below.

2. Introduction to Debye Model of Lattice Heat Capacity

To overcome the difficulties of Einstein's model, we must consider the distribution of the vibrational frequencies of the atoms in the solid and superimpose the contributions of the various vibrational frequencies to the energy and heat capacity. With $g(\nu)d\nu$ denoting the number of vibration-freedom degrees in the frequency intermission of $(\nu, \nu + d\nu)$, $g(\nu)$ is the frequency distribution function; once $g(\nu)$ is known, heat capacity can be calculated. In 1912, Debye proposed an approximation model by considering solid as an isotropic elastic medium. He proposed that the medium can propagate two kinds of elastic waves, one being the expansion wave, which is a longitudinal wave with a velocity of C_l; the other being the swing-wave, which is a transverse wave with a propagation velocity of C_t. In terms of frequency, the longitudinal wave has only one vibration mode, that is, the vibration in the propagation direction; the transverse wave has two vibration modes that are two orthogonal vibrations in the plane perpendicular to the propagation direction. From this, we have

$$g(\nu) = 4\pi V\left(\frac{1}{C_l^3 + C_t^3}\right)\nu^2 \tag{3.29}$$

Note that the total freedom number of vibration is 3N, which should be

$$\int_0^\infty g(\nu)d(\nu) = 3N \tag{3.30}$$

Substituting equation (3.29) into (3.30), the resulting integral is obviously divergent. Therefore, it is necessary to assume that there is a maximum value ν_M of the vibration frequency as

$$\int_0^{\nu_M} g(\nu)d(\nu_M) = 3N$$

From this solution, we can derive

$$\nu_M = \frac{\bar{C}}{2\pi}\left[6\pi^2\left(\frac{N}{V}\right)\right]^{1/3}$$

Where $\bar{C}$ can be obtained by

$$\frac{1}{\bar{C}^3} = \frac{1}{3}\left(\frac{1}{C_l^3} + \frac{2}{C_t^3}\right)$$

and the heat capacity is

$$\begin{aligned} C_V^{(T)} &= k\int_0^{\nu_M}\left(\frac{h\nu}{kT}\right)^2 \frac{e^{\frac{h\nu}{kT}}}{\left(e^{\frac{h\nu}{kT}}-1\right)^2} g(\nu)d(\nu) \\ &= 4\pi V\left(\frac{1}{C_l^3} + \frac{2}{C_t^3}\right)\int_0^{\nu_D}\left(\frac{h\nu}{kT}\right)^2 \frac{e^{\frac{h\nu}{kT}}\nu^4}{\left(e^{\frac{h\nu}{kT}}-1\right)^2} d(\nu) \end{aligned}$$

If ν_M is used, the above equation can be rewritten as

$$C_V^{(T)} = 9R\left(\frac{kT}{h\nu_M}\right)^3 \int_0^{h\nu_M/kT} \frac{x^4 e^x}{(e^x-1)^2}dx \tag{3.31}$$

where $R = Nk$ is the gas constant, and $x = \frac{h\nu}{kT}$.

Equation (3.31) contains only one variable of ν_M, and if

$$\Theta_D = \frac{h\nu_M}{k} \tag{3.32}$$

is the temperature unit, the heat capacity C_V is a universal function

$$C_V\left(\frac{T}{\Theta_D}\right) = 9R\left(\frac{T}{\Theta_D}\right)^3 \int_0^{\frac{T}{\Theta_D}} \frac{x^4 e^x}{(e^x-1)^2}dx \tag{3.33}$$

where Θ_D is referred to as the Debye temperature.

In general, the calculated heat capacity of the Debye model agrees well with the experimental results over a wide range of temperature. However this model also has its limitations, for example, the model sets the upper limit of the vibration frequency. In fact, when the frequency is high enough, the atomic structure of the solid appears to be no longer able to make a continuum. Therefore, equation (3.29) cannot be applied for larger frequency v.

3. Electronic heat capacity

Strictly speaking, only lattice heat capacity is discussed above. The heat capacity of a solid should also include electronic heat capacity. Thermal excitation energy of an electron can be derived from Fermi statistics as

$$E_h = \frac{\pi^2}{6} N(E_F^0)(kT)^2 \tag{3.34}$$

where $N(E_F^0)$ is the energy-state-density of low-temperature- limit Fermi level E_F^0 at 0 K.

On differentiation with respect to T to obtain the electronic heat capacity

$$C_V = \frac{\pi^2}{3} kN(E_F^0)(kT) \tag{3.35}$$

This is a result of quantum statistics, which is much smaller than the classical theoretical value of $3k/2$. This is possibly because, according to quantum theory, the initial energy of most electrons is much lower than E_F^0 and they mainly do not participate in thermal excitation according to the Pauli principle. Only the electrons within the range of kT and the vicinity of E_F^0 contribute to heat capacity. In addition, the electronic heat capacity is considerably smaller than the lattice heat capacity; however, at low temperatures, the lattice heat capacity approaches zero in accordance with T^3 as temperature decreases. The electronic heat capacity is also proportional to T and slowly decreases with temperature. In the liquid-nitrogen temperature range, lattice heat capacity and electronic heat capacity can be compared.

The discussion of solid heat capacity becomes more complicated when electronic heat capacity is also considered. Fortunately, the operating temperatures of solid-state heat capacity lasers so far are generally higher than 100 K, therefore, the influence of electronic heat capacity can be ignored completely. Moreover, in this temperature range, even the lattice heat capacity of the Debye model can be ignored. Therefore, the following discussion in this chapter will be based on the results of the Einstein model, i.e.,

$$C_V(T) = 3R\left(\frac{\Theta_D}{T}\right)^2 \frac{e^{\Theta_D/T}}{\left(e^{\Theta_D/T} - 1\right)^2} \tag{3.36}$$

where $R = Nk$ is the gas constant and Θ_D is Debye temperature.

3.6 Heat-capacity operation model of lasers

3.6.1 Heat storage and increase in temperature

1. Heat-storage power and output optical power

Thermal deposition occurs in the laser emission phase of heat capacity laser, resulting in an increase in working medium temperature, which is the basic characteristics of heat capacity lasers. The average thermal power of the deposition can be expressed as

$$P_{heat} = 2P_p \frac{V}{W} \eta_{abs} \frac{\chi}{1+\chi} D_t \tag{3.37}$$

where P_p is the pump-power-density factor; number 2 shows the bilateral pump; V is the working material volume; W is the plate width; η_{abs} is the upper energy level absorption efficiency; and X is the specific heat, which is the ratio of the heat E_{heat} generated in the medium with the stored energy E_s. Take Fig. 3.13 as an example.

E_p is pump energy; to pump one population from E_1 to E_3, $E_p = h\nu_{13}$, at this time, there is useful energy $E_s = h\nu_{32}$ stored in the work material; the energy of laser lower level $E_h = E_l = h\nu_{21}$ is converted into heat energy when the population returns to the ground state E_l through non-radiative transition from that energy level. By definition, this is

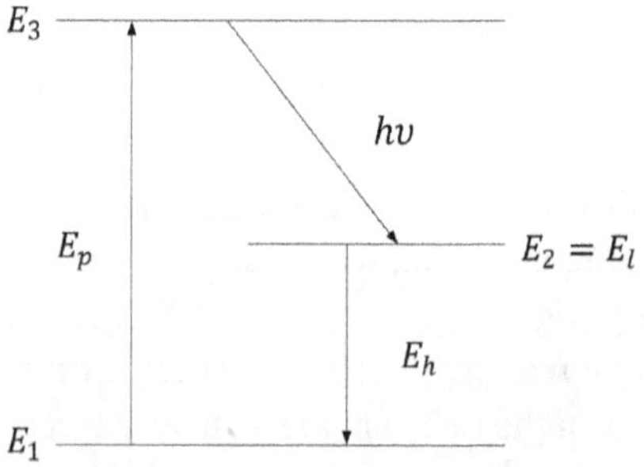

Fig. 3.13: Pump energy, thermal energy, and energy storage diagram.

$$\chi = \frac{h\nu_{21}}{h\nu_{32}} = \frac{E_l}{E_p - E_l} = \frac{1}{\left(E_p/E_l - 1\right)}$$

$$1 + \chi = \frac{E_p/E_l}{E_p/E_l - 1} = \frac{1}{\left(1 - E_l/E_p\right)}$$

and

$$\frac{\chi}{1+\chi} = E_l/E_l = \frac{h\nu_{21}}{h\nu_{13}} \tag{3.38}$$

$$P_{out} = \eta_{extr}(P_s - P_{th}) \tag{3.39}$$

P_{th} refers to the threshold power; and η_{extr} is the extraction efficiency.

Notice

$$P_s = \frac{P_h}{\chi}$$

and

$$P_{th} = \frac{n_l V h\nu}{\tau_l}$$

substituting it into equation (3.39) gives the expression

$$P_{out} = \left(\frac{P_h}{\chi} - \frac{n_0 V h\nu e^{-E_l/kT}}{\tau_l}\right)\eta_{extr} \tag{3.40}$$

where τ_l and $n_l = n_0 e^{-E_l/kT}$ are, respectively, the lifetime and population density of the lower energy level; and n_0 is the density of the total population at every level.

2. Increase in temperature

The temperature increase in the working substance during the time interval t due to the above-described thermal deposition can be expressed as

$$\Delta T = \frac{P_h t}{\rho V C_V(T)} \tag{3.41}$$

where ρ is the density of the medium and $C_V(T)$ changes with the temperature according to equation (3.16).

However, for the resulting analytic expression of ΔT, C_V must be approximated as a constant, i.e., it is assumed to be equal to a certain average of the value $C_V(T_{start})$ at initial temperature T_{start} and the value $C_V(T_{Max})$ at the maximum temperature T_{max}.

Here, the maximum temperature is defined as the temperature at which the output power is reduced to 1/f of the temperature at T_{start}. Then, from equation,

$$\left(\frac{P_h}{\chi}-\frac{n_0 Vhe^{-E_l/kT_{Max}}}{\tau_l}\right)\eta_{extr}=f\left(\frac{P_h}{\chi}-\frac{n_0 Vhe^{-E_l/kT_{start}}}{\tau_l}\right)\eta_{extr} \tag{3.41}$$

Namely,

$$\frac{n_0 Vh\nu}{\tau_l}\left[e^{-E_l/kT_{Max}}-fe^{-E_l/kT_{start}}\right]=\frac{P_h}{\chi}(1-f)=\frac{2P_p\eta_{abs}V}{W(1+\chi)}(1-f) \tag{3.42}$$

When $T_{\max}$ is far greater than T_{start}, the second term of the square bracket can be ignored, therefore

$$e^{-E_l/kT_{Max}}=\frac{2P_p\tau_l\eta_{abs}(1-f)}{n_0 h\nu W(1+\chi)}$$

or

$$T_{Max}=\frac{E_l}{-k\ln\left[\frac{2P_p\tau_l\eta_{abs}(1-f)}{n_0 h\nu W(1+\chi)}\right]}=\frac{E_l}{-k\ln\left[F\left(1-E_l/E_p\right)\right]} \tag{3.43}$$

where

$$F=\frac{2P_p\tau_l\eta_{abs}}{n_0 h\nu W}(1-f) \tag{3.44}$$

The temperature rise caused by thermal deposition is finally obtained as

$$\Delta T=T_{Max}-T_{start}=-\left\{\frac{E_l/kT_{start}}{\ln\left[F\left(1-E_l/E_p\right)\right]}+1\right\}T_{start} \tag{3.45}$$

3. Energy output

During increase in the temperature from T_{start} to T_{Max}, the laser energy output is

$$E_{out}=\int P_{out}dt=\int_{T_{start}}^{T_{Max}}\frac{P_{out}}{\left(\frac{dT}{dt}\right)}dT$$

Substitutions equation (3.40) and (3.41) into it, we obtain

$$E_{out} = \int_{T_{start}}^{T_{Max}} \left(\frac{P_h}{X} - \frac{n_0 Vhe^{-E_l/kT}}{\tau_l} \right) \eta_{extr} \frac{\rho V C_V}{P_h} dT$$

For situations of interest, there are usually $E_l \gg kT$. Therefore, the integral of $e^{-E_l/kT}$ can be neglected. Thus, the equation above can be rewritten as

$$E_{out} = \frac{P_h}{X} \eta_{extr} \frac{\rho V C_V}{P_h} \Delta T = \eta_{extr} V \frac{\rho C_V}{X} \Delta T$$

$$= C_V \rho V T_{start} \eta_{extr} \left(E_p/E_l - 1 \right) \left\{ \frac{E_l/kT_{start}}{-ln\left[F\left(1 - E_l/E_p\right)\right]} - 1 \right\} \tag{3.46}$$

It is clear from equation (3.46) that, in order to ensure that $E_{out} > 0$, is met

$$E_l > -kT_{start} ln\left[F\left(1 - E_l/E_p\right)\right] > kT_{start}(-lnF)$$

Thus, the minimum laser lower level is

$$E_{lmin} = kT_{start}(-lnF) \tag{3.47}$$

Thus, the greater the F, the smaller the allowed E_{lmin} and the equation would more easily be met. The equation (3.44) shows that large F implies that absorbed pump power per unit volume of the working substance is high or f is small.

Equation (3.43) indicates that E_l/E_p should not be too close to one, otherwise it will lead to E_{out} being too small. That is, E_l should neither be too close to the ground state nor too close to the pump energy level (the laser upper level is shown in Fig. 3.8). It is generally believed that the lower level of the ideal heat capacity laser lies somewhere around 1/3 to 1/2 between the ground state and the upper level. Corresponding to 1.06 μm radiation, the upper and lower energy levels of Nd^{3+} need to higher than the ground state level by approximately 1.4 eV and 0.26 eV, respectively.

In the first part of equation (3.46)

$$E_{out} = \eta_{extr} \frac{\rho C_V}{X} \Delta T V \tag{3.48}$$

The right-hand side of equation (3.48) contains two merit factors. One of them is given by $\left(\frac{\rho C_V}{X}\right)$ containing only material parameters besides parameter X, which can be estimated from the spectroscopy of the lasing ion for the zero-order approximation. Another factor is $\eta_{extr} \frac{\rho C_V}{X} \Delta T$ depending on the temperature range over which the material is lased. These factors help in selecting the working substance that is most suitable for the heat capacity lasers.

3.6.2 Temperature distribution and thermal stress

1. Temperature difference between the surface and the center

Assuming that the symmetrical and stable pumping from the two end faces of the slab along the Z axis (in the thickness direction with Z = 0 at the center), under adiabatic boundary conditions temperature distribution can be obtained as

$$T(Z,t)=\frac{2}{t_s}\frac{\chi}{1+\chi}\frac{P}{\rho c_v}[1-exp(-\alpha t_s)]t+T_0-\cos\left(\frac{2\pi}{t_s}Z\right)\frac{1}{\pi^2}\frac{Pt_s\chi}{\lambda(1+\chi)}\cdot$$
$$[1-exp(-\alpha t_s)]\frac{(\alpha t_s)^2}{(\alpha t_s)^2+4\pi^2}\left[1-exp\left(-\gamma\frac{4\pi^2}{t_s^2}t\right)\right] \tag{3.49}$$

where P is the irradiance $(W\cdot m^{-2})$ of the pumping light; α is the absorption coefficient (m^{-1}); t_s is the slab thickness (m); λ is the thermal conductivity $(W\cdot m^{-1}K^{-1})$; γ for the thermal diffusion; $(m^2\cdot s^{-1})$; t is the time; and T_0 is temperature at $t=0$.

From equation (3.49), the temperature difference between the slab surface and the centerline can be expressed as

$$T_s-T_c=T_{Z=\pm\frac{t_s}{2}}-T_{Z=0}=\frac{2}{\pi^2}\frac{Pt_s\chi}{\lambda(1+\chi)}[1-exp(-\alpha t_s)]\frac{(\alpha t_s)^2}{(\alpha t_s)^2+4\pi^2}\left[1-exp\left(-\gamma\frac{4\pi^2}{t_s^2}t\right)\right]$$
$$=\frac{2t_s}{\pi^2\lambda}\Phi\frac{(\alpha t_s)^2}{(\alpha t_s)^2+4\pi^2}\left[1-exp\left(-\gamma\frac{4\pi^2}{t_s^2}t\right)\right] \tag{3.50}$$

where

$$\Phi=P[1-exp(-\alpha t_s)]\frac{\chi}{1+\chi} \tag{3.51}$$

is the heat flow through the slab surface.

Equation (3.50) shows that the temperature at the surface is always higher than that at the center and the temperature difference between the surface and the center can determined by the depth over which the pump power is absorbed. In addition, for time to meet the condition of $t\geq\frac{t_s^2}{\gamma}$, we have

$$T_S-T_C\approx\frac{2}{\pi^2}\frac{Pt_s\chi}{\lambda(1+\chi)}[1-exp(-\alpha t_s)]\frac{(\alpha t_s)^2}{(\alpha t_s)^2+4\pi^2}$$

This implies that the temperature difference no longer changes over time, although the temperatures either on the surface or at the center increase with time, for example

$$T(0,t)=\frac{2}{t_s}\frac{\chi}{1+\chi}\frac{P}{\rho c_v}[1-exp(-\alpha t_s)]t+T_0-\frac{1}{\pi^2}\frac{Pt_s\chi}{\lambda(1+\chi)}[1-exp(-\alpha t_s)]\frac{(\alpha t_s)^2}{(\alpha t_s)^2+4\pi^2}$$

2. Stress compared with traditional working mode

The stress on the slab surface caused by the temperature difference given by equation (3.50) is

$$\sigma_{hc} = \frac{aE}{2}(T_s - T_c) = \frac{aE\Phi t_s}{\pi^2\lambda}\frac{(\alpha t_s)^2}{(\alpha t_s)^2 + 4\pi^2}\left[1 - \exp\left(-\gamma\frac{4\pi^2}{t_s^2}t\right)\right]$$

Under the condition of $t \geq \frac{t_s^2}{\gamma}$,

$$\sigma_{hc} = \frac{aE\Phi t_s}{\pi^2\lambda}\frac{(\alpha t_s)^2}{(\alpha t_s)^2 + 4\pi^2} \tag{3.52}$$

This result can be compared with the tensile stress that occurs on the surface during steady-state average power operation. For the latter we have

$$\sigma_{ssap} \approx \frac{aE}{6\lambda(1-\nu)}\Phi\frac{6[e^{\alpha t_s}(\alpha t_s - 2) + (\alpha t_s + 2)]}{\alpha \cdot \alpha t_s(e^{\alpha t_s} - 1)} \tag{3.53}$$

where ν is the Poisson ratio.

Dividing Equation (3.52) with (3.53), we obtain

$$\frac{\sigma_{hc}}{\sigma_{ssap}} = \frac{(\alpha t_s)^4(e^{\alpha t_s} - 1)}{\pi^2\left[(\alpha t_s)^2 + 4\pi^2\right][e^{\alpha t_s}(\alpha t_s - 2) + (\alpha t_s + 2)]}(1-\nu) \tag{3.54}$$

An estimate of this ratio can be made using Taylor expansion for small αt_s:

$$\begin{aligned}
&\frac{6[e^{\alpha t_s}(\alpha t_s - 2) + (\alpha t_s + 2)]}{\alpha \cdot \alpha t_s(e^{\alpha t_s} - 1)} \\
&= \frac{6\left\{\left[1 + \alpha t_s + \frac{1}{2}(\alpha t_s)^2 + \frac{1}{6}(\alpha t_s)^3 + \frac{1}{24}(\alpha t_s)^4 + \frac{1}{120}(\alpha t_s)^5 + \ldots\right](\alpha t_s - 2) + (\alpha t_s + 2)\right\}t_s}{(\alpha t_s)^2\left[\alpha t_s + \frac{1}{2}(\alpha t_s)^2 + \frac{1}{6}(\alpha t_s)^3 + \frac{1}{24}(\alpha t_s)^4 + \frac{1}{120}(\alpha t_s)^5 + \ldots\right]} \\
&= t_s\frac{1 + \frac{1}{2}(\alpha t_s) + \frac{3}{20}(\alpha t_s)^2 + \frac{1}{30}(\alpha t_s)^3 + \ldots}{1 + \frac{1}{2}(\alpha t_s) + \frac{1}{6}(\alpha t_s)^2 + \frac{1}{24}(\alpha t_s)^3 + \ldots} \\
&\ldots \approx t_s\left[1 - \frac{(\alpha t_s)^2}{60} + \ldots\right]
\end{aligned} \tag{3.55}$$

Substituting equation (3.55) into (3.54), we obtain

$$\frac{\sigma_{hc}}{\sigma_{ssap}} = \frac{(\alpha t_s)^2}{\pi^2\left[(\alpha t_s)^2 + 4\pi^2\right]} \cdot \frac{6(1-\nu)}{\left[1 - \frac{(\alpha t_s)^2}{60}\right]} = \frac{6(1-\nu)(\alpha t_s)^2}{4\pi^2\left[1 + \frac{(\alpha t_s)^2}{4\pi^2}\right]\left[1 - \frac{(\alpha t_s)^2}{60}\right]}$$

$$\approx \frac{6(1-\nu)(\alpha t_s)^2}{4\pi^2\left[1+\frac{(15-\pi^2)}{60\pi^2}(\alpha t_s)^2\right]} \approx \frac{3(1-\nu)}{2\pi^4}(\alpha t_s)^2 - \frac{(1-\nu)(15-\pi^2)}{40\pi^6}(\alpha t_s)^4 \tag{3.56}$$

For a nominal Poisson ratio of 0.3, this is approximately

$$\frac{\sigma_{hc}}{\sigma_{ssap}} \approx 0.01(\alpha t_s)^2 - 0.1\times 10^{-3}(\alpha t_s)^4 \tag{3.57}$$

It can be seen from equation (3.57) that, even when (αt_s) reaches 5, $\frac{\sigma_{hc}}{\sigma_{ssap}}$ is only 0.25. Therefore, keeping other conditions constant, the tension on the surface of the laser medium operating in the heat capacity mode is only one-quarter of that of the laser medium operating at steady-state average power. This is the main advantage of heat capacity lasers, which allow heat capacity lasers to operate at higher power and deliver a better beam quality.

3.6.3 Beam distortion

Although the heat capacity laser was designed to minimize thermal gradients and thermal beam distortion, gradients still exist in a direction transverse to the propagation as a result of non-uniformities in pump illumination.

A complete description of beam distortions requires knowledge thermodynamics, elastic mechanics, and many other areas, which is beyond the scope of this book.

Consider a slab medium where the z axis is through the slab thickness, the x axis is along the slab width, and the y axis is along the slab length. If the surface of the plate of thickness t_s is given by (x, y), the optical path difference across the beam aperture can be expressed as

$$OP = OP_0\left[1-\frac{\frac{\partial s}{\partial y}}{n^3}-\frac{\left(\frac{\partial s}{\partial y}\right)^2(n^4-n^2-3)-\left(\frac{\partial s}{\partial x}\right)^2 n^2}{2n^6}\right] \tag{3.58}$$

where n is the refractive index of the material; OP_0 is the uniform wave front displacement given by the perfectly flat plate of thickness t_s under Brewster's angle θ_B. As shown in Figure 3.14, then

$$OP_0 = nt_s/sin\theta_B \tag{3.59}$$

The surface shape of this slab with linear temperature gradient through its thickness is expressed as

$$S(x,y) = \frac{\alpha\delta T}{2t_s}(x^2+y^2) \tag{3.60}$$

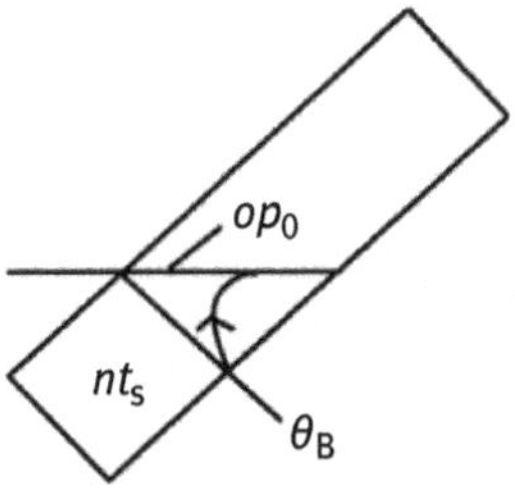

Fig. 3.14: Relationship between OP_0 and θ_B.

where α is the thermal expansion coefficient of the material (K^{-1}); and δT is the temperature rise in the thickness direction. It can be seen from equation (3.60) that $\frac{\partial s}{\partial y}$ is proportional to $\alpha\delta T$. In this book, the typical value of $\alpha\delta T$ has an order of magnitude of 10–, and hence $\left(\frac{\partial s}{\partial y}\right)^2$ is negligible compared to $\frac{\partial s}{\partial y}$. Substituting equation (3.60) into equation (3.58) results in

$$OP \approx OP_0\left(1 - \frac{\alpha\delta T}{n^3} \cdot \frac{y}{t_s}\right) \tag{3.61}$$

For Nd^{3+}:YAG laser working substance, $n \approx 1.82$, $\theta_B \approx 57°$ at λ=1.0 μm, set $t_s = 1cm$, substituting them into equation (3.59) obtained

$$OP_0 \approx 2.1cm$$

The optical path difference between the optical path and the ideal optical path across the beam aperture can be expressed as

$$\frac{\alpha\delta T}{n^2 sin\theta_B} y$$

with the order of magnitude of μm. Or, for 1 μm wavelength, phase difference is approximately 6 rad. Such a phase difference will have a significant impact on the beam quality, which requires the adaptive optical system to be corrected.

3.6.4 Heat capacity laser example

In January 2006, Lawrence Liverpool National Laboratory (LLNL) developed a heat-capacity laser with the best property at that time. Figure 3.15 shows the latest configuration of the heat-capacity laser used at LLNL.

The basic building block of the heat capacity laser is the laser gain module, which consists of a single slab pumped by four high-powered diode arrays, two on either side of the slab. Each diode array pumps the slab's adjacent face at a defined angle, ceramic path surface at an angle to provide uniform pump intensity across the

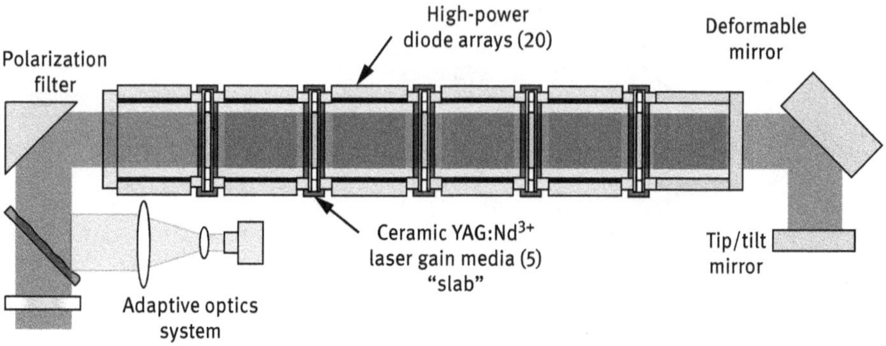

Fig. 3.15: LLNL heat capacity laser structure (from High-Power Laser Handbook).

entire surface. In this example, the laser gain medium is five pieces of Nd^{3+}:YAG transparent ceramic, the edge of which is coated with cobalt-doped GGG to suppress the amplified spontaneous emission. Figure 3.15 shows the five gain modules that are locked to form a compact cavity. The wave front sensor, deformable mirror, tilting mirror, and controller of the cavity adaptive optical system maintain uniformity of the wave front phase.

Two key components make up the gain module of the heat capacity laser. The first is a high-power LD array used to pump the laser gain medium. Each diode array contains hundreds of relatively small high-power diode bars that are carefully arranged to form a diode array. The second key component of heat capacity lasers is the laser gain medium. The emergence of large volume transparent ceramics as a laser gain medium is a key development in heat capacity lasers, mainly because the output power of the heat capacity lasers increases linearly with the gain medium volume.

4 Optical fiber lasers

Because of their numerous advantages and potential applications, fiber lasers have attracted considerable attention since they first appeared in the 1960s. However, the output power of the early device was small, which restricted its application to a great extent. In the 1990s, with the invention of the double-clad fiber, the output of fiber lasers was greatly improved, which was comparable to that of the bulk lasers. Moreover, the characteristics of fiber lasers such as compact structure, high efficiency, and high beam quality are far superior than that of the traditional rod lasers. Therefore, in recent years, fiber laser technology and its applications have developed rapidly, becoming a new hot spot in the laser field.

In this chapter, after a brief introduction, we will introduce the working principle of fiber lasers. In Section 4.3, we will discuss the modes of the fibers and the conditions for single-mode operation. In Section 4.4, we will describe the double-clad technology that plays an important role in the development of fiber lasers. Finally, in Section 4.5, we will introduce the stimulated scattering fiber lasers.

4.1 Introduction

For the first time in 1961, Snitor at the American Optical Company observed the stimulated emission of radiation in Nd^{3+}-doped glass fiber. Soon after, he developed the first optical lasers with the transversely pumped coaxial flashlamp. Although the pumping efficiency was very low, the optical fiber had a high gain level. Over the next 10 years, the laser pumping system made significant progress. First, laser pumping was changed from transverse modes to longitudinal modes; second, the laser diode replaced the flashlamp as the pumping source. This new technology greatly reduced the threshold of the fiber laser and further improved the working efficiency.

Third, over the next 10 years, the working media also improved. As a result of the invention of the silicon-based optical fiber and the rare earth ion-doped optical element, the output of Er^{3+}-doped silica fiber lasers could completely cover the optical communication window of 1525~1565 nm C-band. Since then, the research of erbium-doped fiber lasers and amplifiers (EDFLA) have attracted great interest. At the same time, with the development of the heavy metal fluoride fiber, the compact visible light source which radiates blue, green, and red light has also shown good prospects. The narrow-line width fiber lasers, chirped fiber lasers, and lasers without any reflectors have also emerged successively.

With renewed focus on improving the power of fiber lasers, in the early 1990s, the double-clad fiber and laser diode array pump technology were introduced, which considerably improved the output power of the fiber laser. At present, the output power of the single-mode fiber laser is up to 2000 W, and the output of multimode

https://doi.org/10.1515/9783110500608-004

fiber lasers can reach up to 50 kW. An output of 105 kW can be obtained by utilizing coherent combination techniques.

In recent years, the output power and beam quality of fiber lasers have been further improved by excellent properties of some dopants such as Yb^{3+}.

4.2 Energy levels and spectra of several rare earth ions

In this section, we focus on several rare earth ions which have been widely used in the fiber laser field and have shown better performances. We mainly introduce the structure of their energy levels and the absorption and emission spectra in silicon- and fluorine-based fiber.

4.2.1 Introduction

It is clear from the previous section that the working medium of the fiber laser is a glass or crystal waveguide fiber doped with transition elements or rare earth ions. Therefore, it is conceivable that the working mechanism of the fiber laser is similar to that of the bulk laser. For example, the necessary and sufficient conditions for generating the laser can be discussed in reference to relevant contents discussed in Chapter 2. In particular, the gain coefficient can be expressed as a function of the energy level and the absorption and emission cross-sections. Here, we will only introduce some unique properties of fiber lasers. Because the working medium of the optical fiber is very long, the energy-level population will likely vary significantly with the position, on the Z-axis along the fiber axis. Gain coefficient at point Z can be expressed as

$$\mathrm{G}(Z) = N_u(Z)\sigma_{ul}(\upsilon) - N_L(Z)\sigma_{ul}(\upsilon) \tag{4.1}$$

The total gain coefficient is

$$\int_0^L Gdz = \int_0^L [N_u(Z)\sigma_{ul}(\upsilon) - N_L(Z)\sigma_{lu}(\upsilon)]dz \tag{4.2}$$

where $\sigma_{ul}(\upsilon)$ and $\sigma_{lu}(\upsilon)$ are scattering and absorption cross-sections at a frequency υ, and L is the total length of the fiber.

Because the fiber is extremely long, the gain obtained by equation (4.2) is also very high, which is one of the major advantages of fiber lasers.

First, even if the pumping level is not very high, a high gain can be obtained, implying that the fiber laser can easily pumped by the diode laser whose advantages over the flashlamp were described in Chapter 3.

Second, the higher gain allows for higher loss and lower quantum efficiency. Allowable higher loss provides greater freedom for people to design the laser resonator, whereas allowable lower efficiency enables gain at a frequency far away from the center of the emission line that is sufficient to produce the laser. Therefore, it is easy to achieve a wide wavelength tuning range. In fact, the tuning range of the fiber laser is not limited by the gain but by the radiation spectra of the emitting ions.

The above phenomena leads to a noteworthy result, that is, the fiber laser cannot be simply referred to as three- or four-energy level systems, with its performance lying in between. We will take an example shown in Fig. 4.1 to illustrate this point. In the region close to the end of the laser pumping where the pumping signal is strong, the particles on the ground state of E_0 are pumped to the level of E_2, and then relaxed to E_1 by nonradiative transition, and the radiation wave with a wavelength of λ_0 is generated by transition from E_1 to E_0. Because the particles returning to level E_0 are pumped to level E_2, the level E_0 remains almost empty, which is similar to that of the three-energy level system of the solid laser, as shown in Fig. 4.1(a).

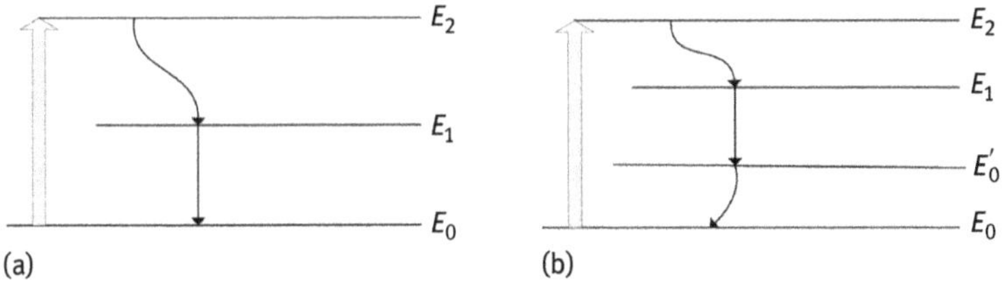

Fig. 4.1: An example of the energy-level diagram of the fiber laser. (a) three-energy level system without energy level splitting (b) the ground state splitting.

In the region far away from the end of the laser pumping, the pumping signal is too weak to make the level E_0 almost empty. However, the ground state is split into a new energy level $E_{0'}$ due to Stark effect, the latter is located above the level E_0, and the particles on it tend to be relaxed to level E_0, which enables it to remain empty. Thus, it is easy for particles to be transited from level E_1 to level E_0, generating waves with a wavelength λ' (larger than λ_0). Here, the operation mode is similar to that of the four-energy level system. Therefore, fiber lasers with different structures of the energy levels are referred to as a quasi three- or quasi four-energy level system.

A large number of energy levels of trivalent rare earth ions seem to offer a broad selection for potential laser transitions. However, in reality, many energy levels are unsuitable for laser energy because of their short lifetime. This is because particles on an energy level with a short lifetime will be relaxed into the low energy level by the non-radiative transition of multiphonon emission.

The multiphonon relaxation also has an important influence on the up-conversion laser, which is generated when particles are pumped to a high energy level by successive absorption of two or more photons first, followed by the radiation of a high-energy photon. To ensure that the above process proceeds smoothly, the high rate multiphonon relaxation process is not allowed in the intermediate energy level and the upper laser level.

Table 4.1 shows the transition spectra of dopants in the working media of some representative fiber lasers. Some of them will be further discussed below.

Table 4.1: Transition spectra of the working media of some representative fiber lasers.

Dopant	Energy level	Wavelength/ μm
Er	$^4I_{11/2} - {}^4I_{13/2}$	2.75(2.70~2.83)
	$^4I_{13/2} - {}^4I_{15/2}$	1.55(1.52~1.62)
	$^4S_{3/2} - {}^4I_{15/2}$	0.55
Ho	$^5I_6 - {}^5I_7$	2.9(2.83~2.95)
	$^5I_7 - {}^5I_8$	2.04
	$^5S_2 - {}^5I_8$	1.55
Nd	$^4F_{3/2} - {}^4I_{13/2}$	1.34
	$^4F_{3/2} - {}^4I_{11/2}$	1.06(1.05~1.14)
	$^4F_{3/2} - {}^4I_{9/2}$	0.94(0.9~0.95)
Tm	$^3F_4 - {}^3H_5$	2.3(2.25~2.5)
	$^3H_4 - {}^3H_6$	1.9(1.65~2.05)
	$^3F_4 - {}^3H_4$	1.47
	$^1G_4 - {}^3H_6$	0.48
Yb	$^2F_{5/2} - {}^2F_{7/2}$	1.04(1.01~1.17)

4.2.2 Laser energy levels and spectra of several rare earth ions in silicon optical fiber

The first Tm^{3+}-doped fiber laser had a silicon optical fiber as the substrate. The energy levels of Tm^{3+} in an aluminosilicate fiber is shown in Figs. 4.2(a), and 4.2(b) shows the corresponding absorption and emission spectra.

The typical transition of Tm^{3+} in silicon fiber occurs between $^3H_4$ and $^3H_6{}^3$, and the radiative central wavelength is 1.9 μm. Corresponding to this transition, radiation with low threshold, high efficiency, and a wide tuning range has been realized.

Pumping the particle to the upper-level $^3H_4$ can be either a single photon absorption process or a multiphoton absorption process, the latter is also called an upconversion

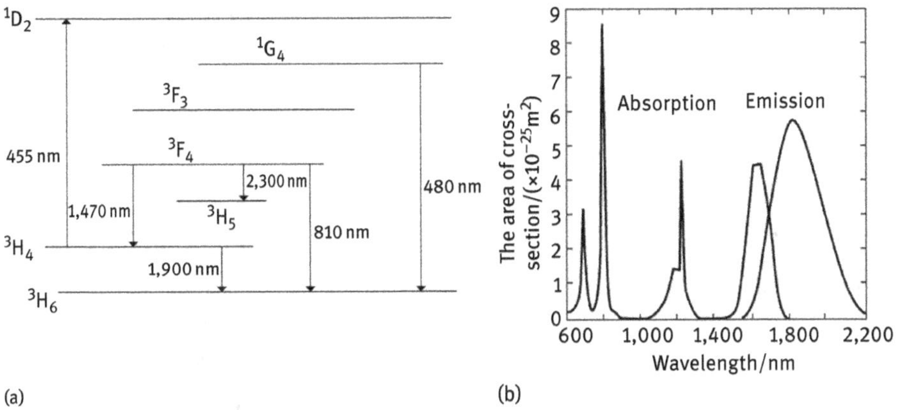

Fig. 4.2: Energy level diagram of Tm^{3+} in an aluminosilicate fiber (a) and typical emission and absorption cross-sections (b).

laser process. In single photon absorption, the radiation at a wavelength of 0.79 μm pumps the particle to the $^3F_4$ level. The rapid multiphonon relaxation results in an accumulation of a large number of particles on the upper level $^3H_4$ while the lower level $^3H_6$ remains almost empty. Thus, radiation transition with low threshold and high efficiency will be achieved, with the laser emitting at 1.9 μm.

To some extent, the lifetime of particles on the $^3H_4$ level is effected by the nonradiative multiphonon relaxation, which reduces from 6 ms (pure radiation) to hundreds of microseconds. The reduction in lifetime also increases pumping threshold, which does not impact the slope efficiency. In fact, the slope efficiency can be significantly larger than the radiation quantum efficiency because a part of energy of Tm^{3+} pumped into $^3F_4$ level is passed to nearby particles on the $^3H_6$ level, which enables the two to reach the $^3H_4$ energy level. The above-mentioned case often occurs in low energy phonon substrates such as zirconium fluoride (ZBLAN, ZrF_4–BaF_2. LaF_3–AlF_3–NaF) with large dopants.

Because of the reduced relaxation rate of the excited level, the Tm^{3+} particles show more laser transitions in ZBLAN than in silicon. For example, after Tm^{3+} particles are pumped to $^3F_4$, in silicon, through multiphonon relaxation the Tm^{3+} particles reach the $^3H_4$ level and then jump to the $^3H_6$ level, with the laser emitting at 1.9 μm. While in ZBLAN, the particles in the $^3F_4$ level can also jump to the $^3H_5$, $^3H_4$, and $^3H_6$ levels, with the laser emitting at 2.3 μm ($^3F_4$ to $^3H_5$), 1.47 μm ($^3F_4$ to $^3H_4$) and 810 nm ($^3F_4$ to $^3H_6$), respectively, resulting in the reduction of pump quantum efficiency during the transition between ($^3F_4$ to $^3H_4$ and to $^3H_6$). To obtain the optimum working state of laser, such as the minimum threshold and the maximum efficiency, ideally, the ground state should have the largest phonon energy between ZBLAN and Si. In this manner, on one hand, the efficiency of nonradiative transitions from $^3H_4$ to $^3H_4$ is higher; on the other hand, the lifetime of $^3H_4$ level can basically be retained at a pure radiation value. This has been achieved in Tm^{3+}-doped germanate glass fiber, which shows a lower threshold at 1.9 μm than that for ZBLAN or Si glasses.

A typical example of multiphoton absorption process is three-photon absorption. In this, the pump source is a Nd^{3+}:YAG laser operated at 1.064 μm, through which blue light can be emitted. The process is as follows: first, on absorbing a pump photon, particles jump from the $^3H_6$

level to the $^3H_5$ level then into the $^3H_4$ level on relaxation; second, on absorbing another pump photon, the particles jump from the $^3H_4$ level into the $^3H_{2,3}$ level, then into the $^3H_4$ level by relaxation, and finally, on absorbing a pump photon again, the particles jump from the $^3F_4$ level into the $^1G_4$ level, and blue light is emitted during the transition from $^1G_4$ to $^3H_6$. The first upconversion fiber lasers used the Tm^{3+}:ZBLAN fiber whose ion pump was realized by two-photon absorption of red light. After that, it was noted that the Tm^{3+}:ZBLAN upconversion fiber laser has very good performance when the ion pump was realized by absorption of three photons of infrared radiation. Longer wavelength (~1.12 μm) pump radiation can increase the absorption cross-section of the pump. The upconversion fiber lasers have attracted considerable interest, and their overall efficiency (from pumping photons to outputting photons) can be as high as 30%.

Er^{3+}-doped silica fiber laser is another type of laser that has received considerable attention and plays an important role in the development of fiber laser technology. The main reason for this is that its radiation spectrum just covers the general C-band communication window of 1525~1565 nm. The partial energy levels and absorption and emission spectra of Er^{3+} in Si are shown in Fig. 4.3.

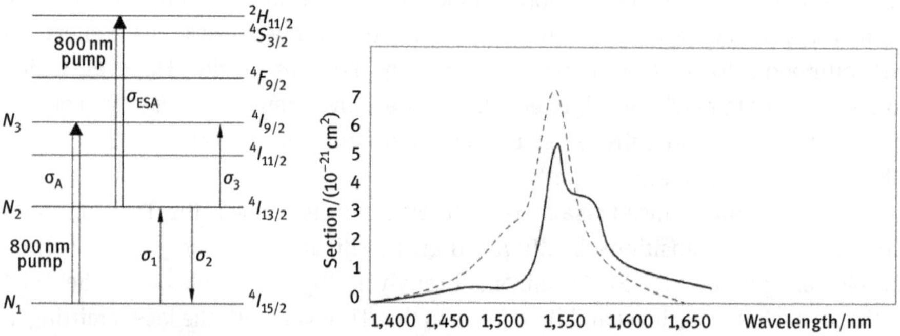

(a) Energy level diagram (b) Absorption and emission spectra

Fig. 4.3: The partial energy levels of Er^{3+} in Sib (a), the absorption (imaginary line) and emission spectra(solid line) of Er^{3+} in Si (b).

It is not difficult to see from Fig. 4.3 that the radiation at 1480 nm is more suitable to be a pumping source, which has a large absorption cross section and a small emission cross section. In addition, since the wavelength of an ideal light source is near the 800 nm, 800 nm is also a potential pump band which can be seen from the energy level diagram (a).

Yb^{3+}-doped fiber laser has been widely used in recent years. Figure 4.4(a) shows a simplified two-energy level diagram of Yb^{3+} ions in silicon optical fiber, and Fig. 4.4(b) shows a typical emission and absorption spectrum of aluminum silicate and phosphorus silicate fibers, including the formation of subenergy levels induced by Stark splitting.

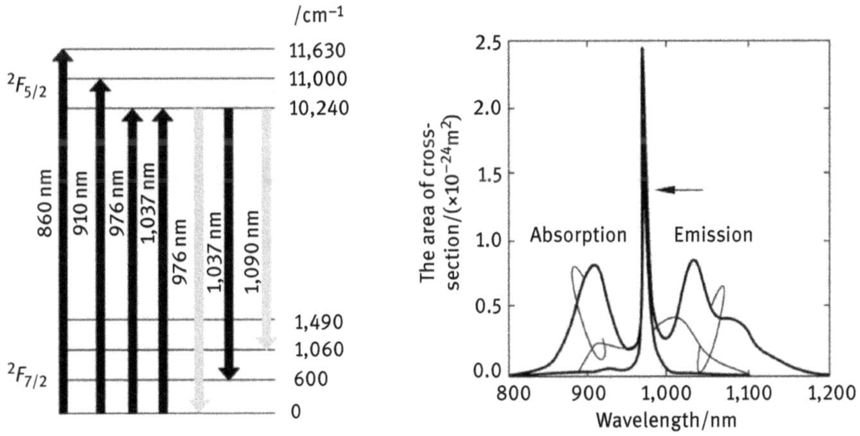

(a) Two-energy level system (b) Absorption and emission spectra

Fig. 4.4: Simplified two-energy level diagram of Yb^{3+} ions in silicon optical fiber (a) and a typical emission and absorption spectrum of aluminum silicate and phosphorus silicate fibers (b).

4.2.3 Laser energy levels and spectra of several rare earth ions in fluoride optical fiber

Figures 4.5 to 4.8 show the laser energy levels and the corresponding absorption and emission spectra of several rare earth ions in fluoride optical fiber.

Figure 4.8(a) shows a way to emit 480 nm blue light on excitation by 1140 nm laser. In ZBLAN glass optical fiber, the fluorescence spectrum of Yb^{3+} was found to have a single emission band centered at 980 nm, and the absorption spectrum of Yb^{3+} consisted of a single band centered at 974 nm.

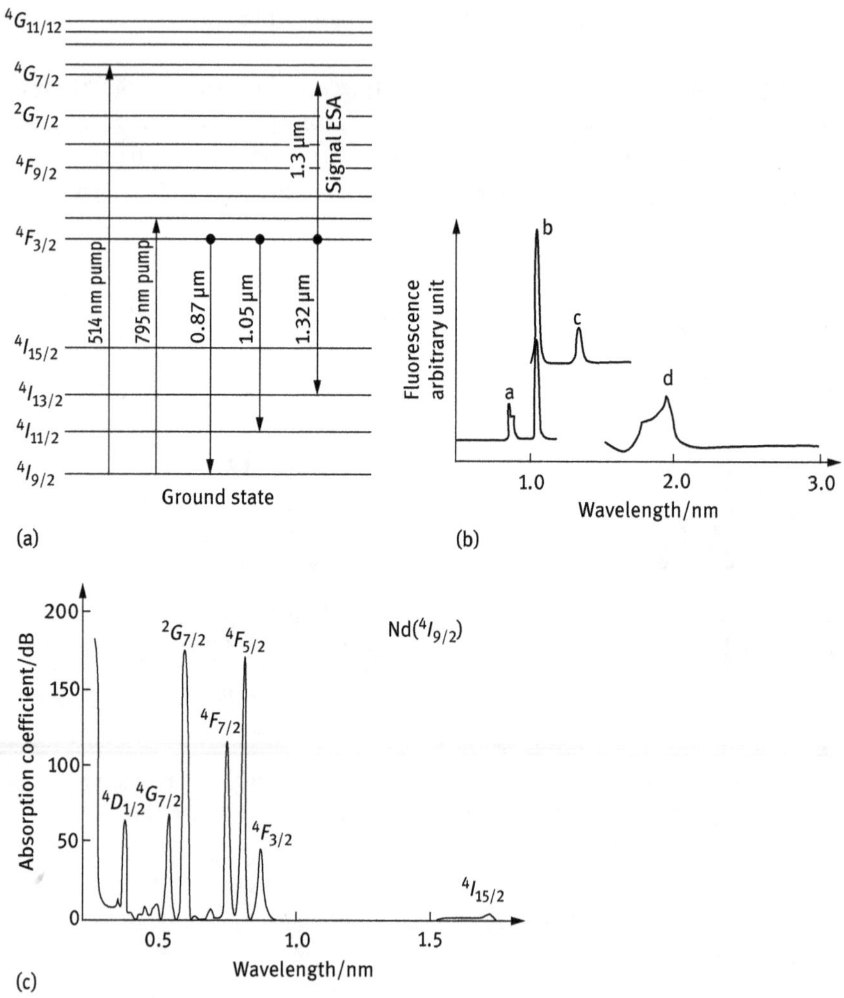

Fig. 4.5: Energy level diagram for Nd^{3+} (a) and typical emission (b) and absorption (c) cross-sections in ZBLAN glass optical fiber.

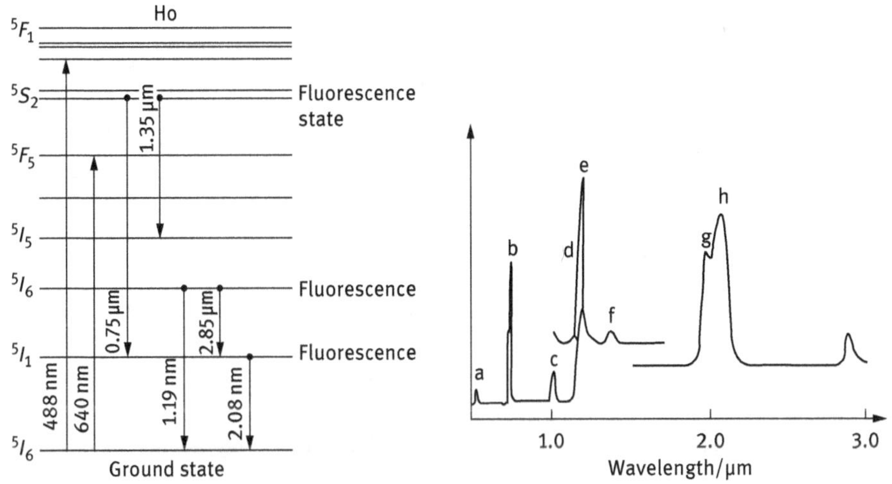

(a) Energy level diagram

(b) Emission spectra

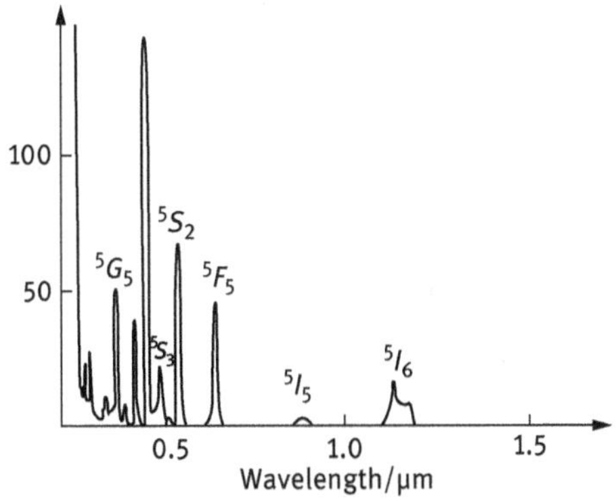

(c) Absorption spectra

Fig. 4.6: Energy level diagram for Ho^{3+} (a) and typical emission (b) and absorption (c) cross-sections in ZBLAN glass optical fiber.

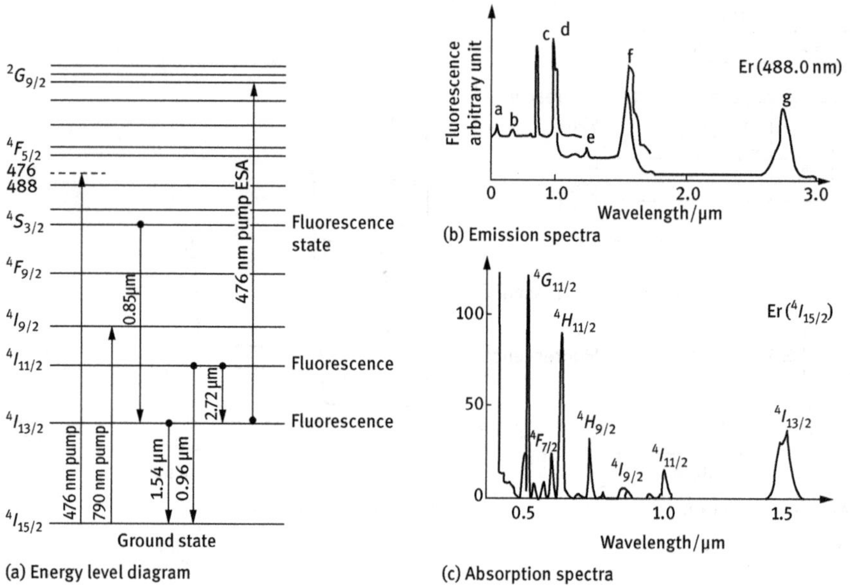

Fig. 4.7: Energy level diagram for Er^{3+} (a) and typical emission (b) and absorption (c) cross-sections in ZBLAN glass optical fiber.

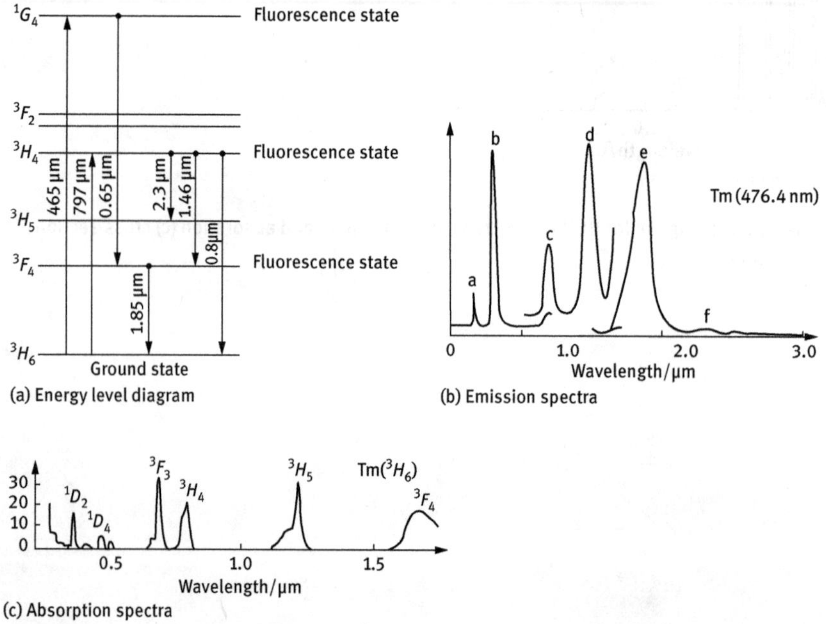

Fig. 4.8: Energy level diagram for Tm^{3+} (a) and typical emission and absorption cross-sections in ZBLAN glass optical fiber.

4.3 Mode and conditions for single-mode operation

In this section, we will derive the mode, cutoff frequency, and conditions for single-mode operation in optical fiber media. For convenience, the related concepts of bulk media will be introduced first.

4.3.1 Bulk media

One of the important advantages of optical fiber laser is that its threshold pump power is significantly lower than that of bulk devices, which will be discussed in this subsection.

Figure 4.9 shows a simplified diagram of the longitudinal pumping operation of a bulk laser. The laser beam from the pump source is focused on the focal plane in the working medium, with a beam waist size of ω_0, which is transmitted forward with a divergence angle of $2\Delta\theta$. A relationship exists between $\Delta\theta$ and ω_0. If the mode of pump beam is TEM_{00}, then

$$\Delta\theta = \frac{\lambda}{\pi\omega_0} \tag{4.3}$$

However, if the pump light source is multimode, the spot diameter in one plane is M times that of the diffraction limited spot, and the divergence angle after the focal plane is approximately M times that of the diffraction limited beam having the same waist spot, that is,

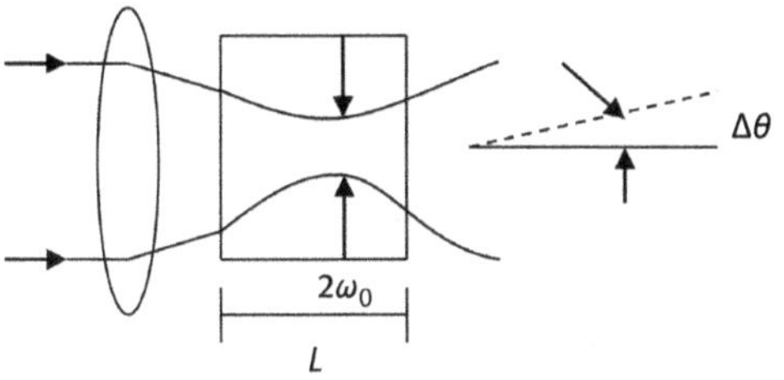

Fig. 4.9: Schematic diagram of longitudinal pumping operation of a bulk laser.

$$\Delta\theta = M \times \frac{\lambda}{\pi\omega_0} \tag{4.4}$$

For a four-level laser system without ground-state depletion, the gain coefficient will increase linearly with the pumping intensity. However, in general, when the ground-state depletion exists, the gain coefficient will only monotonically increase with the pump light intensity. Thus, the total pumping intensity through the gain medium once is L, where L is the length of the operating medium and is the average pumping intensity along the length of the gain medium. Obviously, the stronger the convergence of the pump beam to the gain medium, the smaller the spot, but the greater the divergence. For a given gain medium, the optimum focusing condition is given approximately by Equation (4.5), which can result in the largest L.

$$L_{opt}\Delta\theta \approx 3\omega_0 \tag{4.5}$$

At TEM_{00} mode laser pumping, substituting Equation (4.3) into (4.5), we get

$$L_{opt} \approx \frac{3\pi\omega_0^2}{\lambda} \tag{4.6}$$

The choice of L depends on the maximum pump power required to be absorbed by the working medium. If only 10% of the pump power is absorbed by the gain medium, the quantum efficiency of the laser will not exceed 10%. In rare earth ion-doped materials, the effective length of the working medium L_{opt} can be reduced by increasing the concentration of active ions. However, if the concentration of rare earth ions in materials is high and the distance between adjacent ions is small, then the ion–ion interaction may affect particle dynamics in two ways described below, which can further affect laser performance.

One way is shown in Fig. 4.10(a): two adjacent ions are initially at the same level E_f which is equidistant from the laser lower energy level E_e and the upper level E_h. Thus, when it is excited, an ion may jump up into the E_h level and another may jump down to the E_e level. This will undoubtedly lead to an increase in pump threshold. However, if ΔE_{he} is greater than the energy of the pump photon, a useful synergistic upconversion laser will be obtained, that is, the laser output wavelength will be shorter than the pump wavelength.

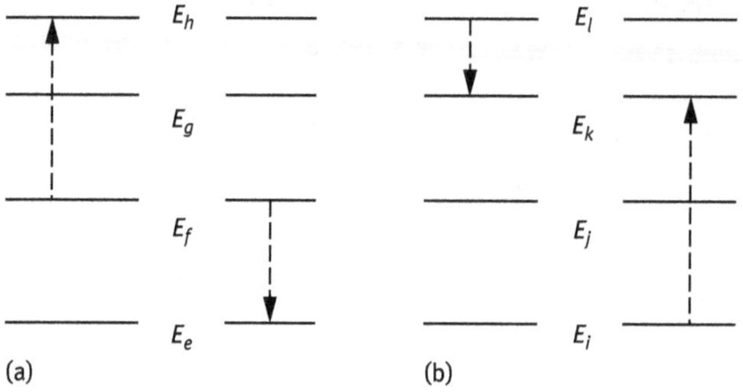

Fig. 4.10: Schematic diagram of adjacent ion transition in high doping medium.

Another way is simply the opposite: the two adjacent ions that are initially, respectively, in the high-energy state E_l and the ground state E_i reach the same final state E_k after transition, as shown in Fig. 4.10(b). For example, in Nd^{3+}, E_l is the upper level of 1.06 μm laser radiation, the population on the upper level will be consumed, resulting in an increase in threshold pump power.

4.3.2 Optical fiber working material

Fiber lasers are different from bulk devices (see Fig. 4.11). No matter how big the fiber size is, once the pump power enters into the core, it will be transmitted continuously

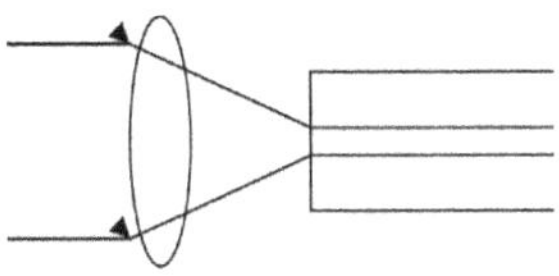

Fig. 4.11: Schematic of fiber laser pumping.

until completely absorbed. The diameter of a core is only a few microns, which makes the pumping intensity in fiber to be 100 times as large as that in bulk devices, or even higher in some cases. Under the same conditions, the threshold power of fiber laser is expected to decrease proportionally. For a four-level system, this threshold reduction may not be the most important because the threshold pump power accounts for only a small fraction of the maximum pump power. However, for a three-level system, a high pumping intensity is required only to achieve the threshold. Another advantage of fiber lasers is that the pump spot size and the length of the operation material are independent of each other. Thus, if the rare earth ion dopants in the material are reduced by half, the length of the optical fiber can be simply doubled to maintain the excitation efficiency. This property is especially important when ion–ion interactions do not affect system performance when rare earth ion doping is required to maintain sufficiently low concentration. In addition, long working material enables weak absorption to excite laser transition, which becomes important when the weak pumping band coincides with the wavelength of the commercial LD.

When we design fiber laser oscillators and amplifiers, the efficiency of the pumped laser entering into the fiber core is an important issue that needs to be considered. The efficiency depends on two parameters: the allowed number of transverse modes and the spatial distribution of the pumped beam. When the TEM00 mode beam enters into the single-mode fiber, the efficiency is expected to be 50% in typical cases. However, when it enters into the multimode waveguide, this efficiency may be higher. If the pumped laser is multi-transverse mode, such as an LD array, but the fiber is single-mode, then the efficiency of the pumped laser entering into the optical fiber is low. It is still assumed that the divergence angle of multimode-pumped beam is M times that of the single-mode beam, and the efficiency is approximately one-Mth of 50%, or $1/M \times 50\%$. Although the single-mode fiber is preferred in many cases, the multimode fiber is still superior to bulk material in reducing the threshold power as long as the fiber length satisfies the condition of $L \gg 3\pi a^2/\lambda$, where a is the radius of the core.

4.3.3 Mode property and cutoff frequency

As discussed in the previous subsection, the threshold pump power of the fiber laser is closely related to the allowed modes in the fiber. This subsection will discuss the mode property of the optical fiber.

4.3.3.1 Wave equations in cylindrical coordinates

It is convenient to express the wave equations in cylindrical coordinates (γ, φ, z) because the refractive indexes of most fibers are column symmetric, and the field components are E_γ, E_φ, E_z; H_γ, H_φ, and H_z. Because unit vectors in γ, φ direction are not constant anymore, the wave equations which include the transverse components are very complicated, however, the Z-component that satisfies the Helmholtz equation is relatively simple.

$$(\nabla^2 + k^2)\begin{pmatrix} E_z \\ H_z \end{pmatrix} = 0$$

Therefore, we usually solve E_z and H_z first, and then get all transverse components by means of $E_z H_z$. Because E_z and H_z satisfy the same form of the equation, only one of them need to be solved, such as E_z.

In cylindrical coordinates, Laplace operator is expressed by equation (4.7)

$$\nabla^2 = \frac{1}{r}\frac{\partial}{\partial r}\left(r\frac{\partial}{\partial r}\right) + \frac{1}{r^2}\frac{\partial^2}{\partial \phi^2} + \frac{\partial^2}{\partial Z^2} \tag{4.7}$$

where

$$k^2 = \omega^2 n^2 / c^2$$

where ω is the angular frequency of light; n is the refractive index of light; and c is the velocity of light in vacuum.

Equation (4.7) can be solved using variables separation method, and the solutions are

$$E_z = \varphi(r)e^{\pm il\varphi}, l = 0, 1, 2, \ldots$$

The above solution is a single-valued function of φ, so equation (4.7) changes to L-order Bessel equation

$$\left[\frac{\partial^2}{\partial r^2} + \frac{1}{r}\frac{\partial}{\partial r} + \left(k^2 - k_2^2 - \frac{l^2}{r^2}\right)\right]\varphi(r) = 0 \tag{4.8}$$

The general solutions of equation (4.8) are as follows

$$\varphi(r) = \begin{Bmatrix} C_1 J_l(hr) + C_2 Y_l(hr), h^2 = k_1^2 - k_z^2 > 0 \\ C_1 I_l(qr) + C_2 K_l(qr), q^2 = k_z^2 - k_2^2 > 0 \end{Bmatrix} \tag{4.9}$$

where C_1 and C_2 are constants; J_l and Y_l are L-order Bessel functions of the first kind and second kind, respectively; and I_l and K_l are their deformations, respectively. If the original function form are used, then the arguments are purely imaginary; thus, the variants of Bessel functions are also called the imaginary order Bessel functions.

Next taking $J_l(x)$ as an example, we will further discuss equation (4.9). For any value of x, the integer-order Bessel function

$$J_l(x) = \sum_{m=0}^{\infty} \frac{(-1)^m}{m!} \frac{1}{(l+m)!} \left(\frac{x}{2}\right)^{2m+l}$$

can be written as

$$J_l(x) = \frac{1}{l!} \left(\frac{x}{2}\right)^l (1+\theta)$$

where

$$\theta \le \frac{1}{l+1} \left\{ exp\left(\frac{|x|^2}{4} - 1\right) \right\}$$

When $x \ll 1$

$$\theta \le \frac{1}{l+1} \left\{ \frac{|x|^2}{4} + \frac{1}{2}\left(\frac{|x|^2}{4}\right)^2 + \ldots \right\} \ll 1$$

Then, we get the asymptotic expression

$$J_l(x) \to \frac{1}{l!} \left(\frac{x}{2}\right)^l, x \ll 1$$

However, when $x \to \infty$, we get

$$J_l(x) \sim \frac{1}{\sqrt{2\pi x}} \left\{ exp\left[i\left(x - \frac{l\pi}{2} - \frac{\pi}{4}\right)\right] + c.c. \right\} = \sqrt{\frac{2}{\pi x}} cos\left(x - \frac{l\pi}{2} - \frac{\pi}{4}\right)$$

Then, the asymptotic expression is as follows

$$J_l(x) \to \sqrt{\frac{2}{\pi x}} \cos\left(x - \frac{l\pi}{2} - \frac{\pi}{4}\right), x \to \infty$$

For other functions, a similar method is used to derive the asymptotic expression when the argument is very small or very large, and for further simplification, only the first or second term of the expansion are retained; finally, we can obtain

For $x \ll 1$,

$$I_l(x), J_l(x) \to \frac{1}{l!} \left(\frac{x}{2}\right)^l$$

$$Y_0(x) \to \frac{2}{\pi}(x + 0.88 + \ldots) \tag{4.10}$$

$$Y_l(x), K_l(x) \to \frac{(l-1)!}{\pi} \left(\frac{2}{x}\right)^l$$

$$K_0(x) \to -(x + 0.88 + \ldots)$$

For $x \gg Max\{1, l\}$

$$J_l(x) \to \sqrt{\frac{2}{\pi x}}\cos\left(x - \frac{l\pi}{2} - \frac{\pi}{4}\right)$$

$$Y_l(x) \to \sqrt{\frac{2}{\pi x}}\sin\left(x - \frac{l\pi}{2} - \frac{\pi}{4}\right) \tag{4.11}$$

$$I_l(x) \to \frac{1}{\sqrt{2\pi x}}e^x$$

$$K_l(x) \to \sqrt{\frac{\pi}{2x}}e^{-x}$$

When x changes from small to large, the corresponding asymptotic expressions exist in the area of $x{\sim}l$, where l is a non-negative integer.

Substituting equations (4.10) and (4.11) into (4.9), we can get, $\Psi(r)$ and obtain E_z, after getting E_z and H_z, we can use them to express the transverse components of E and H. For example, we can write the following equations by Maxwell curl equation.

$$i\omega\varepsilon E_r = ik_z H_\phi + \frac{1}{r}\frac{\partial}{\partial\phi}H_z$$

$$i\omega\mu H_\phi = ik_z E_r + \frac{1}{r}\frac{\partial}{\partial\phi}E_z$$

We can get E_r or H_ϕ by solving the above equations simultaneously, and E_r or H_ϕ can be got by the similar way, so we can obtain all field components in column symmetrical refractive index optical fiber, that is,

$$E_r = \beta\left(\frac{\partial E_z}{\partial r} + \frac{\omega\mu}{rk_z}\frac{\partial H_z}{\partial\phi}\right)$$

$$E_\phi = \beta\left(\frac{1}{r}\frac{\partial E_\phi}{\partial\phi} - \frac{\omega\mu}{k_z}\frac{\partial H_z}{\partial r}\right) \quad H_r = \beta\left(\frac{\partial H_z}{\partial r} - \frac{\omega\varepsilon}{rk_z}\frac{\partial E_z}{\partial\phi}\right) \tag{4.12}$$

$$H_\phi = \beta\left(\frac{1}{r}\frac{\partial E_\phi}{\partial\phi} + \frac{\omega\varepsilon}{k_z}\frac{\partial E}{\partial r}\right)$$

where

$$\beta = \frac{ik_z}{k_z^2 - \omega^2\mu\varepsilon}$$

4.3.3.2 Gradient refractive index circular waveguide

As a special case, the gradient index circular waveguide is considered, as shown in Fig. 4.12, it consists of two parts: one is a core with the refractive index of n_1 and the radius of r_1, the other is a cladding with a refractive index of n_2 and a radius of r_2.

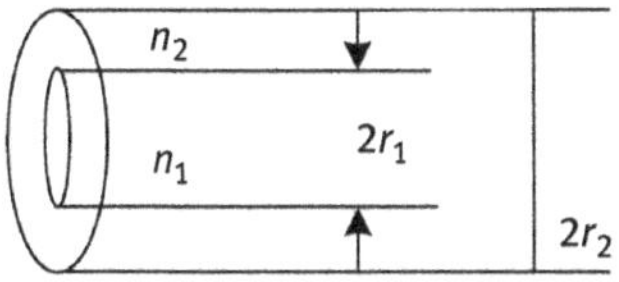

Fig. 4.12: Gradient refractive index circular waveguide.

Usually, $n_1 > n_2$ and r_2 is large enough, therefore, in fact, the field of constrained mode is 0 at $r = r_2$ and we usually make $r_2 = \infty$.
The dependence of E_z and H_z on r can be expressed by equation (4.9) for constraint propagation. If $k_z > n_2\omega/c$, the wave will fade away in the cladding. Therefore, in the region of $r_1 < r \le r_2$, in equation (4.9), $c_1 = 0$

$$E_z(\mathbf{r}, t) = Ck_1(qr)exp[i(\omega t + l\phi - k_z z)]$$
$$H_z(r, t) = Dk_1(qr)exp[-i(\omega t + l\phi - k_z z)], r_1 < r \le r_2 \tag{4.13}$$

where C and D are both constants, but

$$q = \left(k_z^2 - n_2^2\omega^2/c^2\right)^{1/2}$$

For the field in the fiber core, it is necessary to consider the asymptotic behavior of the field when $r \to 0$. Because the field must be limited at $r = 0$, but Y_l and K_l are both divergent at $r \to 0$, it is necessary to make $c_2 = 0$. Taking note of the fact the tangential field components in the core and the cladding are matching, so we can get

$$E_z(\mathbf{r}, t) = AJ_l(hr)exp[i(\omega t + l\phi - k_z z)]$$
$$H_z(r, t) = BJ_l(hr)exp[-i(\omega t + l\phi - k_z z)], r < r_1 \tag{4.14}$$

where A and B are both constants, but

$$h = \left(n_1^2\omega^2/c^2 - k_z^2\right)^{1/2}$$

In equations (4.13) and (4.14), we take positive sign before l; if the sign is negative, then we can get another independent solution which has the same radiation dependence. In physics, the meaning of l is similar to the quantum number of the Z component of photon orbital angular momentum in column symmetry potential field. The positive sign means clockwise rotation around Z-axis, whereas the negative sign means anticlockwise rotation. Because the fiber does not rotate in a particular direction, the above two stags are degenerate.

Substituting equations (4.13) and (4.14) into (4.12), we can calculate all field components in fiber core and cladding. These fields should meet the boundary conditions, that is, E_ϕ, E_z, H_ϕ, and H_z must keep continuous at $r = r_1$, in the region of $\left[\frac{n_1\omega}{c}, \frac{n_2\omega}{c}\right]$, only finite eigenvalues K_z can meet the above condition, once we find these K_z, the coefficients should satisfy the following equations,

$$\frac{C}{A}=\frac{J_l(hr_1)}{K_l(qr_1)},\ \frac{D}{A}=\frac{J_l(hr_1)}{K_l(qr_1)}\frac{B}{A}$$

$$\frac{B}{A}=\frac{ik_z l}{\omega\mu}\left(\frac{1}{q^2r_1^2}+\frac{1}{h^2r_1^2}\right)\left[\frac{J_l'(hr_1)}{hr_1J_l(hr_1)}+\frac{K_l'(qr_1)}{qr_1K_l(qr_1)}\right]^{-1} \tag{4.15}$$

where J_l and K_l are the derivatives of J_l and K_l with respect to their arguments, respectively.

In this chapter, we are particularly interested in B/A because it is the measurement of relative value between E_z and H_z of the mode propagating in the optical fiber, that is,

$$\frac{B}{A}=\frac{H_z}{E_z}$$

4.3.3.3 Mode property and cutoff frequency

The slab guide modes can be divided into two kinds, namely, TE and TM, and the cylindrical waveguide mode are also divided into two kinds, namely, EH and HE, which can be written as

$$\frac{J_{l\pm1}(hr_1)}{hr_1J_l(hr_1)}=\pm\left(\frac{n_1^2+n_2^2}{2n_1^2}\right)\frac{K_l'(qr_1)}{qr_1K_l(qr_1)}+\left(\frac{l}{h^2r_1^2}-R\right) \tag{4.16}$$

When the sign in equation (4.16) is positive or negative, its solution corresponds to the EH or HE mode. Where

$$R=\left[\left(\frac{n_1^2-n_2^2}{2n_1^2}\right)\left(\frac{K_l'(qr_1)}{qr_1K_l(qr_1)}\right)^2+\left(\frac{lk_z}{n_1k_0}\right)^2\left(\frac{1}{q^2r_1^2}+\frac{1}{h^2r_1^2}\right)^2\right]^{1/2} \tag{4.17}$$

Obviously,

$$q^2+h^2=k_0^2\left(n_1^2-n_2^2\right)$$

The normalized frequency is adopted,

$$V=k_0r_1\left(n_1^2-n_2^2\right)^{\frac{1}{2}} \tag{4.18}$$

Then we can get

$$q^2r_1^2=V^2-h^2r_1^2 \tag{4.19}$$

When $l=0$, taking note of $K_0'(x)=-K_1(x), J_{-1}(x)=-J_1(x)$ we obtain

$$R=\frac{n_1^2-n_2^2}{2n_1^2}\left[\frac{-K_1(qr_1)}{qr_1K_{0(qr_1)}}\right]$$

For HE mode, the right-hand side of equation (4.16) is

$$\text{right side} = -\left(\frac{n_1^2+n_2^2}{2n_1^2}\right)\left[\frac{-K_1(qr_1)}{qr_1K_0(qr_1)}\right] - \frac{n_1^2-n_2^2}{2n_1^2}\left[\frac{-K_1(qr_1)}{qr_1K_0(qr_1)}\right]$$
$$= \frac{K_1(qr_1)}{qr_1K_0(qr_1)} \approx -\frac{1}{(qr_1)^2}\frac{2}{ln\left[(qr_1)^2\right]}$$

Then

$$\frac{J_1(hr_1)}{hr_1J_0(hr_1)} = \frac{-K_1(qr_1)}{qr_1K_0(hr_1)} = \frac{2}{(V^2-h^2r_1^2)ln(V^2-h^2r_1^2)},\quad \text{when } hr_1 \to V \tag{4.20}$$

However, for EH mode, equation (4.21) can be deduced similarly,

$$\frac{J_1(hr_1)}{hr_1J_0(hr_1)} = -\frac{n_2^2}{n_1^2}\frac{2}{(V^2-h^2r_1^2)ln(V^2-h^2r_1^2)} \tag{4.21}$$

Equations (4.20) and (4.21) are both transcendental equations, which can be solved by the graph method. First, to guarantee the field in fiber cladding decays exponentially, q required by the constraint mode should be real, so the field of $hr_1 \in [0, V]$ needs to be considered only. The right-hand side of equation (4.21) is always negative, at the starting point, $hr_1 = 0$, $qr_1 = V$, then, we can get

$$\frac{-K_1(qr_1)}{qr_1K_0(qr_1)} = -\frac{K_1(V)}{VK_0(V)}$$

It is a number slightly less than 0. With the increase in hr_1 and the decrease in qr_1, $\frac{-K_1(qr_1)}{qr_1K_0(qr_1)}$ drops to $-\infty(hr_1 \to V)$ monotonically (see the dotted lines in Fig. 4.13). At the starting point of $hr_1 = 0$ the left hand of the above equation is

$$\frac{J_1(hr_1)}{hr_1J_0(hr_1)} = \frac{hr_1}{2}\frac{1}{hr_1} = \frac{1}{2}$$

After leaving $hr_1 = 0$, both J_0 and J_1 are oscillatory attenuation functions, and the value of $\frac{J_1(hr_1)}{hr_1J_0(hr_1)}$ rises monotonically and tends to $+\infty$ at the first zero point of J_0, that is $hr_1 \approx 2.045$. Thereafter, it changes from $-\infty$ to $+\infty$ among the two zero points of $J_0(hr_1)$, and the asymptotic line is $J_0 = 0$ which is shown as the solid line cluster in Fig. 4.13.

If the maximum value of hr_1 is expressed as $(hr_1)_{MAX} = V < 2.405$, if V is between the first and second zero point, then the curve has an intersection, that is, a single mode operation is allowed. If V is between the second and third zero point, the curve has two intersection points. Therefore, we let the mth zero point of J_0 be x_{om}, then for TEom (or TMom) mode, the cutoff frequency can be determined by

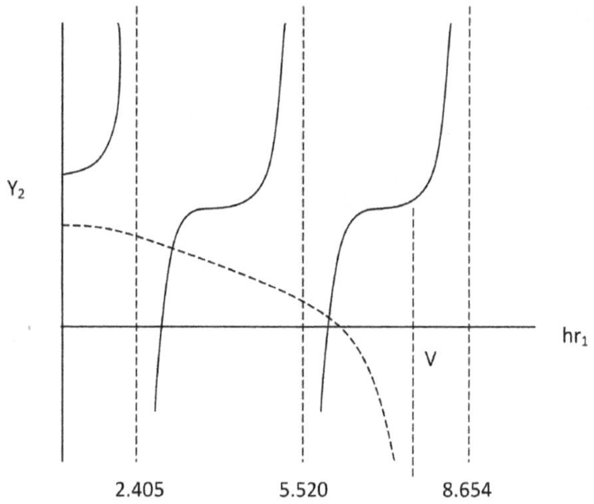

Fig. 4.13: Gradient index fiber mode

$$x_{0m} = V = \frac{2\pi r_1}{\lambda}\left(n_1^2 - n_2^2\right)^{\frac{1}{2}}$$

Then, we can get

$$\left(\frac{r_1}{\lambda}\right)_{0m} = \frac{x_{0m}}{2\pi\sqrt{n_1^2 - n_2^2}} \tag{4.22}$$

where the zero points are $x_{01} = 2.405$; $x_{02} = 5.520$; $x_{03} = 8.654$

For the larger m, there is an approximate recurrence relationship

$$x_{0m} \approx x_{0(m-1)} + \pi \approx 2.4 + (m-1)\pi$$

That is, the distance between the adjacent zero points is π.

4.3.4 The basic structure of optical fiber lasers

The basic structure of optical fiber lasers is shown as Fig. 4.14. A fiber works as the working medium, and an optical resonator consists of two reflectors. The laser is completely reflected and the pump light is completely transmitted by the input

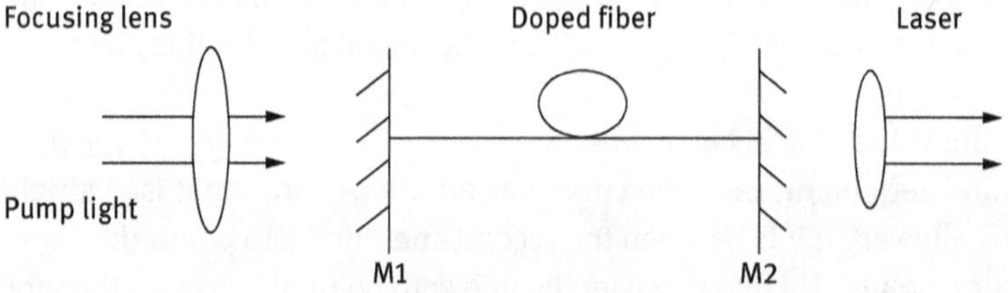

Fig. 4.14: The basic schematic diagram of doped fiber lasers.

reflective mirror M1, which makes the pump light to be used effectively and prevent the output light from becoming unstable owing to the resonance; the laser is partly transmitted by the output reflective mirror M2, and its transmittance is determined by the working state and output power requirements and other factors.

The selection of pump light source is determined by the characteristics of dopants or the desired operating wavelength. LD or its array or Ti:sapphire or other fiber lasers laser can be used as the pumping source.

In this subsection, we only introduce the basic structure of fiber lasers. In recent years, the development of fiber laser is very rapid. To obtain different characteristics to meet the requirements of various applications, a wide variety of devices have been introduced. Due to space limitations, they are not all described here.

4.4 Double-clad fiber laser

4.4.1 Limitation of the single-clad fiber

In many practical applications, the average power of the laser is required to be high. However, the early single-clad, single-mode fiber laser is hard to achieve high power output. In fact, the maximum continuous power that the fiber laser can output can be estimated by the cross-section or the diameter $2r_1$ of the fiber core. We assume that the cross-section of the fiber core is represented by μm^2, and at the current level, $1w/\mu m^2$ is the accepted safety value, which will not cause fiber damage. In addition, the highest value is no more than $1.5w/\mu m^2$. Thus, to improve the output power, we can only use the fiber with a thicker core. However, we want the fiber laser to work in the condition of fundamental transverse mode to achieve a high-quality beam. According to the discussion in the above section, this limits that the cross-section of the fiber core cannot be too large. Specifically, according to equation (4.24), the diameter of the fiber core cannot exceed

$$2r_1 = \frac{2.405\lambda}{\pi\sqrt{n_1^2 - n_2^2}}$$

If the typical value of the numerical aperture is obtained, that is $\left(n_1^2 - n_2^2\right)^{1/2} = 0.15$, then we can get

$$2r_1 \approx 5\lambda$$

where λ is the wavelength of the output laser. Supposing that $\lambda = 1.5\mu m$, then we get

$$2r_1 \approx 7.5\mu m$$

If this fiber is used as the gain medium, the safety value of the output power of the laser is only approximately 44 W and the maximal value is no more than 66 W. There is no doubt that this output power is far from enough in many practical applications.

Therefore, since the first fiber laser was introduced 30 years ago, its development has been gradual. Not until the double-clad fiber was used as the gain medium of the laser and the high-power high-quality laser output was achieved in the late 1980s of the last century, did the fiber laser technology start to develop rapidly.

4.4.2 Double-clad fiber laser

Double-clad fiber laser nearly has all the advantages of the traditional single-mode fiber laser, but not liking the latter, it does not have the limitation of only being pumped by the low-power, single-mode LD. Thus, double-clad fiber laser can use multimode high-power LD or its array to pump to gain the high efficient and high-power, single-mode laser output.

4.4.2.1 Double-clad structure

The double-clad fiber has three layers. The innermost layer is the fiber core whose refractive index is n_1 and is surrounded by the inner cladding whose refractive index is n_2. Then, there is an outer cladding whose refractive index is n_3 outside the inner cladding. Moreover, the relationship between the refractive indexes of each layer is $n_1 > n_2 > n_3$. The cross-section and refractive index of a standard double-clad fiber are shown in Fig. 4.15(a). For contrast, the cross-section and partitioned refractive index diagram of the standard single-mode fiber are shown in Fig. 4.15(b).

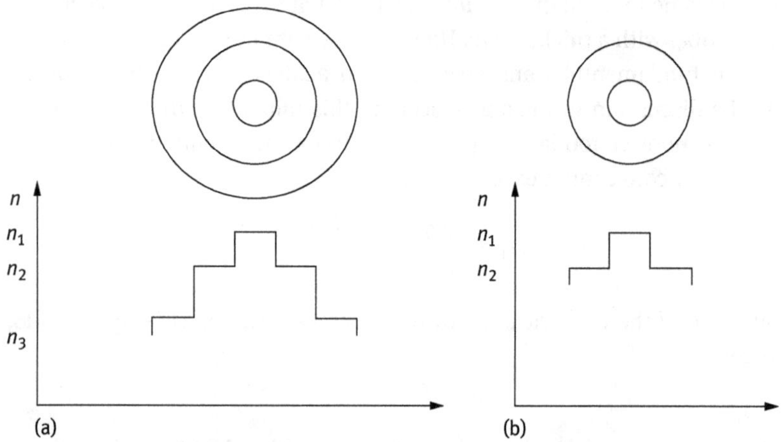

Fig. 4.15: Schematic diagrams of cross-section and refractive index type of the double-clad fiber (a) and single-clad fiber (b).

Rare earth ions, which are used to produce lasers, only exist in the fiber core. The core is very thin, usually the diameter of the fiber core typically ranges only from a few microns to more than ten microns so that the laser can finally output high-,quality single-mode beam. The diameter of the inner cladding is 10 times or larger than that of the fiber core, which makes it easy to accept the high-power, multimode pumping light. The outer cladding is used to limit the pumping light in the fiber and to prevent it from escaping from the inner cladding. Thus, when the pumping light propagates inside the inner cladding, it can repeatedly pass through the fiber core and fully excite the ions inside the core (Fig. 4.16).

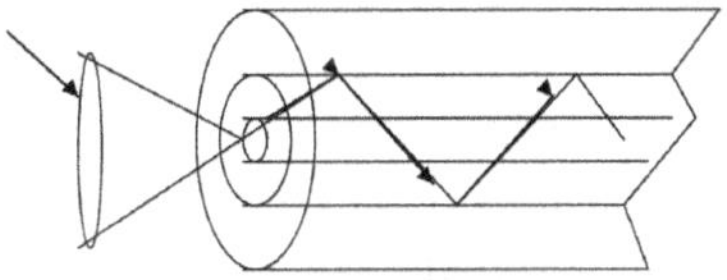

Fig. 4.16: Pumping in double-clad fiber.

Because the pumping light propagates inside the inner cladding, double-clad fiber is also called the cladding-pumped structure. Furthermore, the pump methods can be roughly divided into two different kinds, namely, end pump and side pump. Figure 4.17 (a)–(c) show the end pump technologies mainly used in fiber laser whereas Fig. 4.17 (d)–(f) most frequently used show side pump technologies.

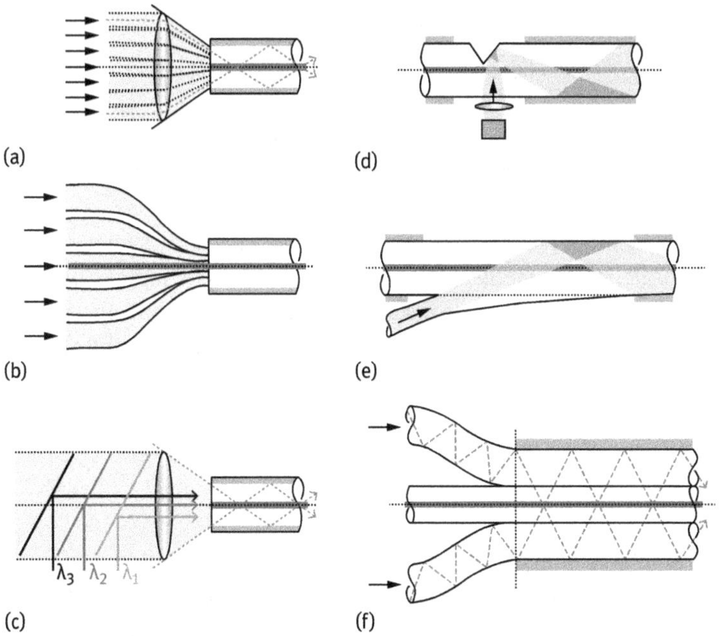

Fig. 4.17: Main cladding-pumped structures: (a)-(c) are the end pumps whereas (d)-(f) are the side pumps.

4.4.2.2 Absorption of the pumping light

As mentioned above, dopants only exist in the fiber core. The waveband of the pumping light is determined according to the absorption band of these ions, so the absorption coefficient of the fiber core α_{co} is far larger than that of the cladding α_{c1}.

Moreover, there is a close relationship between α_{co} and the section shape of the inner cladding. In circular inner cladding, the low order modes, which have substantial overlap with the doped region, are absorbed quickly so that they cannot propagate forward anymore. On the other hand, the overlap between the forward-propagating hollow tubular modes and the doped region is close to zero, and therefore, these modes hardly participate in the pumping process. Consequently, obvious absorptions no longer exist after these modes leave the pump end of the fiber for a very short distance (in typical condition, the distance is only tens of centimeters). Therefore, people are more willing to use inner cladding structure with slightly complicated section shape. Some common section shapes are shown in Fig. 4.18(a)–(e) as square, rectangle, regular hexagon, quincunx, and D-section, respectively.

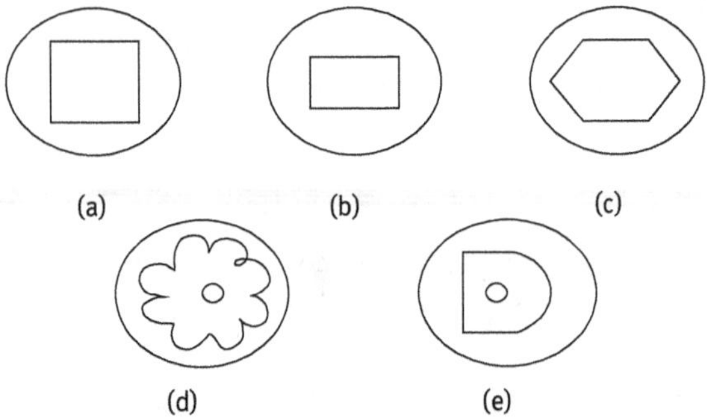

Fig. 4.18: Sectional views of the typical double-clad fibers.

When these fibers whose inner cladding cross section is not circularly symmetric are used as the gain medium, all the modes have complicated structure and the absorption along the fiber is approximately a constant. In this condition, all the modes equally participate in the pumping process and the relationship between the absorption coefficient and the sectional area of the fiber core and the inner cladding can be expressed as follows:

$$\frac{\alpha_{c1}}{\alpha_{c0}} = \gamma \approx \frac{S_{c0}}{S_{c1}} \tag{4.23}$$

where S_{c0} and S_{c1} are the sectional area of the fiber core and the cladding, respectively, and γ is called the multimode overlap factor. Furthermore, the larger the γ is the easier it is for the pumping to be absorbed. In this sense, for a given S_{c0}, it is advantageous for S_{c1} to be small. On the other hand, the smaller the S_{c1} is, the more difficult it is for the pumping light to be coupled into the fiber. Thus, the selection of

S_{c1} should be considered with tradeoffs. In the case that S_{c1} is not too large, to improve the coupling of the pumping light, the numerical aperture should be larger which is larger than 0.4 in the typical condition.

4.4.2.3 Output power and threshold

In 1998, A. Bertoni published a coupled equation group which can be used to describe the transmission of light in the double-clad fiber laser. The equation group can be expressed as follows:

$$\begin{aligned} \frac{dI_S^+}{dz} &= \sigma_i n_i \frac{I_p}{I_p^{sat}} \frac{(I_S^+ + I_n)}{1 + \frac{I_S^+ + I_n^-}{I_S^{sat}}} - \alpha_S I_S^+ \\ \frac{dI_S^-}{dz} &= -\sigma_i n_i \frac{I_p}{I_p^{sat}} \frac{(I_S^- + I_n)}{1 - \frac{I_S^+ + I_S^-}{I_S^{sat}}} + \alpha_S I_S^- \end{aligned} \tag{4.24}$$

where I_S^+ and I_S^- denote the intensity of the forward-propagating laser beam and backward-propagating laser beam, respectively. σ_i is the radiation cross-section of the ions in the fiber core. Furthermore, n_i is the ion number density determined by the doping concentration I_n is used to represent the contribution of the spontaneous radiation to the laser intensity, whereas I_S^{sat} and I_p^{sat} denote the saturation intensity of the laser and the pumping light, respectively. Moreover, α_S is the attenuation coefficient of the laser.

In equation (4.24), the intensities of laser and pumping light are all functions of three-dimensional space coordinates. If normalized functions of intensity $\delta_s(x, y)$ and $\delta_p(x, y)$ are introduced, then I_s and I_p can be expressed as follows:

$$I_s(x, y, z) = P_s(z)\delta_s(x, y),\ I_p(x, y, z) = P_p(z)\delta_p(x, y)$$

If equation (4.24) is integrated over the horizontal ordinate, and effective cross-sections σ_S^{eff} and σ_p^{eff} are introduced into the result, then the transmission equations which are similar to those of the single-mode fiber laser can be expressed as follows

$$\begin{aligned} \frac{dP_S^+}{dz} &= \sigma_i n_i \frac{P_p}{P_p^{sateff}} \frac{(P_S^+ + P_n)}{1 + \frac{P_S^+ + P_S^-}{P_S^{sateff}}} - \alpha_S P_S^+ \\ \frac{dP_S^-}{dz} &= -\sigma_i n_i \frac{P_p}{P_p^{sateff}} \frac{(P_S^- + P_n)}{1 + \frac{P_S^+ + P_S^-}{P_S^{sateff}}} + \alpha_S P_S^- \end{aligned} \tag{4.25}$$

Where,

$$P_S^{sateff} = \frac{h\nu_S}{\sigma_S^{eff}\tau},\ P_p^{sateff} = \frac{h\nu_p}{\sigma_p^{eff}\tau}$$

P_S^{sateff} and P_p^{sateff} denote the effective saturation power of the laser and the pumping light, respectively.

In 2002, P. Even and D. Pureur established the physical model and got the expression of the laser photon flow, that is:

$$F_S^+ = \left[F_p(1 - e^{g_p}) - \frac{n_i S_{cO} \Delta N L}{\tau}\right] \frac{1}{1+R} \tag{4.26}$$

where, F_p is the photon flow of the pump and g_p is the multimode absorption rate of the pumping light. τ is the average life of the ion at excited state and ΔN is the mean number of the particle inversion. L is the length of the fiber, and

$$R = \left(\frac{1 - R_1}{1 - R_2}\right) \sqrt{\frac{R_2}{R_1}}$$

where R_1 and R_2 are the intensity reflectivity of the mirror. To rewrite equation (4.4) as a form of power, we introduce the ratio of $\frac{v_s}{v_p}$ to represent the quantum efficiency of the photon energy changing form hv_p to hv_s. Then we can get

$$P_S^+ = \left[P_p(1 - e^{g_p}) - \frac{n_i S_{cO} \Delta N L h v_p}{\tau}\right] \frac{v_s}{(1+R) v_p} \tag{4.27}$$

where P_p is the pumping power entered into the inner cladding of the fiber. Thus, we can get the pumping threshold power, that is:

$$P_p^{th} = \frac{n_i S_{cO} \Delta N L h \gamma_p}{\tau(1 - e^{g_p})} \tag{4.28}$$

Equation (4.28) shows that pumping threshold through g_p, which is relevant to the multimode overlap, depends on the fiber type which is double-clad or standard single-mode. Figure 4.19 is an example of the pumping threshold of the Y_b^{3+} -doped fiber laser. For comparison, corresponding values of the single-mode fiber are indicated by dotted line in Fig. 4.19. The values of the relevant parameters are

$$\lambda_p = 972nm, S_{c1} = 2.1 \times 10^4 \mu m^2, S_{cO} = 13.8 \mu m^2$$

$$n_i = 76 \times 10^{24} m^{-3}, \tau = 7.6 \times 10^{-4} s, R_1 = 0.99, R_2 = 0.03$$

Figure 4.19 shows that when the length of the double-clad fiber is very short, the pumping threshold power of the corresponding laser needs to be very high. The reason why double-clad devices are more applicable for the four-level system rather than the three-level system which requires the fiber to be very short. Figure 4.19 also shows that when the fiber is long enough, the requirements of the pumping threshold power of the double-clad devices tend to be the same as those of the single-clad devices. However, further research shows that there are still differences between them even if the fiber is long. Especially, at the pumping input end, standard single-

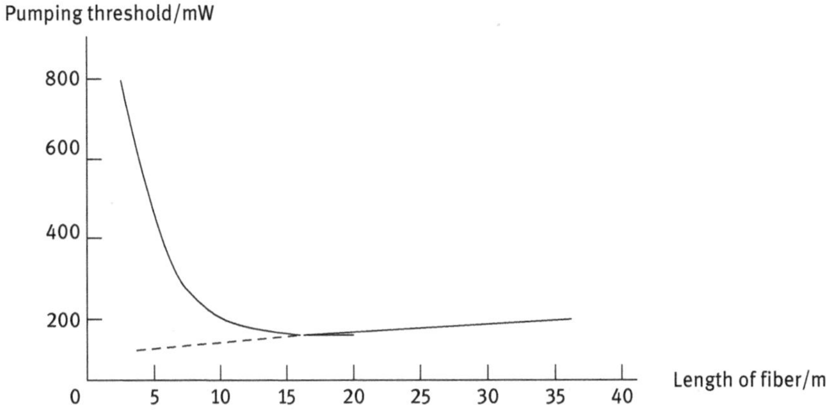

Fig. 4.19: Example of the pumping threshold of the Y_b^{3+}-doped fiber laser.

mode fiber always has a high inversion population density and with the increase of distance away from the input end, the inversion population density decreases rapidly. Nevertheless, the situation in double-clad fiber laser is different. The cladding pump technology is used such that pumping light can propagate inside the inner cladding, which can keep the inversion population density more balanced along the fiber core.

4.4.3 Introduction of the photonic crystal fiber laser

The appearance of the photonic crystal fiber (PCF) at the end of the 20th century provides a new choice for the double-clad fiber laser. PCF uses the air-hole ring as the inner cladding (Fig. 4.20), which greatly reduce the effective refractive index of the cladding. Moreover, the propagation of light waves in this kind of fiber is determined by the geometry and size of the air hole rather than the materials of the fiber.
The working condition of the transverse mode in the common fiber is given in the previous discussion of this chapter, expressed as

$$dNA = 2.405\lambda/\pi \tag{4.29}$$

where d is the diameter of the fiber core while NA is the numerical aperture. Thus, the core of the fiber, which works in the condition of fundamental transverse mode, is very thin, and in typical condition, its diameter is only several microns. Moreover, at the current level, $1w/\mu m^2$ is the accepted safety value, which will not cause fiber damage. The highest value is no more than $1.5w/\mu m^2$. Thus, the output power of the fiber laser which operates in the condition of the fundamental transverse mode is very low.

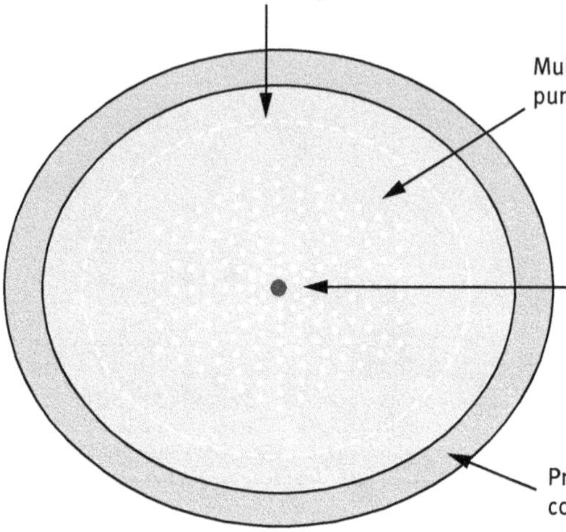

Fig. 4.20: Diagram of the photonic crystal fiber.

Using double-clad structure can substantially increase the transmission cross-section of the pumping light, which will benefit the coupling of light. In addition, the larger the cross-sectional area of the inner cladding S_{c1} is, the more favorable it is for the pumping light to be coupled into the fiber. However, equation (4.23) shows that the larger the S_{c1} is, the smaller the absorption coefficient of the fiber core will be. Thus, the selection of S_{c1} should be considered with tradeoffs. In the situation that S_{c1} is not too large to increase the coupling of the pumping light, the numerical aperture should be larger. It would be best to let $NA = \left(n_1^2 - n_2^2\right)^{1/2} \geq \frac{1}{\sqrt{\pi}} \approx 0.55$. However, nowadays, NA of the common double-clad fiber is typically hard to exceed 0.48 because of the limitation of the cladding refractive index.

In summary, the cross-sectional area of the mode and the numerical aperture of the fiber are the important factors that limit the increase in fiber laser output power. Although the double-clad structure is used to improve the output power, it is not ideal enough.

The appearance of photonic crystal fiber laser (PCFL) changes the situation entirely. Researches show that all-silica PCF which is made by pure silicon can be single mode for any wavelength. If we notice equation (4.24), in other words, fiber still remains in the single mode when the diameter of the fiber core is arbitrarily large, which leads to the appearance of the large mode area (LMA) PCFL.

In addition, because there is an air hole structure in the inner cladding of the fiber, its refractive index differs considerably from that of the fiber core. Thus, a large numerical aperture can be achieved. At present, NA of 0.9 has been reported.

Thus, it can be seen that PCFL solves two key problems, i.e., sectional area of the fundamental transverse mode in fiber and a small numerical aperture of fiber. Therefore, it provides an effective way to gain the high power and high beam quality laser.

4.5 Stimulated scattering fiber lasers

Scattering is a phenomenon that a medium absorbs the light with certain frequency and emits the light that has a certain frequency shift relative to the incident light. This section focuses on the stimulated Raman scattering fiber lasers and stimulated Brillouin scattering fiber lasers.

4.5.1 Raman scattering fiber lasers

4.5.1.1 Raman scattering

It has been nearly 90 years since Raman scattering was first observed. In 1928, the Indian scholar C. V. Raman and his collaborators first observed the significant frequency shift of light in the media. It is then found that this frequency shift is the vibration frequency ω_M of the molecules in the sample material. This phenomenon is called Raman scattering, and the frequency shift in the scattering is called Raman frequency shift.

Raman frequency shift can be either positive or negative, and it is corresponding to the frequency of the incident light being shifted up or down by ω_M. The phenomenon which produces the former is also known as anti-Stokes scattering, and the phenomenon which produces the latter is called Stokes scattering. When the frequency of the incident light is set as ω_L, the frequency of the Stokes spectral line is

$$\omega_S = \omega_L - \omega_M$$

Whereas the frequency of the anti-Stokes spectral line is

$$\omega_{AS} = \omega_L + \omega_M$$

The corresponding transitions of the two processes are shown in Fig. 4.21, where g and r are the molecular vibrational levels; O and P correspond to the lowest excited states of electrons; and v denotes a virtual state.

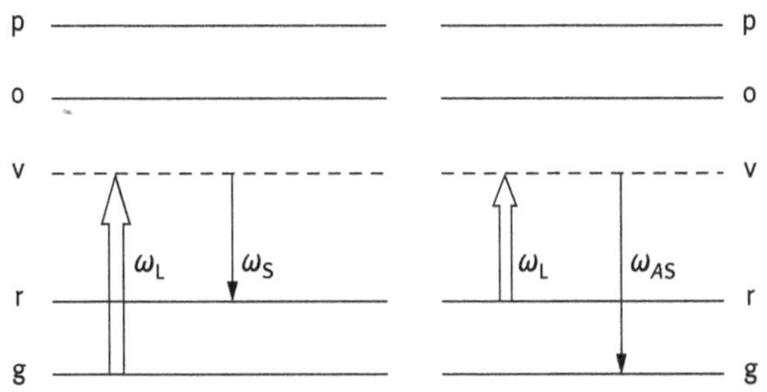

(a) Stokes scattering (b) Anti-Stokes scattering

Fig. 4.21: Diagram of energy level transition of spontaneous Raman scattering.

The above process has been widely used in spectroscopy. However, a severe drawback is that the efficiency of non-elastic scattering is very low and the scattering section is usually only 10^{-12} times the infrared absorption cross-section, thus the scattering light intensity is only 10^{-7}~10^{-15} times the incident light intensity, so it is difficult to monitor it in real time.

With the invention of laser and the development of laser technology, the above-mentioned situation has been completely changed since the 1960s. The incident laser is tuned to respond to the electron absorption of the medium, that is, the electron is excited to the actual level O or P to obtain the so-called resonance Raman scattering. The energy level transition is shown in Fig. 4.22. The efficiency of the resonance Raman scattering is more than 100 times higher than that of non-resonance Raman scattering.

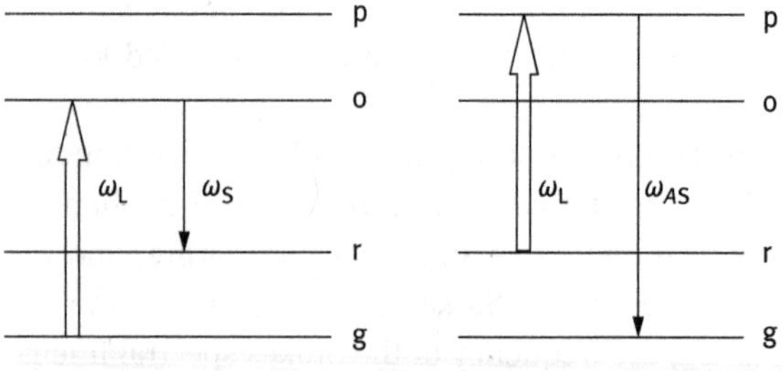

(a) Stokes scattering (b) Anti-Stokes scattering

Fig. 4.22: Diagram of energy level transition of resonance Raman scattering.

In the above two processes, the scattered light is emitted spontaneously, which is called spontaneous Raman scattering. If the laser entering into the sample not only contains the excitation frequency ω_L but also the Stokes frequency ω_S, the latter will excite the scattering of the light with a frequency of ω_S. This process is called the stimulated Raman scattering (SRS), which is much stronger than the spontaneous Raman scattering. Figure 4.23 shows the energy level transition of SRS.

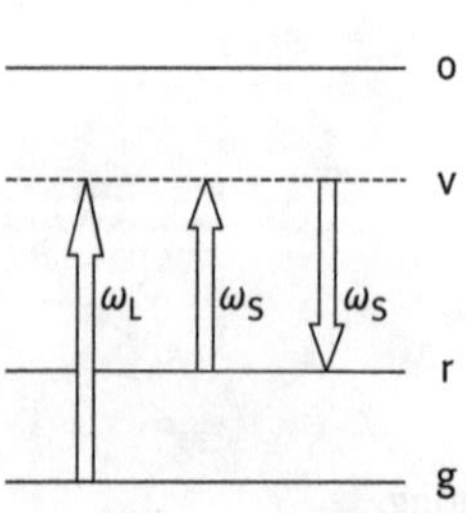

Fig. 4.23: Diagram of energy level transition of stimulated Raman scattering.

4.5.1.2 Raman scattering fiber lasers

Raman scattering described above can also occur in fibers. Considering the positive Stokes scattering, we will find that, when the pump light enters into the core of a fiber, it will act on the lattice. The pump photon converts a part of the energy into the molecular vibrational energy, which will cause molecular transition and the emission of Stokes photon with lower energy or the frequency shifting down by the Raman frequency shift. The frequency shift is mainly determined by the material of the fiber, for example, in the Ge-Si fiber, the typical frequency shift is 13 THz.

Yb^{3+}-doped double-clad fiber lasers are very effective for pumping Raman resonators. The Raman fiber laser cavity consists of a thin single-mode fiber and a Bragg grating reflector. The output light of the Yb^{3+}-doped fiber laser enters into the fiber from the input hand of the Raman cavity and excites the first Stokes radiation at the wavelength of λ_{S1}. Because of the action of the reflector, the first Stokes light is oscillated in the laser cavity and increases gradually. Then, as the second pump light, it excites the optical fiber to emit the second Stokes light at the wavelength of λ_{S2}. This above process will be continued to achieve desired wavelength output, as shown in Fig. 4.24.

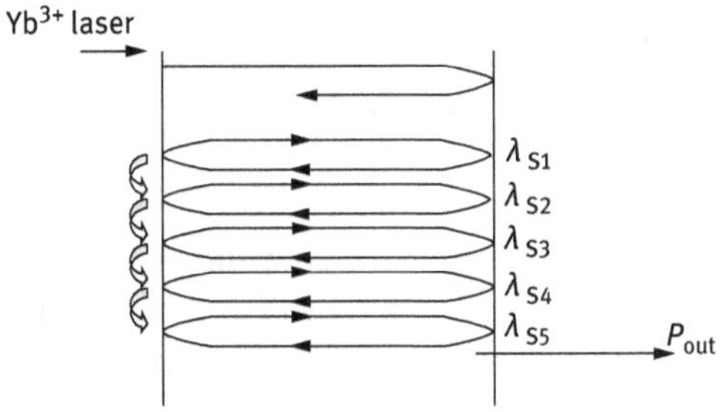

Fig. 4.24: Cascaded Raman cavity and Stokes–Stokes power conversion based on Raman effect.

An important advantage of Raman fiber laser is the availability of relatively high power output and a flexible wavelength output that covers virtually any inherent wavelength in the transparent window of the fiber material. It provides the pump source with superior performance for the 1370–1500 nm Yb^3-doped fiber amplifier used for long-distance communication. In addition, multiple wavelengths output can be obtained from the same laser, which is useful for interference and anti-interference.

4.5.2 Stimulated Brillouin scattering fiber lasers

This subsection describes another fiber laser based on nonlinear optical scattering, namely, the stimulated Brillouin scattering fiber laser (SBSFL). First, we introduce the Brillouin scattering (SBS) process.

4.5.2.1 Stimulated Brillouin scattering (SBS)

The SBS process is basically the same as the SRS process described in the previous subsection. The main difference between them is that the energy of the incident photon is transmitted to the medium by molecular vibrations in the SRS but by phonon in the SBS. Thus, if Fig. 4.23 is used to show the process of SBS, g and r no longer represent the molecular vibrational levels but the phonon energy levels. The Stokes frequency shift of the corresponding SBS process is determined by

$$\nu_B = (2n\nu)/\lambda_p$$

where n is the refractive index of the medium for the pump light; v is the velocity of the sound in the medium; λp is the wavelength of the Brillouin pump light.

Because the pitch of the phonon energy is much smaller than that of the molecular vibration level, the Stokes shift of the SBS process is much smaller than that of the SRS process. For SBS, the typically Stokes shift is 1–10 GHz.

4.5.2.2 Erbium-doped stimulated Brillouin scattering fiber laser

The important characteristic of SBSFL is that its output signal has very good coherence and directivity. However, as the Brillouin gain factor is very low, the laser output power is low, which limits its applications in many areas.

On the other hand, erbium-doped fiber (EDF) has a higher gain, which can produce the high power laser output with single longitudinal mode and narrow linewidth. If we combine the nonlinear Brillouin gain in single-mode fiber (SMOF) with the high linearity gain in EDF to form the Brillouin scattering erbium-doped fiber laser (BEFL), we expect to obtain the laser output with high power and high beam quality. In the normal operating mode, the laser emission still originates from the Brillouin gain, but no longer requires a critical coupling cavity of a conventional laser based on the Brillouin gain. It compensates for cavity loss by EDF amplifiers.

Although BEFL has similarity to BFL and erbium doped fiber lasers (EDFL) in some respects, it is different from the both lasers. The operating wavelength is precisely controlled by the Brillouin pumped Stokes shift, while the high power output is from the EDF, BEFL has the laser output with high power and narrow linewidth.

A typical structure of the BEFL is shown in Fig. 4.25. The laser resonator consists of an EDF fiber ring, a SMOF and an optical isolator. The effect of the optical isolator is to prevent the Brillouin pump light from entering into the EDF, thereby preventing the injection-locking of Brillouin pump.

When the narrow linewidth Brillouin pump is injected into the SMOF, the SBS produces a narrow band gain in clockwise direction, and its wavelength is determined by the Brillouin pumped Stokes shift. For such a loss cavity, only depending on the Brillouin gain may not be enough to get the laser output. While the gain generated by using the light at 980 nm to pump EDF is sufficient to overcome cavity losses, through which we can get high power lasers.

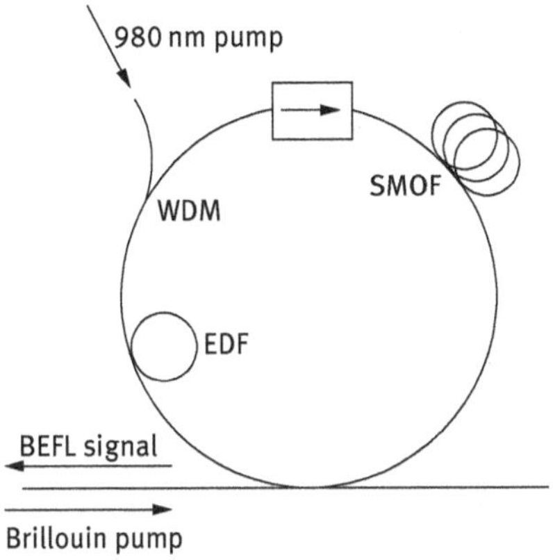

Fig. 4.25: A typical structure of BEFL.

Figure 4.26 shows the case where the broadband gain curve of the EDF is superimposed with the narrowband gain curve from the SBS process. The peak of the former appears at the wavelength of point *a*, and the latter peak appears at the wavelength of point *b*. If point *b* and point *a* are sufficiently close, the broadband gain of EDF still has a higher value at point *b* so that the total gain at point *b* will be greater than the gain at point *a* (as shown in the upper one of Fig. 4.26) and reach the threshold gain, then a laser at the wavelength of point *b* is generated. On the other hand, if the peak of the narrowband gain curve of the SBS process appears at the wavelength of point *c* which is far away from point *a* (as shown in the lower one of Fig. 4.26), the broadband gain of the EDF will drop to a low value. Consequently, even if the value at point *c* is superposed, the broadband gain of the EDF is still lower than the value at point *a* on the broadband gain curve of the EDF and eventually it reaches the threshold gain at point *a*, and then a laser at the wavelength of point *a* is generated.

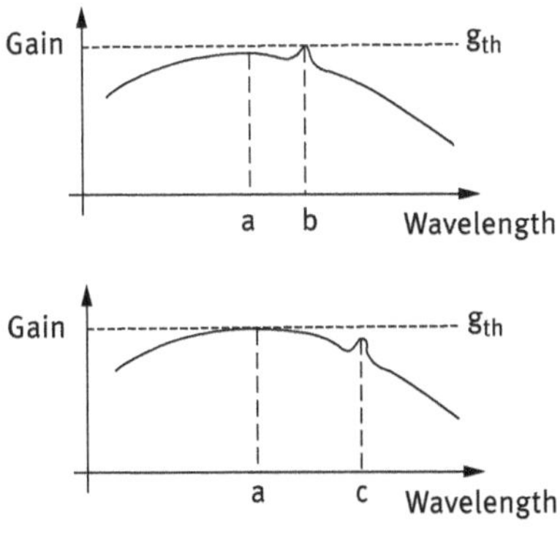

Fig. 4.26: Working wavelength of BEFL.

4.5.2.3 Multi-wavelength operation of the stimulated Brillouin scattering fiber laser

Many applications of stimulated Brillouin scattering fiber laser require a dual wavelength or even multiple-wavelength output, which can be achieved through the cascaded BEFL.

Stimulating the SBS process in SMOF requires two conditions. One is that the line width of Brillouin pump is narrower than the Brillouin gain bandwidth and the other is that the power is large enough to reach the threshold. The output of BEFL has the above characteristics clearly, so two or more BEFL can be cascaded to obtain the dual-wavelength and even multi-wavelength laser output. If the components for each BEFL are the same, especially all SMOFs are the same, the output spectrum lines are equally spaced and the spacing between the adjacent spectrum lines is the Stokes frequency shift.

5 Beam propagation and propagation media

5.1 Beam propagation in homogeneous media and media boundary

5.1.1 Beam propagation in homogeneous media

In the homogenous medium, the reference axis z is chosen arbitrarily, and the beam propagates along a path with a distance of L along the direction with a certain angle of axis z. At the beginning and endpoint of the path, we make the cross-section M_1 and M_2, which are vertical to axis z and the corresponding joints are z_1 and z_2, respectively (Fig. 5.1).

The beam in a cross-section can be characterized with two coordinate parameters: one is the distance r between the beam and reference axis z, and the other is the included angle θ between them. Here, we specify that θ is positive when the propagation direction of beam is above the axis z, otherwise, θ is negative. Hence, the beam can be characterized with r_i, θ_i and r_0, θ_0 in the cross-section M_i and M_0, respegctively. According to their geometric relations, the relationship between the above two sets of parameters can be expressed as follows

$$\begin{cases} r_0 = r_i + L\tan\theta_i \\ \theta_0 = \theta_i \end{cases}$$

Only considering the paraxial ray, then $\tan\theta \approx \theta$, the above equation can be simplified as

$$\begin{cases} r_0 = r_i + L\theta_i \\ \theta_0 = \theta_i \end{cases} \tag{5.1}$$

Equation (5.1) can be expressed as a matrix form as

$$\begin{pmatrix} r_0 \\ \theta_0 \end{pmatrix} = \begin{pmatrix} 1 & L \\ 0 & 1 \end{pmatrix}\begin{pmatrix} r_i \\ \theta_i \end{pmatrix} = T_L\begin{pmatrix} r_i \\ \theta_i \end{pmatrix} \tag{5.2}$$

Where the column matrix represents the beam coordinates in the cross-section, and the square matrix T_L expresses the transformation effect of the media to the beam.

$$T_L = \begin{pmatrix} 1 & L \\ 0 & 1 \end{pmatrix} \tag{5.3}$$

In the following discussion, the above description method is extended to all kinds of situations, that is, when the beam propagates through different media boundaries or some kind of optical element, or the combination of several kinds of situation, the transformation of the beam can be expressed by the square matrix T as

https://doi.org/10.1515/9783110500608-005

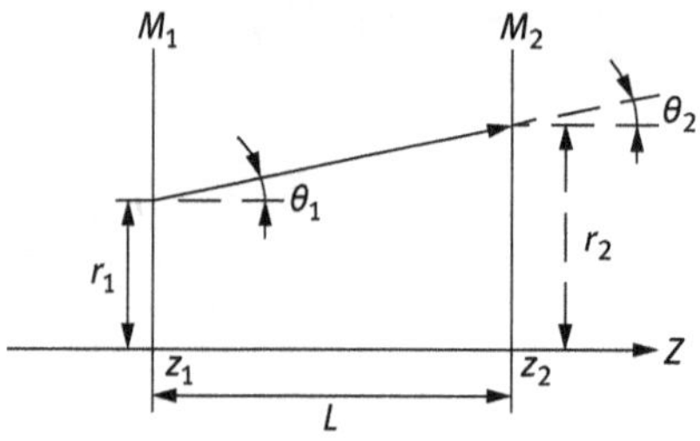

Fig. 5.1: The beam propagates along a path with a distance of L in homogeneous media.

$$T = \begin{pmatrix} A & B \\ C & D \end{pmatrix} \tag{5.4}$$

Equation (5.4) is called as an *ABCD* matrix, and this notation is called the matrix method accordingly. Below, we will discuss other cases based on this method. Moreover, the beam is always supposed as the paraxial ray.

5.1.2 Beam transmission in the media boundary

As shown in Fig. 5.2, we suppose that the two media with the refractive index *of* n_1 and n_2, respectively, have the boundary plane M, the beam propagates in the medium of refractive index n_1 and reaches the point P (at plane M) along the direction with a certain angle θ_i of axis z. Then, the beam enters into the medium of refractive index n_2 through point P, and continues propagating along the direction with a certain angle θ_0 of axis z. In this section, we will examine the transformation matrix of the boundary M to the beam, that is, we will find the relationship between the incident ray and the outgoing ray at point P. In the current situation, apparently, $r_0 = r_i$, so only the relationship between θ_0 and θ_i needs to be found out.

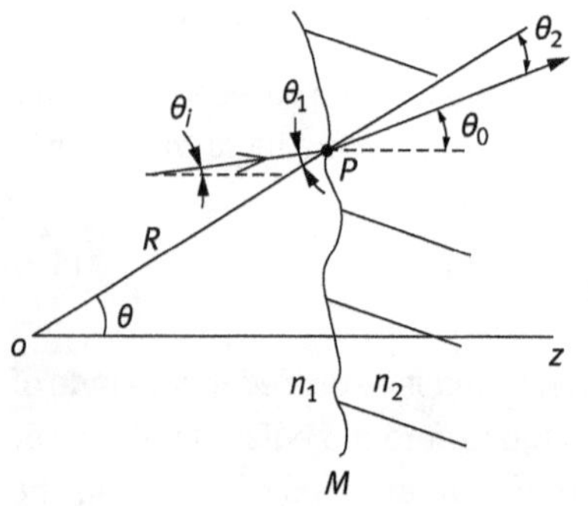

Fig. 5.2: Beam transmission in the media boundary.

We draw a circle with the center of a circle point P and the radius ρ which is the radius of curvature of boundary M at point P, which intersects with the reference axis z at point O. According to the geometrical relationship shown in Fig. 5.2, we can derive

$$\begin{aligned}\theta_0 &= \theta - \theta_2 \\ &= \theta - \frac{n_1}{n_2}\theta_1 \\ &= \theta - \frac{n_1}{n_2}(\theta - \theta_i) \\ &= \frac{n_2 - n_1}{n_2}\frac{r_i}{\rho} + \frac{n_1}{n_2}\theta_i\end{aligned}$$

then,

$$\begin{cases} r_0 = r_i \\ \theta_0 = \frac{n_2 - n_1}{n_2}\frac{1}{\rho} r_i + \frac{n_1}{n_2}\theta_i \end{cases} \tag{5.5}$$

The transformation matrix T_0 of boundary M in point P (the radius of curvature is ρ) to the beam is

$$T_\rho = \begin{pmatrix} A_R & B_R \\ C_R & D_R \end{pmatrix} = \begin{pmatrix} 1 & 0 \\ \frac{1}{\rho}\frac{n_2 - n_1}{n_2} & \frac{n_1}{n_2}\theta_i \end{pmatrix} \tag{5.6}$$

Two exceptions of the above results are very useful, they are

(1) If the boundary M is a sphere with the radius of curvature ρ, T_0 fits any point of M.

(2) If the boundary M is a flat plane, that is $\rho \to \infty$, then $T_\infty = \begin{pmatrix} 1 & 0 \\ 0 & \frac{n_1}{n_2} \end{pmatrix}$.

5.1.3 Beam propagation through a thin lens

For an optical element, naturally, the optical axis can be the reference axis. For a thin lens, the distance between the incidence point and the optical axis is equal to that between the exit point and the optical axis.

$$r_0 = r_i \tag{5.7}$$

The lens will then refract the beam twice.

For simplicity, and in general, we let the two surfaces constituting a lens have the same curvature. In the first refraction, the beam enters into the medium of refractive index n_2 from the medium of refractive index n_1 (shown as Fig. 5.3), for the incident ray, the boundary plane is a convex, according to equation (5.6), the transformation matrix T_1 is

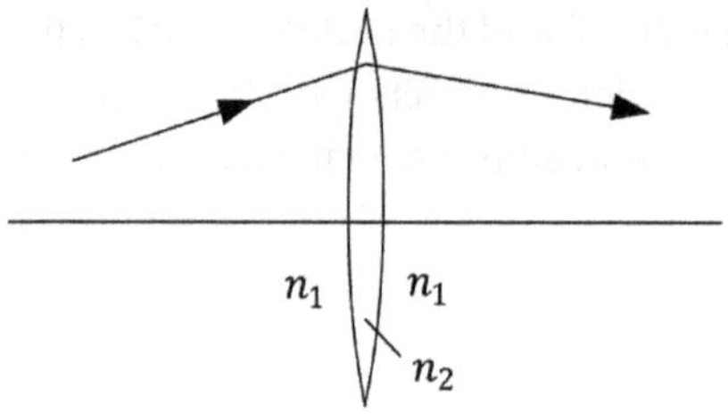

Fig. 5.3: Beam propagation through a thin lens.

$$T_1 = \begin{pmatrix} 1 & 0 \\ -\frac{1}{\rho}\frac{n_2 - n_1}{n_2} & \frac{n_1}{n_2} \end{pmatrix}$$

In the second refraction, the beam enters into the medium of refractive index n_1 from the medium of refractive index n_2, for the incident ray, the boundary plane is a concave, based on equation (5.6) again, the transformation matrix T_2 is

$$T_2 = \begin{pmatrix} 1 & 0 \\ -\frac{1}{\rho}\frac{n_1 - n_2}{n_1} & \frac{n_2}{n_1} \end{pmatrix}$$

The transformation matrix of lens T_f can be got by multiplying T_1 by T_2

$$T_f = T_2 T_1 = \begin{pmatrix} 1 & 0 \\ -\frac{2}{\rho}\left(\frac{n_2 - n_1}{n_2}\right) & 1 \end{pmatrix} = \begin{pmatrix} 1 & 0 \\ -\frac{1}{f} & 1 \end{pmatrix} \tag{5.8}$$

where $f = \frac{\rho}{2}\left(\frac{n_1}{n_2 - n_1}\right)$ is the focal length of the lens.

Equation (5.8) indicates that when the beam propagates through a thin lens, the relationship between exit angle θ_0 and incident angle θ_i is

$$\theta_0 = \theta_i - \frac{r_i}{f} \tag{5.9}$$

In fact, equation (5.9) expresses the law of spherical wave imaging through a thin lens. In Fig. 5.3, the incident ray and exit ray are, respectively, extended along the backward and forward to intersect with the optical axis at point P and P', as shown in Fig. 5.4, the distances between P, P' and the center of the lens O are R and R', respectively. Based on equation (5.7), we can get

$$r_0 = (-R')(-\theta_0) = R\theta_I = r_i$$

Substituting it into equation (5.9), we obtain

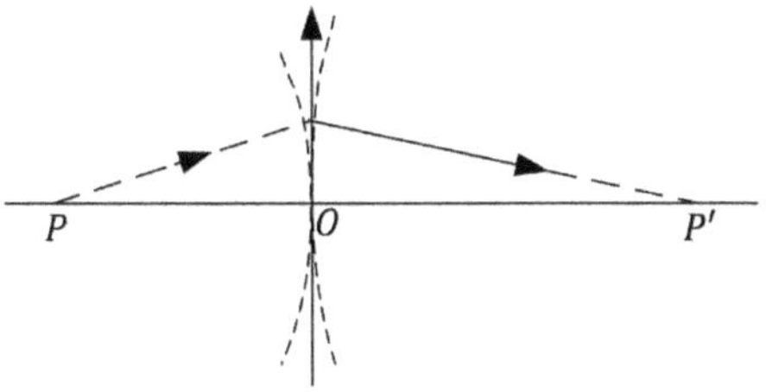

Fig. 5.4: The transformation of thin lens to spherical wave

$$\frac{r_0}{R'} = \frac{r_i}{R} - \frac{r_i}{f}$$

Or

$$\frac{1}{R} - \frac{1}{R'} = \frac{1}{f} \tag{5.10}$$

Equation (5.10) is the familiar formula of spherical wave imaging, and can be expressed as form,

$$R' = \frac{R}{-\frac{1}{f}R + 1}$$

Noticing the transformation matrix of lens as

$$\begin{pmatrix} A & B \\ C & D \end{pmatrix} = \begin{pmatrix} 1 & 0 \\ -\frac{1}{f} & 1 \end{pmatrix}$$

then

$$R' = \frac{AR + B}{CR + D} \tag{5.10)'$$

We can see that the lens will transform the incident spherical wave radiated by object into a new one in image space, however, the incidence spherical wave can be uniform or nonuniform, or the center of curvature is fixed or moving, based on equation (5.10) or (5.10); we can built a relationship between the radii of curvature of these two waves through the focal length of the lens, which has the universal significance. Especially, equation (5.10) not only fits the thin lens but also has more complex optical system. If the transformation matrix is known, the radius of curvature R' can be easily obtained using R.

In general, if R and R' are not known, but R_1 which is the radius of curvature of incident spherical wave in point u of the lens and R_2 which is the radius of the curvature of exit spherical wave in point v of the lens are known, then

$$R = R_1 + u$$

$$-R' = v - R_2$$

Substituting them into equation (5.10), we can get

$$\frac{1}{u+R_1} + \frac{1}{v-R_2} = \frac{1}{f} \tag{5.11}$$

Equation (5.11) is the more general transformation law for thin lens, when $u = v = 0$, equation (5.10) is used. However, when $R_1 = R_2 = 0$, that is, the distance between the center of incident spherical wave and the center of lens is u, and the distance between the center of exit spherical wave and the center of lens is v, from equation (5.11), we obtain

$$\frac{1}{u} + \frac{1}{v} = \frac{1}{f} \tag{5.12}$$

Equation (5.12) shows the common relationship between the object and its image, where u and v are the object distance and image distance, respectively.

5.2 Gaussian beam propagation

In modern optics, we will mostly focus on the propagation of laser beam, which includes laser propagation in free space or in homogeneous media, and the transformation of lens to laser, and so on. In this section, we mainly introduce the issues of laser propagation.

Gaussian beam is emitted when the stable-cavity laser operates in the fundamental transverse mode, and Hermite-Gauss mode beam (rectangular stable spherical cavity) or Laguerre-Gauss mode beam (circular symmetry stable cavity) are emitted when the laser operates in higher order modes. So the study of laser propagation involves the study the propagation of Gaussian beam. For simplicity, we only discuss Gaussian beam confined to the fundamental Gaussian mode, and in the following section, when we express it, the "fundamental mode" will be omitted.

5.2.1 Gaussian beam and its parameters

Gaussian beam propagating along the z axis can be expressed

$$\Psi_{00}(x,y,z) = \frac{C}{\omega(z)} exp\left[-\frac{x^2+y^2}{\omega^2(z)}\right] exp\left\{-i\left[k\left(z+\frac{x^2+y^2}{2R}\right) - tan^{-1}\frac{z}{z_R}\right]\right\} \tag{5.13}$$

where $\omega(z)$ and $R(z)$ are the width of beam at point z and the curvature radius of equiphase surface, respectively. From equation (5.13), we can see that the field amplitude of Gaussian beam is maximal in the center of the cross-section $(x=y=0)$, and falls down from the center based on the law of Gauss function $e^{\omega^2(z)}$. In general, $\omega(z)$ is the radius at which the field amplitudes fall to *1/e* of their axial values. ω_0, the beam waist, is the smallest beam width, the relationship between $\omega(z)$ and ω_0 is

$$\omega(z)=\omega_0\sqrt{1+\left(\frac{z}{z_R}\right)^2} \tag{5.14}$$

where $z_R=\frac{\pi\omega_0^2}{\lambda}$ is called as the Rayleigh length of Gaussian beam.

The last item of equation (5.13) is the phase factor of Gaussian beam.

$$\phi(x,y,z)=k\left(z+\frac{x^2+y^2}{2R}\right)-tan^{-1}\frac{Z}{Z_R} \tag{5.15}$$

At the origin, $\phi(0,0,0)=0$, and $\phi(x,y,z)$ expresses the phase lag of Gaussian beam at point (x,y,z) relative to the origin, where $\frac{k(x^2+y^2)}{2R}$ expresses the phase shifting related with the horizontal coordinate (x,y), which indicates the equiphase surface is a spherical plane with the radius of curvature of R, and

$$R(z)=z+\frac{z_R^2}{z} \tag{5.16}$$

The main parameters of Gaussian beam are shown in Fig. 5.5.

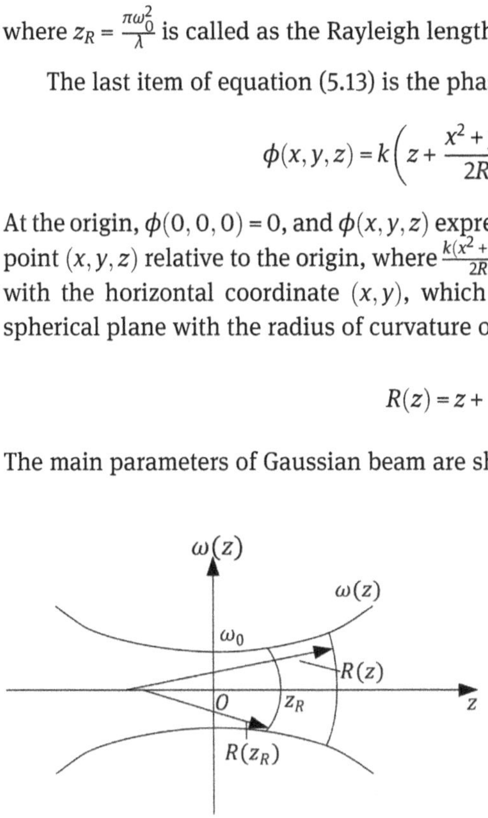

Fig. 5.5: Main parameters of fundamental mode Gaussian beam.

5.2.2 Gaussian beam propagation in free space

From Subsection 5.2.1, we know that the equiphase surface of Gaussian beam is a spherical plane with the radius of curvature of R. Being different from the common spherical wave, its center of curvature will change with axis z when Gaussian beam propagates, which can be seen from equation (5.16), for example, $R(z)=2z_R$ when $z=z_R$, the center of curvature is located at $-z_R$; while $R(z)=z\to\infty$ when $z\to\infty$, the center of curvature is located at the beam waist.

For conveniently describing the Gaussian beam propagation, a new parameter $q(z)$ is recommended and defined as follows

$$\frac{1}{q(z)}=\frac{1}{R(z)}-i\frac{\lambda}{\pi\omega^2(z)} \tag{5.17}$$

Substituting equations (5.15) and (5.16) into (5.17), we can deduce $q(z)=z+iz_R$

$$\begin{aligned}\frac{1}{q(0)}&=\frac{1}{R(0)}-i\frac{\lambda}{\pi\omega^2(0)}\\&=-i\frac{1}{z_R}\end{aligned} \tag{5.18}$$

That is $q_0=iz_R$

then, from equation (5.18), we obtain

$$q(z)=q_0+z \tag{5.19}$$

Considering two points z_1 and z_2 away from the distance of L, the relationship between $q(z_1\ )$ and $q(z_2\ )$ can be easily obtained from equation (5.19) and expressed as

$$q(z_2\ )=q(z_1\ )+L \tag{5.20}$$

Equation (5.20) has the same form with the relational expression of common spherical wave propagation in space, if we replace the radius of curvature with parameter q.

5.2.3 Gaussian beam propagation through a thin lens

From Subsection 5.2.2, we can see that when the Gaussian beam propagates in space, the transformation law of parameter q is the same as that of common spherical wave. In fact, the above conclusion can be applied to other optical elements. In the following section, we will take the thin lens as an example to discuss this phenomenon.

We suppose that, Gaussian beam with the parameter q enters into the thin lens, then, after the transformation of lens, the parameter is changed to q', which is shown in Fig. 5.6, from equation (5.17), we obtain

$$\begin{aligned}\frac{1}{q}&=\frac{1}{R}-i\frac{1}{\pi\omega^2}\\\frac{1}{q'}&=\frac{1}{R'}-i\frac{1}{\pi\omega'^2}\end{aligned} \tag{5.21}$$

In Subsection 5.1.3, we know that equation (5.10) is workable even though the center of curvature of spherical wave is moving, but for a thin lens, $\omega'=\omega$ is required. Bring in the above conditions, we can obtain

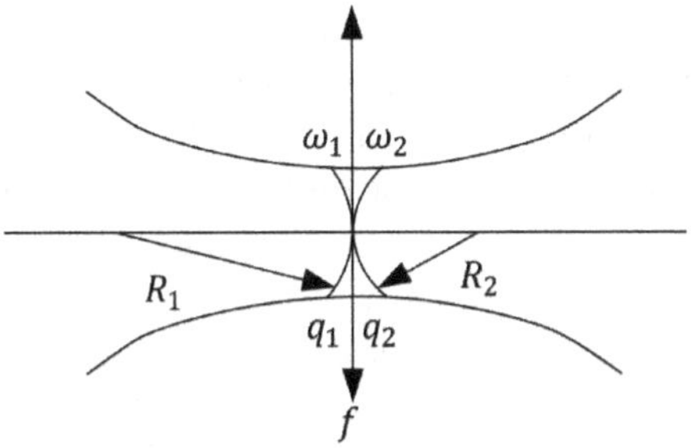

Fig. 5.6: The transformation of thin lens to Gaussian beam.

$$\begin{aligned}\frac{1}{q'} &= \frac{1}{R'} - i\frac{1}{\pi\omega'^2} \\ &= \left(\frac{1}{R} - \frac{1}{f}\right) - i\frac{1}{\pi\omega'^2} \\ &= \left(\frac{1}{R} - i\frac{1}{\pi\omega'^2}\right) - \frac{1}{f} \\ &= \frac{1}{q} - \frac{1}{f}\end{aligned} \tag{5.22}$$

Where f is the focal length of lens. If we replace q with R in equation (5.22), it will have the same form with equation (5.10).

In general, for more complex optical system, if we know its transformation matrix $\begin{pmatrix} A & B \\ C & D \end{pmatrix}$, the relationship between q and q' can be expressed as follows, it has the same form with equation (5.10).

$$q' = \frac{Aq + B}{Cq + D} \tag{5.23}$$

We can simply study Gaussian bean propagation by utilizing equation (5.23), which is the main reason to introduce parameter q to describe Gaussian beam.

5.3 Ray optics theory of planar dielectric optical waveguides

For the first time, the dielectric layer optical waveguide phenomenon was observed in the experiment in the early 1960s soon after a new kind of active and passive optical element based on it was came out. By the late 1960s, guidewave optic elements had indicated the beginning of an advanced technology with broad prospect, and "integrated optics" was appeared, and from then on, guidewave optics have been developing rapidly in terms of theory and application.

In classical theory, electromagnetic field follows Maxwell equations and constitutive relation, or equivalently, we can describe all the optical phenomena with boundary conditions and wave equations. However, in many cases, results from ray optics theory with the advantages of being more simple and intuitive are identical to that from wave optics theory; hence, in this chapter, we will introduce the ray optics image to the dielectric optical waveguides theory.

Optical fiber is the familiar optical waveguide device for us, however, in integrated optics we focus on the planar dielectric optical waveguides such as thin films or narrow strips. In this chapter, we will mainly study the planar dielectric optical waveguides whose fundamental theories fit to all kinds of waveguide structures.

5.3.1 Beam reflection and refraction in media boundary

In this subsection, we will discuss the phenomena of beam reflection, refraction and total internal reflection in media boundary, which is important for us to understand the waveguide.

A beam enters into the semi-infinite medium boundary from the medium of refractive index n_1 along the direction of angle θ_1, which is shown in Fig. 5.7, we suppose that the two media are lossless, uniform, isotropous, and with the refractive index of n_1 and n_2, respectively. A certain fraction of the light is reflected from the boundary and the remainder is refracted. The reflection light returns into the medium of refractive index n_1, and the angle between the normal and reflection light equals the angle θ_1 at which the wave is incident on the surface, while the remainder is refracted into the medium of refractive index n_2 with the refraction angle of θ_2.

According to Snell Law, θ_2 depends on

$$n_1 \sin\theta_1 = n_2 \sin\theta_2 \tag{5.24}$$

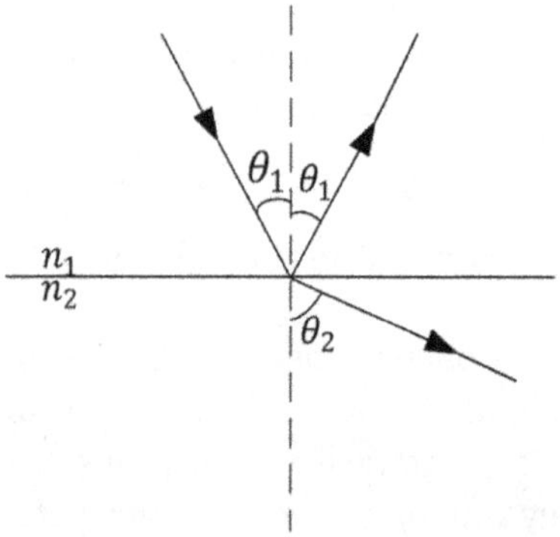

Fig. 5.7: Beam reflection and refraction in media boundary.

the relative amplitude of reflection beam is expressed with the reflection coefficient R, which depends on the angle of the incidence and the polarization direction which can be calculated using Fresnel Formula. For the electric-field component of horizontal polarization TE (i.e., electric field direction is perpendicular to the incident plane wave composed of the wavefront normal and boundary normal) and magnetic-field component TM, we obtain

$$R_{TE} = \frac{n_1 cos\theta_1 - n_2 \cos\theta_2}{n_1 \cos\theta_1 + n_2 \cos\theta_2} = \frac{n_1 \cos\theta_1 - \sqrt{n_2^2 - n_1^2 sin^2\theta_1}}{n_1 \cos\theta_1 + \sqrt{n_2^2 - n_1^2 sin^2\theta_1}} \tag{5.25}$$

$$R_{TM} = \frac{n_2 \cos\theta_1 - n_1 \cos\theta_2}{n_2 \cos\theta_1 + n_1 \cos\theta_2} = \frac{n_2^2 \cos\theta_1 - n_1\sqrt{n_2^2 - n_1^2 sin^2\theta_1}}{n_2^2 \cos\theta_1 + n_1\sqrt{n_2^2 - n_1^2 sin^2\theta_1}} \tag{5.26}$$

Form equation (5.25) and (5.26), when $\theta_1 < \sin^{-1}(n_2/n_1)$, R is real number, that is the part of light is reflected, and the other part is refracted. When $\theta_1 = \sin^{-1}(n_2/n_1)$, $\theta_2 = \pi/2$ is got by equation (5.24),which means no light enters into the medium 2, "transmitted wave" only propagates along the boundary between these two media (shown in Fig. 5.8), and the incident angle θ_1 is called as the critical angle, which is expressed by θ_c in equation (5.27).

$$\theta_c = \sin^{-1}\left(\frac{n_2}{n_1}\right) \tag{5.27}$$

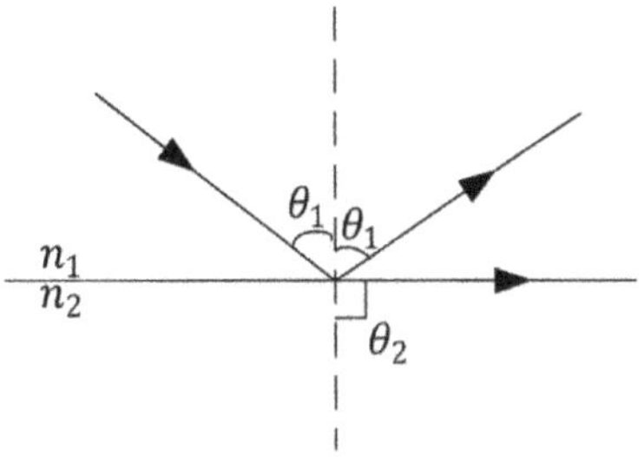

Fig. 5.8: Beam enters from the medium of refractive index n_1 into the medium of refractive index n_2 with critical angle.

With the increase of incident angle, when $\theta_1 > \theta_c$, θ_2 is not a real number, there is no transmitted light, which is called as the total internal reflection shown in Fig. 5.9. The total internal reflection is the theoretical base of light guide.

When the total internal reflection happens, R is the complex number whose module is 1, and can be expressed

$$R = e^{2j\varphi} \tag{5.28}$$

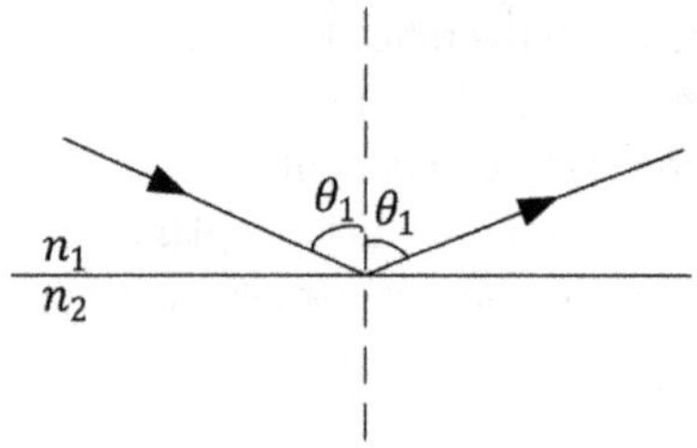

Fig. 5.9: The total internal reflection.

where

$$\phi_{TE} = tg^{-1} \frac{\sqrt{n_1^2 sin^2\theta_1 - n_2^2}}{n_1 \cos\theta_1} \tag{5.29}$$

$$\phi_{TM} = tg^{-1} \frac{n_1^2}{n_2^2} \frac{\sqrt{n_1^2 sin^2\theta_1 - n_2^2}}{n_1 \cos\theta_1} \tag{5.30}$$

Figure 5.10 shows how ϕ_{TE} depends on the incident angle θ_1, and n_2/n_1 is the parameter.

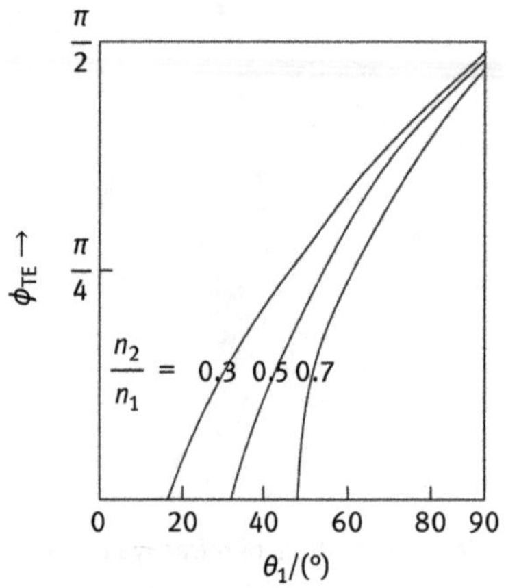

Fig. 5.10: The relationship between ϕ_{TE} and the incident angle θ_1.

5.3.2 The beam propagation in planar waveguide

In this subsection, the ray-optics model of beam propagation in planar waveguide will be discussed. The most simple planar waveguide has three-layer shown in Fig. 5.11, the planar dielectric film with the refractive index of n_1 painted on the substrate with the refractive index of n_2, and the coverage with the

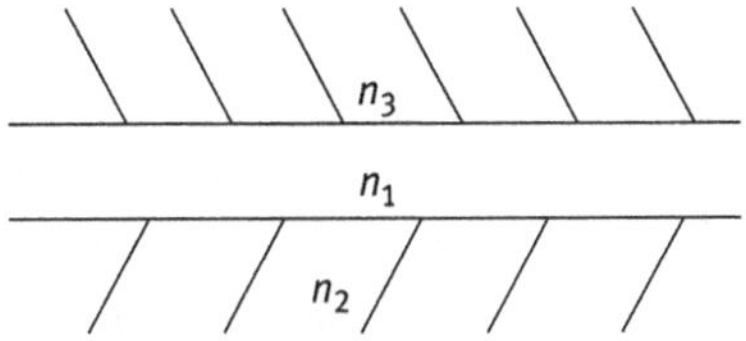

Fig. 5.11: Planar dielectric waveguide with three-layer.

refractive index of n_3 is on the planar dielectric film, which is shown in Fig. 5.11. The typical thickness of intermediate planar dielectric film is only μm order of magnitude, which is also called as the thin film waveguide.

In the planar dielectric waveguide with three-layer, the index of refraction n_i (i=1,2,3) meets the following relationship

$$n_1 > n_2 \geq n_3 \tag{5.31}$$

When $n_1 > n_2 = n_3$, the planar dielectric waveguide is called as the symmetric planar waveguide, otherwise, when $n_1 > n_2 > n_3$, it is called as the asymmetric planar waveguide. The order of magnitude of $n_1 - n_2$ is approximately $10^{-3} - 10^{-1}$, moreover, in many conditions the cladding layer is the air, so $n_3 \approx 1$. Consequently, the common waveguide belongs to the asymmetric type. Under different wavelengths, the refractive indexes of several optical waveguide materials are shown in Table 5.1.

Next we will discuss the several existing kinds of waves in the planar waveguide. Supposing that the waveguide belongs to the asymmetric type, then we know

$$n_1 > n_2 > n_3 \tag{5.32}$$

Table 5.1: The refractive indexes of common optical waveguide materials.

Medium variety	λ /μm	n
Silica glass(SiO_2)	0.633	1.46
Standard microcrystalline glass (ZERODUR)	0.633	1.51
Sputtering Kang Ning 7059 glass	0.633	1.62
Epitaxial $LiNbO_3$ (n_o)	0.80	2.28
(n_e)	0.80	2.19
Epitaxial $LiTaO_3$ (n_o)	0.80	2.15
(n_e)	0.80	2.16
GaAs	0.90	5.6
InP	1.51	5.17

Accordingly there are two critical angles, one is θ_{c12} which makes the beam to have total internal reflection at the boundary of the film and substrate; the other is θ_{c13} which makes the beam has the total internal reflection at the boundary of film and cladding layer. For the smaller angle of incidence θ_i, when $\theta_i < \theta_{c12}$, θ_{c13}, according to Neil's law, the light entering into the substrate will go through the boundary reflection twice and then escape from the cladding layer. With the increase in θ_i, until it meets $\theta_{c13} < \theta_i < \theta_{c12}$, the light entering into the substrate will be first be refracted at the boundary of film and substrate, and then total reflected at the boundary of film and cladding layer, lastly it will come back to the substrate by being refracted again at the boundary of film and substrate. The above two situations are referred to as the radiation wave and the substrate radiation wave, respectively, which are shown as Fig. 5.12(a) and (b).

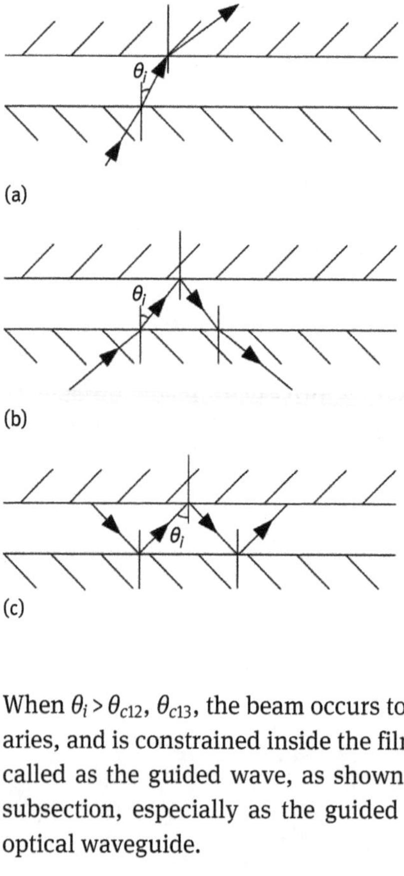

Fig. 5.12: The existing kinds of waves in the planar waveguide: (a) radiation wave, (b) substrate radiation wave, (c) guided wave.

When $\theta_i > \theta_{c12}$, θ_{c13}, the beam occurs total internal reflection at the above two boundaries, and is constrained inside the film to propagate along the zigzag path, which is called as the guided wave, as shown in Fig. 5.12(c). We will discuss it in the next subsection, especially as the guided wave has the self-evident importance in the optical waveguide.

5.3.3 Guided wave in planar dielectric waveguide

Figure 5.13 shows the side view of the planar waveguide. We suppose that the guided wave propagates along the z–axis, and is lateral constrained in x-direction; moreover, both waveguide structure and light wave are homogeneous in y-direction.

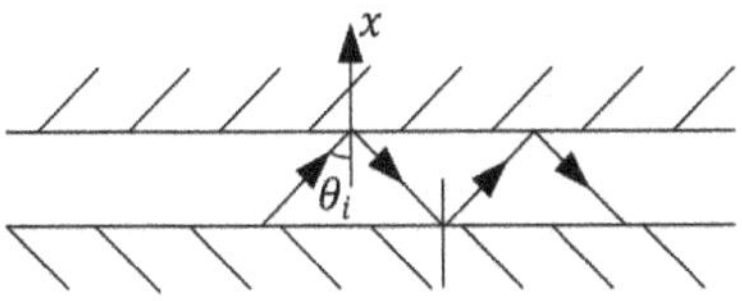

Fig. 5.13: Guided wave in planar dielectric waveguide.

When $\theta_i < \theta_{c12}$, θ_{c13}, the beam will be constrained inside the film to propagate along the zigzag path.

Introducing the propagation constant along z-direction, then we obtain

$$\beta = k_z = k_1 \sin\theta_i \tag{5.33}$$

Apparently $\beta < k_1 = n_1 k_0$

where k_0 is the wavenumber in vacuum. On the other hand, because

$$\frac{\beta}{k_1} = \sin\theta_i > \sin\theta_{c12} = \frac{n_2}{n_1} = \frac{k_2}{k_1}$$

So $\beta > k_2$

Then, there are

$$k_2 < \beta < k_1 \tag{5.34}$$

Or we can write $n_2 k_0 < \beta < n_1 k_0$

If the guided wave effective refractive index is defined as

$$N = \frac{\beta}{k_0} \tag{5.35}$$

Then from Equation (5.34), we can get

$$n_2 < N < n_1 \tag{5.36}$$

From Equation (5.33), the maximum propagation constant along the z- direction is

$$\beta_M = k_1 = n_1 k_0 \tag{5.37}$$

There is no guided waves in the region of $\beta < n_1 k_0$, which is called the cutoff region of guided wave.

5.3.4 Goos–Hänchen displacement and effective depth of waveguides

In the above subsection, we discussed the light path in the planar waveguide based on the viewpoint of beam propagation, we will consider the light energy stream of planar waveguide in this subsection.

From the above discussion, we know that the modulus of reflection coefficient under the condition of total reflection is 1, that is $|R| = 1$, which implies that the incident light energy changing to reflection energy to be retained in the waveguide without flowing into the substrate at all. However, when with the incident angle of θ_i, the beam coming from the medium of refractive index n_1 (here is the waveguide layer) reaches A point at the boundary of medium of refractive index n_1 and medium of refractive index n_2 it will enter into the medium of refractive index n_2 with the incident angle of θ_i, and then be total internal reflected by permeating a certain depth x_s, the reflected beam coming from the medium of refractive index n_2 will reach B point at the boundary of medium of refractive index n_1 and medium of refractive index n_2 and then come back to the medium of refractive index n_1 (shown in Fig. 5.14). The above optical phenomenon was observed and explained by Goos–Hänchen in 1947, so half of the distance between B and A is called as Goos–Hänchen displacement, which is expressed by z_s.

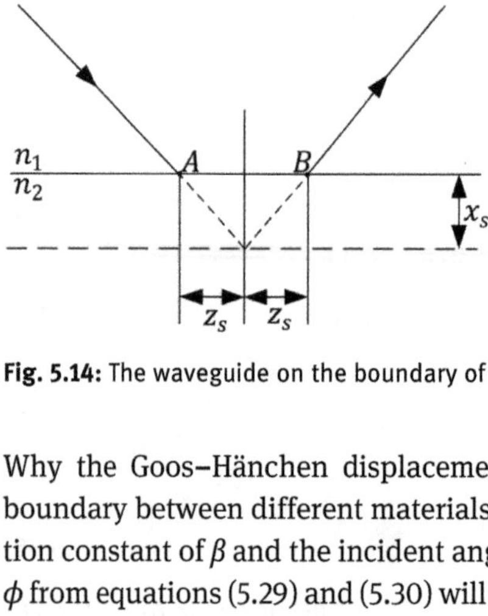

Fig. 5.14: The waveguide on the boundary of planar waveguide.

Why the Goos–Hänchen displacement exists? In the anisotropic medium or the boundary between different materials with different refractive indices, the propagation constant of β and the incident angle of θ_i will be drifted, accordingly, the phase ϕ from equations (5.29) and (5.30) will be drifted. If the drifting distance is small, then we can obtain

$$\phi(\beta + \Delta\beta) = \phi(\beta) + \frac{d\phi}{d\beta}\Delta\beta = \phi(\beta) + z_s\Delta\beta \tag{5.38}$$

and where

$$z_s = \frac{d\phi}{d\beta}$$

z_s is the Goos–Hänchen displacement.

In the same manner, the beam will permeate the cladding layer with a certain depth of x_c before being total internally reflected, which will create the beam lateral displacement of z_c.

For calculating the value of z_s quantitatively, equation (5.29) is written as

$$\text{tg}\phi_{TE} = \frac{\left(n_1^2 sin^2\theta_i - n_2^2\right)^{1/2}}{n_1 \sin\theta_i}$$

Then, we will take the derivative of the above equation with respect to β on both sides of the equal sign, for the left side, the result of the derivative is

$$\frac{d}{d\beta}[tg\phi_{TE}] = \frac{d}{d\phi_{TE}}(\phi_{TE}) = \frac{\left(N^2 - n_2^2\right)tg^2\theta_i + N^2}{N^2} z_s \tag{5.39}$$

For the right side, the result is

$$\begin{aligned}\frac{d}{d\beta}\left[\frac{\left(n_1^2 sin^2\theta_i - n_2^2\right)^{\frac{1}{2}}}{n_1 \sin\theta_i}\right] &= \frac{d}{d\theta_i}\left[\frac{\left(n_1^2 sin^2\theta_i - n_2^2\right)^{\frac{1}{2}}}{n_1 \cos\theta_i}\right]\frac{d\theta_i}{d\beta} \\ &= \frac{\left(N^2 - n^2\right)^{-1/2}\left[N^2 + \left(N^2 - n^2\right)tg^2\theta_i\right]}{N}\frac{tg\theta_i}{kN} \\ &= \frac{\left(N^2 - n^2\right)^{-1/2}\left[N^2 + \left(N^2 - n^2\right)tg^2\theta_i\right]}{kN^2} tg\theta_i\end{aligned} \tag{5.40}$$

Making equation (5.39) equal to equation (5.40), then we will get

$$z_s = k^{-1}\left(N^2 - n_2^2\right)^{-1/2} \text{tg}\,\theta_i \tag{5.41}$$

Similarly, for TM mode,

$$z_s = k^{-1}N^{-2}\left(N^2 - n_2^2\right)^{-1/2}\left(\frac{1}{n_1^2} + \frac{1}{n_2^2} - \frac{1}{N^2}\right)^{-1} \text{tg}\theta_i \tag{5.42}$$

Based on the geometrical relationship shown in Fig. 5.14, the depth of beam permeating the medium of refractive index n_2 is

$$x_s = \frac{z_s}{tg\theta_i} \tag{5.43}$$

Substituting equation (5.41) and (5.42) into (5.43), we can get

$$x_s = k^{-1}\left(N^2 - n_2^2\right)^{-1/2} \tag{5.44}$$

and

$$x_s = k^{-1}N^{-2}\left(N^2 - n_2^2\right)^{-1/2}\left(\frac{1}{n_1^2} + \frac{1}{n_2^2} - \frac{1}{N^2}\right)^{-1} \tag{5.45}$$

The above results can be applied to the boundary between the waveguide layer and the cladding layer.

For TE mode,

$$z_c = k^{-1}\left(N^2 - n_3^2\right)^{-1/2} \text{tg}\,\theta_i \tag{5.46}$$

For TM mode,

$$z_c = k^{-1}N^{-2}(N^2 - n_3^2)^{-1/2}\left(\frac{1}{n_1^2} + \frac{1}{n_3^2} - \frac{1}{N^2}\right)^{-1} \mathrm{tg}\theta_i \tag{5.47}$$

And for the above two modes, the depths of penetration are

$$x_c = k^{-1}(N^2 - n_3^2)^{-1/2} \tag{5.48}$$

and

$$x_c = k^{-1}N^{-2}(N^2 - n_3^2)^{-1/2}\left(\frac{1}{n_1^2} + \frac{1}{n_3^2} - \frac{1}{N^2}\right)^{-1} \tag{5.49}$$

Supposing the original depth of waveguide layer is d, the effective depth of waveguide can be expressed as equation (5.50) if we consider the penetration of beam in substrate and cladding layer (shown in Fig. 5.15).

$$d_{eff} = d + x_s + x_c \tag{5.50}$$

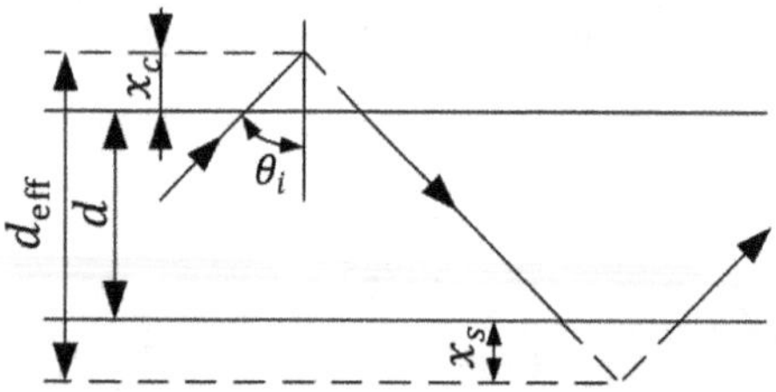

Fig. 5.15: The effective depth of waveguide layer.

And for TE mode,

$$d_{eff} = d + k^{-1}\left[(N^2 - n_2^2)^{-1/2} + (N^2 - n_3^2)^{-1/2}\right] \tag{5.51}$$

For TM mode,

$$d_{eff} = d + k^{-1}N^{-2}\left[(N^2 - n_2^2)^{-1/2}\left(\frac{1}{n_1^2} + \frac{1}{n_2^2} - \frac{1}{N^2}\right)^{-1} + (N^2 - n_3^2)^{-1/2}\left(\frac{1}{n_1^2} + \frac{1}{n_3^2} - \frac{1}{N^2}\right)^{-1}\right] \tag{5.52}$$

Sometimes the regularization depth and effective regularization depth are introduced, which are expressed by D and D_{eff}, respectively,

$$D = kd(n_1^2 - n_2^2)^{1/2}$$

$$D_{eff} = kd_{eff}(n_1^2 - n_2^2)^{1/2}$$

We can get the index of the waveguide

$$a = \frac{n_2^2 - n_3^2}{n_1^2 - n_2^2}, b = \frac{N^2 - n_2^2}{n_1^2 - n_2^2}$$

Then

$$N^2 = n_2^2 + b\left(n_1^2 - n_2^2\right),\ D_{eff} = D + b^{-1/2} + (a+b)^{-1/2}$$

5.4 The electromagnetic theories foundation of planar waveguide

Although the ray optics has the advantages of simplicity and direct-viewing, the optical phenomenon should be perfectly described by electromagnetic theories. In this subsection, we will make a further discussion on the planar dielectric waveguide based on a certain understanding.

5.4.1 The general form of Maxwell's equation

Under the condition of no free charge and conduction current, the two inhomogeneous equations in Maxwell's equations are as follows

$$\nabla \times \boldsymbol{E}(x, y, z; t) = -\frac{\partial}{\partial t}\boldsymbol{B}(x, y, z; t) \tag{5.53a}$$

and

$$\nabla \times \boldsymbol{H}(x, y, z; t) = \frac{\partial}{\partial t}\boldsymbol{D}(x, y, z; t) \tag{5.53b}$$

Where $\boldsymbol{E}$ is electrical field intensity vector, $\boldsymbol{H}$ is magnetic field intensity vector, $\boldsymbol{B}$ is magnetic flux density, and $\boldsymbol{D}$ is electric flux density.

In the system of rectangular coordinates, the Hamilton operator ∇ is written as

$$\nabla = \left(\frac{\partial}{\partial x}, \frac{\partial}{\partial y}, \frac{\partial}{\partial z}\right)$$

We suppose that equation (5.53) has the solutions of periodic change over time, that is

$$\boldsymbol{E}(x, y, z; t) = \boldsymbol{E}(x, y, z)e^{j\omega t} + c.c. \tag{5.54a}$$

$$\boldsymbol{H}(x, y, z; t) = \boldsymbol{H}(x, y, z)e^{j\omega t} + c.c. \tag{5.54b}$$

where ω is the angular frequency with periodic change, and $c.c.$ denotes the complex conjugate of $\boldsymbol{H}(x, y, z)e^{j\omega t}$. In the following discussion, we will use $\boldsymbol{E}$ and $\boldsymbol{H}$ to express the spatial part of the field.

Supposing there is no loss inside the medium, ε is the permittivity of the medium, and μ is the permeability of the medium, the constitutive equations can be written as

$$\boldsymbol{D} = \varepsilon \boldsymbol{E} \tag{5.55}$$

And

$$\boldsymbol{B} = \mu \boldsymbol{H} \tag{5.56}$$

Substituting equations (5.55), (5.56), and (5.54) into (5.53), we can get

$$\nabla \times Ee^{j\omega t} + c.c. = -\boldsymbol{B}(j\omega)e^{j\omega t} + c.c. = -j\omega\mu\boldsymbol{H}e^{j\omega t} + c.c.$$

That is

$$\nabla \times \boldsymbol{E} = -j\omega\mu\boldsymbol{H} \tag{5.57}$$

Similarly

$$\nabla \times \boldsymbol{H} = j\omega\varepsilon\boldsymbol{E} \tag{5.58}$$

Equations (5.57) and (5.58) are Maxwell's equations followed by electric and magnetic field vectors under the condition of no free charge and conduction current.

The left side of equation (5.57) can be expanded as follows

$$\nabla \times \boldsymbol{E} = \left(\frac{\partial E_z}{\partial y} - \frac{\partial E_y}{\partial z}\right)e_x + \left(\frac{\partial E_x}{\partial z} - \frac{\partial E_z}{\partial x}\right)e_y + \left(\frac{\partial E_y}{\partial x} - \frac{\partial E_x}{\partial y}\right)e_z$$

And the right side of equation (5.57) is

$$-j\omega\mu\boldsymbol{H} = -j\omega\mu\left(H_x e_x + H_y e_y + H_z e_z\right)$$

Making the above sides equal, we can get

$$\begin{aligned}
\frac{\partial E_z}{\partial y} - \frac{\partial E_y}{\partial z} &= -j\omega\mu H_x \\
\frac{\partial E_x}{\partial z} - \frac{\partial E_z}{\partial x} &= -j\omega\mu H_y \\
\frac{\partial E_y}{\partial x} - \frac{\partial E_x}{\partial y} &= -j\omega\mu H_z
\end{aligned} \tag{5.57a}$$

where E_x, E_y, , H_z are components of $\boldsymbol{E}$ and $\boldsymbol{H}$ in the direction of x, y, z, respectively, and e_x, e_y, e_z are unit vectors in x, y, z-direction, respectively.

In the same manner, from equation (5.58), we can get

$$\begin{aligned}
\frac{\partial H_z}{\partial y} - \frac{\partial H_y}{\partial z} &= -j\omega\varepsilon E_x \\
\frac{\partial H_x}{\partial z} - \frac{\partial H_z}{\partial x} &= -j\omega\varepsilon E_y
\end{aligned} \tag{5.58a}$$

$$\frac{\partial H_y}{\partial x} - \frac{\partial H_x}{\partial y} = -j\omega\varepsilon E_z$$

5.4.2 Maxwell's equations for planar waveguide

Maxwell's equations (5.57) and (5.58), or (5.57a), (5.58a) is universal, that is, under the condition of no free charge and conduction current, the above equations are true.

Especially, for the planar dielectric waveguide with three-layer, as is pointed out in subsection 5.3, the wave propagates in z-direction, then, we know,

$$E = E_t e^{-j\beta z}, H = H_T e^{-j\beta z} \tag{5.59}$$

where t denotes the lateral component of field.

Then, we take the derivative of the first formula of equation (5.59) with respect to z, and get

$$\frac{\partial E}{\partial z} = E_t(-j\beta)e^{-j\beta z} = -j\beta E$$

Thus,

$$\frac{\partial}{\partial z} = -j\beta \tag{5.60}$$

Moreover, for waveguide film, its scale in y-direction can be treated as infinite according to its depth, so $\boldsymbol{E}$ and $\boldsymbol{H}$ are not changed in y-direction, that is

$$\frac{\partial E}{\partial y} = \frac{\partial H}{\partial y} = 0 \tag{5.61}$$

Substituting equations (5.60) and (5.61) into (5.57a), we get

$$\beta E_y = -\omega\mu H_x \tag{5.62a}$$

$$j\beta E_x + \frac{\partial E_z}{\partial x} = j\omega\mu H_y \tag{5.62b}$$

$$\frac{\partial E_y}{\partial x} = -j\omega\mu H_z \tag{5.62c}$$

While substituting them into (5.58a), we get

$$\beta H_y = \omega\varepsilon E_x \tag{5.63a}$$

$$j\beta H_x + \frac{\partial H_z}{\partial x} = -j\omega\varepsilon E_y \tag{5.63b}$$

$$\frac{\partial H_y}{\partial x} = j\omega\varepsilon E_z \tag{5.63c}$$

After taking the derivative of equation (5.62) with respect to x, we substitute it into equation (5.63), and get

$$\frac{\partial^2 E_y}{\partial x^2} = -j\omega\mu\frac{\partial H_z}{\partial x} = -j\omega\mu\left(-j\omega\varepsilon E_y - j\beta H_x\right)$$

Then, equation (5.62) is substituted into the above equation, and finally we obtain

$$\frac{\partial^2 E_y}{\partial x^2} = \left(\beta^2 - n^2k^2\right) \quad E_y \tag{5.64}$$

where $k = \omega\sqrt{\omega_0\mu_0}$ is the wavenumber in vacuum, μ_0 is the permeability in vacuum, whose value is $4\pi\times 10^{-1}mkg/C^2$ measured by the electrostatic experiment; ε_0 is the permittivity in vacuum, whose experimental value is $8.85\times 10^{-12}s^2C^2/m^3kg$

For TM mode, in the same manner, utilizing the other three equations of equation (5.62) and (5.63), we can deduce the similar relationship expressed as

$$\frac{\partial^2 H_y}{\partial x^2} = \left(\beta^2 - n^2k^2\right)H_y \tag{5.65}$$

E_y and H_y can be solved by equations (5.64) and (5.65), in principle, we can get other field components.

5.4.3 Solutions of TE wave equations

As mentioned above, for TE mode, the transverse components of the electric field vectors exist but the longitudinal component of the electric filed should be zero, that is

$$E_z = 0$$

Substituting the above formula into equation (5.63c), we get

$$\frac{\partial H_y}{\partial x}$$

This means that H_y is a constant that does not depend on x. For simplicity and universality, we let $H_y = 0$, and substitute it into equation (5.63a) and obtain

$$E_x = 0$$

That is to say, for six components of the electromagnetic field vectors, there are no more than three zeroes, which are E_y, H_z and H_x, moreover, H_z and H_x can be expressed by E_y

$$H_x = -\frac{\beta}{\omega\mu}E_y, H_z = \frac{j}{\omega\mu}\frac{\partial E_y}{\partial x} \tag{5.66}$$

So for TE mode, nonzero components of the electromagnetic field vectors can be determined by solving equation (5.66) as long as we get E_y by solving equation (5.64) and substitute it into equation (5.66).

Equation (5.64) is a second-order differential equation with constant coefficients, its general solution form is

$$E_y(x) = \alpha e^{j\gamma x} + \alpha' e^{-j\gamma x} \tag{5.67}$$

where α and α' are integration constants.

Equation (5.64) and its general solution (5.67) fit to describe the wave propagating in the waveguide layer, the substrate and the cladding layer.

For waveguide, γ is expressed as

$$\gamma = \left(n_1^2k^2 - \beta^2\right)^{1/2}$$

For substrate and cladding layer,

$$\gamma_2 = \left(n_2^2k^2 - \beta^2\right)^{1/2}, \gamma_3 = \left(n_3^2k^2 - \beta^2\right)^{1/2}$$

To let the wave propagate inside the waveguide layer, $E_y(x)$ should has the solution with periodic variation in waveguide layer, whereas it is the evanescent wave in the substrate and cladding layer, which implies that γ is real number, but γ_2 and γ_3 are both imaginary numbers.

$$n_2^2k^2, n_3^2k^2 < \beta < n_1^2k^2 \tag{5.68}$$

Supposing the waveguide layer with the depth of d is located between the x=-d and x=0, then the solutions of $E_y(x)$ in waveguide layer, substrate and cladding layer are

$$E_y^{(2)}(x) = A_2e^{\gamma_2 x} + B_2e^{-\gamma_2 x}, \ x < -d,$$

$$E_y(x) = A\cos\gamma x + B\sin\gamma x, \ -d < x < 0$$

$$E_y^{(3)}(x) = A_3e^{-\gamma_3 x} + B_3e^{\gamma_3 x}, \ x > 0$$

When x → ±∞, the electric field is impossible infinite, apparently,

$$B_2 = B_3 = 0$$

Then $E_y(x)$ can be simplified as

$$E_y^{(2)}(x) = A_2e^{\gamma_2 x}, \ x < -d,$$

$$E_y(x) = A\cos\gamma x + B\sin\gamma x, \ -d < x < 0 \tag{5.69}$$

$$E_y^{(3)}(x) = A_3e^{-\gamma_3 x}, \ x > 0,$$

Substituting equation (5.69) into (5.66), we can get

$$H_x^{(2)}(x) = \frac{\beta A_2}{\omega\mu} e^{\gamma_2 x}, \ x < -d$$

$$H_x(x) = -\frac{\beta}{\omega\mu}(A\cos\gamma x + B\sin\gamma x), \ -d < x < 0 \tag{5.70}$$

$$H_x^{(3)}(x) = -\frac{\beta A_3}{\omega\mu} e^{-\gamma_3 x}, \ x > 0$$

and

$$H_z^{(2)}(x) = \frac{jA_2\gamma_2}{\omega\mu} e^{\gamma_2 x}, \ x - d$$

$$H_x(x) = \frac{j\gamma}{\omega\mu}(-A\sin\gamma x + B cos\gamma x), \ -d < x < 0 \tag{5.71}$$

$$H_x^{(3)}(x) = \frac{-jA_3\gamma_3}{\omega\mu} e^{-\gamma_3 x}, \ x > 0$$

Now for TE mode, we have the expressions for three nonzero components of the electromagnetic field vectors, and among them, there are four arbitrary constants A, B, A_2, and A_3, whose relationships can be deduced by the boundary conditions. On the boundary between $x=0$ and $x=-d$, the tangential components of electromagnetic field vectors should be continuous. First, we consider the components of electric field vectors, at $x=0$,

$$E_y(0) = E_y^{(3)}(0)$$

According to equation (5.69), we can obtain

$$A_3 = A \tag{5.72}$$

At x=-d

$$E_y(-d) = E_y^{(2)}(-d)$$

Thus

$$A_2 = (A\cos\gamma d + B\sin\gamma d)e^{\gamma_2 d} \tag{5.73}$$

Then, considering the components of magnetic field vectors, at $x=0$

$$H_z^{(2)}(0) = H_z(0)$$

Substituting the above formula into equation (5.71), we obtain

$$\gamma B = -A_3\gamma_3$$

From equation (5.72), we can get

$$B = -\frac{\gamma_3}{\gamma}A \tag{5.74}$$

Substituting equation (5.74) into (5.73), then

$$A_2 = A\left(\cos\gamma d + \frac{\gamma_3}{\gamma}\sin\gamma d\right)e^{\gamma_2 d} \tag{5.75}$$

So B, A_2, and A_3 can be expressed by A based on equations (5.74), (5.75), and (5.72). If we suppose $E_y(0) = E_0$ at $x=0$, then

$$A = E_0 \tag{5.76}$$

So all constants are known, the nonzero components of the electromagnetic field vectors in waveguide layer, substrate, and cladding layer are as follows

Substrate ($x<-d$)

$$\begin{aligned} E_y^{(2)}(x) &= E_0\left(\cos\gamma d + \frac{\gamma_3}{\gamma}\sin\gamma d\right)e^{\gamma_2(x+d)} \\ H_x^{(2)}(x) &= -\frac{\beta}{\omega\mu}E_0\left(\cos\gamma d + \frac{\gamma_3}{\gamma}\sin\gamma d\right)e^{\gamma_2(x+d)} \\ H_z^{(2)}(x) &= \frac{j\gamma_2}{\omega\mu}E_0\left(\cos\gamma d + \frac{\gamma_3}{\gamma}\sin\gamma d\right)e^{\gamma_2(x+d)} \end{aligned} \tag{5.77}$$

Waveguide layer ($-d<x<0$)

$$\begin{aligned} E_y(x) &= E_0\left(\cos\gamma x - \frac{\gamma_3}{\gamma}\sin\gamma x\right) \\ H_x(x) &= -\frac{\beta}{\omega\mu}E_0\left(\cos\gamma x - \frac{\gamma_3}{\gamma}\sin\gamma x\right) \\ H_z(x) &= \frac{-j\gamma}{\omega\mu}E_0\left(\sin\gamma x + \frac{\gamma_3}{\gamma}\cos\gamma x\right) \end{aligned} \tag{5.78}$$

Cladding layer （$x>0$）

$$\begin{aligned} E_y^{(3)}(x) &= E_0 e^{-\gamma_3 x} \\ H_x^{(3)}(x) &= -\frac{\beta}{\omega\mu}E_0 e^{\gamma_3 x} \\ H_z^{(3)}(x) &= -\frac{\gamma_3}{\omega\mu}E_0 e^{-\gamma_3 x} \end{aligned} \tag{5.79}$$

5.4.4 The modes of TE wave and cutoff condition

In the above subsection, we have derived the relationships between four arbitrary constant A, B, A_2, and A_3 based on the boundary conditions of $E_y(x)$ being tangential continuous at $x=0$, $-d$, and $H_z(x)$ being tangential continuous at $x=0$. Next, in this subsection, we will discuss some results when the boundary condition of $H_z(x)$ being tangential continuous at $x=-d$ are applied.

Let

$$H_z^{(2)}(-d) = H_z(-d)$$

Substituting the above formula into equation (5.77) and (5.78), we obtain

$$\gamma_2\left(\cos \gamma d + \frac{\gamma_3}{\gamma}\sin \gamma d\right) = \gamma \sin \gamma d - \gamma_3 \cos \gamma d$$

That is,

$$tg\gamma d = \frac{\gamma(\gamma_2 + \gamma_3)}{\gamma^2 - \gamma_2\gamma_3} \tag{5.80}$$

Equation (5.80) is the eigenvalue equation of dielectric waveguide. Because γ, γ_2, γ_3 are functions of propagation constant β, equation (5.80) is a equation of β. (5.80) is a transcendental equation, which is solved only by graphical method or numerical solution but not by elementary method, for each solution, there is a waveguide called as an eigenmode of TE wave. Only a set of discrete eigenmodes exist in the waveguide.

The right side of equation (5.80) can be written as

$$tg\gamma d = \frac{\gamma_2/\gamma + \gamma_3/\gamma}{1 - (\gamma_2/\gamma)\cdot(\gamma_3/\gamma)} \tag{5.80a}$$

Then

$$tg\gamma d = tg\left(tg^{-1}\frac{\gamma_2}{\gamma} + tg^{-1}\frac{\gamma_3}{\gamma}\right)$$

$$\gamma d - \left(tg^{-1}\frac{\gamma_2}{\gamma} + tg^{-1}\frac{\gamma_3}{\gamma}\right) = \nu\pi \tag{5.81}$$

Equation (5.81) is another form of eigenvalue equation, where ν is a positive integer, denoting the mode number.

For symmetric waveguide, $n_2 = n_3$, $\gamma_2 = \gamma_3 = \gamma$ equation (5.80) can be simplified as

$$tg\gamma d = \frac{2\gamma\gamma'}{\gamma^2 - \gamma'^2} \tag{5.80b}$$

For equation (5.80b), we divide the numerator and denominator by γ^2, then obtain

$$tg\gamma d = \frac{2\gamma'/\gamma}{1-(\gamma'/\gamma)^2}$$

Apparently

$$tg(\gamma d/2) = \gamma'/\gamma \tag{5.82}$$

Equation (5.82) is followed by all symmetric waveguides.

If we divide the numerator and denominator of equation (5.80b) by γ'^2, then we will obtain

$$tg\gamma d = \frac{2\gamma/\gamma'}{1-(\gamma/\gamma')^2}$$

Apparently

$$tg(\gamma d/2) = -\gamma/\gamma' \tag{5.83}$$

Equations (5.82) and (5.83) are much easier to be solved by graphical or numerical methods than equation (5.80). After having the numerical solutions, we can make an expected conclusion that only single mode waveguide exists at the low frequency (λ is longer, k is smaller), the electromagnetic field distribution is symmetrical of x=-$d/2$, which changes with cosine style inside the waveguide layer (-d<x<0), but has a exponential decay outside the waveguide layer (-d>x or x>0) (Fig. 5.16).

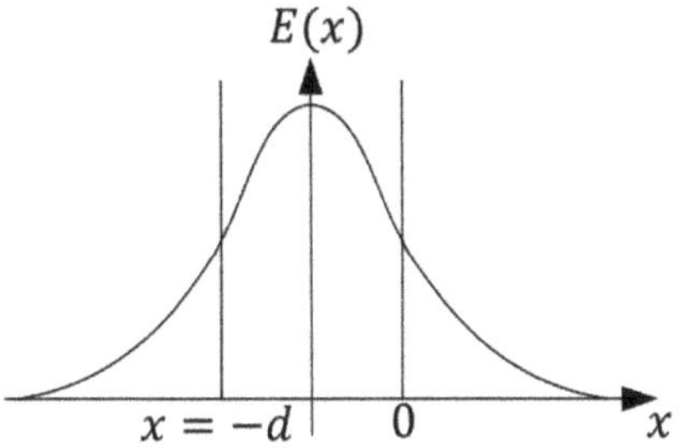

Fig. 5.16: The lowest order symmetric mode of planar waveguide.

With the increase of light wave frequency, the field will be concentrated at the center of the waveguide layer. When the frequency increases to a certain extent, there possibly exists antisymmetric solution and dual-mode waveguide propagates, as shown in Fig. 5.17.

As mentioned above, at the critical state, $\theta_i = \theta_c$, $n_1 \sin\theta_i = n_2$, that is $\beta \to n_2 k$, $\gamma_2 \to 0$, for the planar waveguide with symmetrical structure, $\gamma_3 \to 0$, that is, $\gamma' \to 0$, the cutoff condition of symmetric mode can be obtained by equation (5.82)

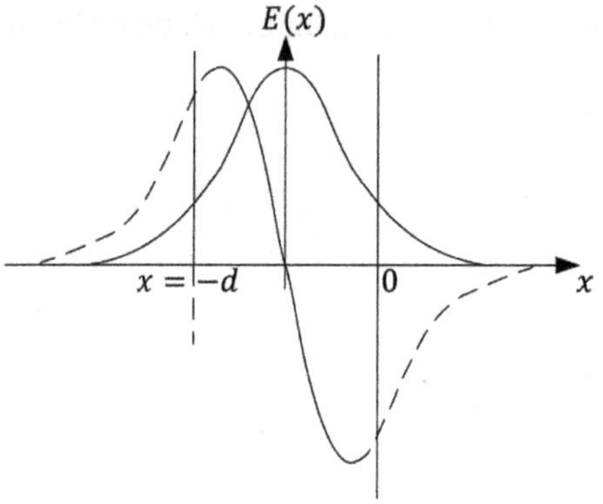

Fig. 5.17: Dual-mode field distribution of planar waveguide.

$$tg(\gamma d/2) = 0 \tag{5.84}$$

Or

$$\gamma d/2 = 0, \pi, 2\pi, \ldots \tag{5.84a}$$

While the cutoff condition of antisymmetric mode can be obtained by equation (5.83)

$$tg(\gamma d/2) \to \infty \tag{5.85}$$

Or

$$\gamma d/2 = \pi/2, 3\pi/2, \ldots \tag{5.85a}$$

Combining equations (5.84a) and (5.85a), we can obtain

$$\gamma d/2 = \nu\pi/2 \tag{5.86}$$

where ν is mode number. Equation (5.86) offers the cutoff conditions of all symmetric waveguide modes.

The above results can be obtained by equation (5.80) directly. In fact, when $\gamma_2 = \gamma_3 \to 0$, from (5.80), we obtain

$$tg\gamma d = 0$$

That is

$$\gamma d = \nu\pi$$

At the cutoff condition,

$$\gamma = k\sqrt{n_1^2 - n_2^2}$$

Hence,

$$(\gamma d/2) = (kd/2)\sqrt{n_1^2 - n_2^2} \tag{5.87}$$

It is a dimension-less parameter, which is called as "V" of waveguide. When $V \geq 1$, the multi-mode waveguide propagates. Making equation (5.86) be equal to (5.87), we obtain

$$(kd/2)\sqrt{n_1^2 - n_2^2} = v\pi/2 \tag{5.88}$$

Equation (5.88) offers the cutoff conditions of all possible waveguide modes in the waveguide layer with the depth of d being composed of dielectric layers with the refractive index of n_1 and n_2, respectively

$$v = \frac{2d}{\lambda}\sqrt{n_1^2 - n_2^2} \tag{5.88a}$$

If n_1, n_2, d, and λ are known, we can estimate that how many waveguide modes will be appeared at most. Next, we will take an example utilizing equation (5.88(a)).

Example 5.1 Supposing the refractive indexes of symmetric waveguide are $n_1 \approx n_2 \approx 2$, respectively, and $n_1 - n_2 = 0.01$, the depth of middle waveguide layer is $d = 2\mu m$, when $\lambda = 0.8\mu m$, how many modes are there in wave propagation?

Solution: Substituting the above data into equation (5.88(a)), we can obtain

$$v = \frac{2 \times 2}{0.8} \times 0.2 = 1$$

Hence, there are two modes corresponding to v=0 and 1.

Equation (5.88a) can be applied to determine the depth of waveguide layer when we need to limit the guided wave modes in a certain range. Therefore, equation (5.88a) is rewritten as

$$d = \frac{v\lambda}{2\sqrt{n_1^2 - n_2^2}} \tag{5.88b}$$

When equation (5.88b) is founded, there are (v+1) modes. If the maximum mode number is v, the depth of waveguide layer should meet the following in equation

$$d < \frac{v\lambda}{2\sqrt{n_1^2 - n_2^2}} \tag{5.89}$$

Especially, if a single-mode waveguide propagates, then

$$d < \frac{\lambda}{2\sqrt{n_1^2 - n_2^2}} \tag{5.90}$$

Example 5.2 Supposing the refractive indexes of symmetric waveguide are $n_1 \approx n_2 \approx 1.8$ respectively, and $n_1 - n_2 = 0.001$, when $\lambda = 0.6\mu m$ to realize the single-mode waveguide propagating, what is the depth of the waveguide layer?

Solution: substituting the above data into Equation (5.90), we can obtain

$$d < \frac{0.6}{2 \times 0.06} = 5(\mu m)$$

So the depth of waveguide layer should be less than 5 μm.

It is worth pointing out that,

Provided $n_1 - n_2 \neq 0$,

Then

$$(kd/2)\sqrt{n_1^2 - n_2^2} = 0$$

That is to say, for the lowest-order mode (corresponding to $v=0$), no matter what the wavelength and the depth of the waveguide layer are, the term cutoff for the symmetric waveguide does not occur, which is one of properties for planar waveguide.

5.4.5 Properties of waveguide mode

Waveguide modes have many important properties, in this subsection, we intend to present the main results without proof in the simplest case. Readers who wish to have a deeper understanding need to read other related literature.

First, different modes are orthogonal. Using $E_{vy}(x)$ and $E_{v'y}$ to express component of electric field in y-direction of two differential TE modes ($v' \neq v$), we can prove

$$\int_{-\infty}^{\infty} E_{vy}(x) E'_{v'y}(x) dx = 0 \tag{5.91}$$

That is to say, the integral of the product of two different modes in the waveguide cross section (the integral along y-direction has been omitted, so two-dimensional integral can be simplified as one-dimensional integral in x-direction) is zero. An example is shown in Fig. 5.17, and the more simple form of equation (5.91) is expressed by

$$< E_{vy}(\mathrm{x}), E'_{v'y}(x) > = 0 \tag{5.91a}$$

Where angle brackets denotes the inner product of two functions, which has been used in theoretical physics widely.

Second, the component of field of waveguide modes can be normalized, that is, for the same mode ($v' = v$), the integral on the left side of equation (5.91) is equal to 1 as long as the component of optical field is multiplied by an appropriate coefficient (normalization factor)

The above properties make arbitrary transverse field be expanded into a form of mode component superposition. For example, in the case of only discrete mode existing, we find

$$E_y(x) = \sum_v a_v E_{vy}(\mathrm{x}) \tag{5.92}$$

where a_v is the expansion coefficient,

$$a_v = \frac{\int_{-\infty}^{\infty} E_y(x)E_{vy}(x)dx}{\int_{-\infty}^{\infty} E_{vy}^2(x)dx} \tag{5.93}$$

Or a_v can be expressed in a form of simple

$$a_v = \frac{\langle E_y(\mathrm{x}), E_{vy}(\mathrm{x})\rangle}{\langle E_{vy}(\mathrm{x}), E_{vy}(\mathrm{x})\rangle} \tag{5.93a}$$

Finally, utilizing the orthogonality, we can obtain the energy density of all waveguide modes propagating along z -direction,

$$P_z(x) = \sum_v a_v a_v^* P_v \tag{5.94}$$

where P_v is the energy density of vth waveguide mode, which is expressed as follows

$$P_v = (\beta/2\omega\mu) \int_{-\infty}^{\infty} E_{vy}^2(x)dx = (\beta/2\omega\mu) < E_{vy}(\mathrm{x}), E_{vy}(\mathrm{x}) > \tag{5.95}$$

The planar waveguide which we have discussed above has the simplest structure. Another waveguide structure called as channel waveguide has been widely used in many integrated optical active and passive devices, including lasers, modulators, switches, and directional coupler. We will make a brief introduction on channel waveguide in the next subsection.

5.5 Channel waveguide introduction

As we know, the planar waveguide confine the optical field in x-direction (i.e., the direction of depth of waveguide). The size of waveguide in y or z direction is considered as infinite relative to the depth of waveguide layer, so for planar waveguide, the optical field is unconfined in the plane of y-z.

In this subsection, we will discuss another waveguide structure, namely channel waveguide. Being different from the planar waveguide, its size in y-direction can not be regarded as infinite, so it is arranged to confine the optical field in two directions, x and y.

First, we will introduce the types of channel waveguides, then the vector wave equations are derived to describe the channel waveguide. However, we do not intend to solve the problem, but give approximate scalar equations to describe the channel waveguide and find some solutions.

5.5.1 Channel waveguide types

Figure 5.18(a) shows the cross-section through the basic geometry, and Figs. 5.18 (b)–5.18(e) show several kinds of common structure form. For simplicity, the refractive

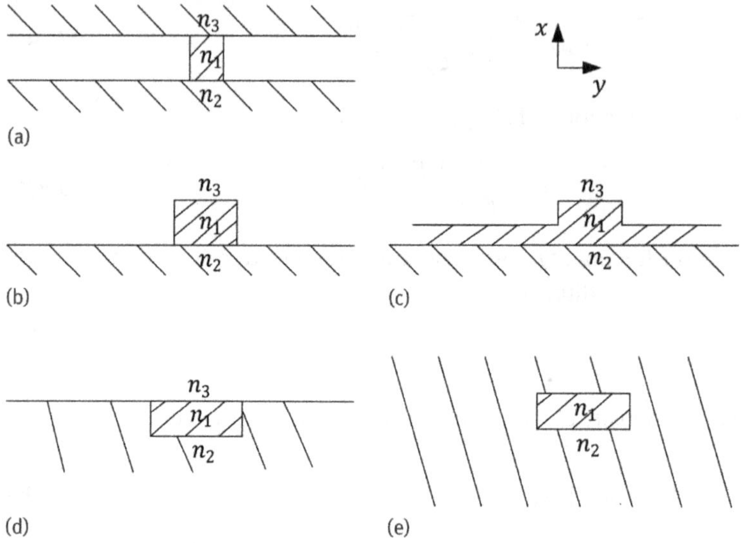

Fig. 5.18: Several common kinds of cross–sections of channel waveguide: (a) general, (b) ridge, (c) strip-loaded, (d) buried, (e) recessed.

indexes shown in the above figures change in step mode, but in fact, the refractive indexes change with a certain gradient because of diffusion fabrication process. In Fig. 5.18, the wave is mainly confined inside the film waveguide layer with the refractive index of n_1. Figure 5.18(b) shows a ridge waveguide, which is usually fabricated from a planar guide that has been patterned and etched to leave a ridge; Fig. 5.18(c) shows a strip-loaded waveguide having the similar fabrication process as ridge waveguide, but the film around the waveguide being not completely removed. Buried waveguide shown in Fig. 5.18(d) and recessed striped waveguide shown in Fig. 5.18(e) are made by modifying the properties of the substrate material so that a higher refractive index is obtained locally.

5.5.2 Vector wave equation

For channel waveguide, $n = n(x, y)$ is the function of two horizontal coordinates, so the analysis of waveguide modes is more complicated than that of planar waveguides. But we still start from the general Maxwell equations

$$\nabla \times \boldsymbol{E} = -j\omega\mu\boldsymbol{H} \tag{5.96a}$$

$$\nabla \times \boldsymbol{H} = j\omega\varepsilon\boldsymbol{E} \tag{5.96b}$$

We apply the curl to equation (5.96a), and by substituting equation (5.96), we obtain

$$\nabla \times \nabla \times \boldsymbol{E} = \omega^2 \varepsilon \mu \boldsymbol{H} \tag{5.97}$$

While applying the curl to equation (5.96b), and utilizing the following identity

$$\nabla \times (\boldsymbol{ab}) \equiv a \nabla \times \boldsymbol{b} + (\nabla a) \times \boldsymbol{b}$$

We get

$$\begin{aligned} \nabla \times \nabla \times \boldsymbol{H} &= j\omega \nabla \times (\varepsilon \boldsymbol{E}) \\ &= j\omega [\varepsilon \nabla \times \boldsymbol{E} + (\nabla \varepsilon) \times \boldsymbol{E}] \end{aligned}$$

By substituting equation (5.96a), we note

$$\nabla \ln \varepsilon = \frac{\nabla \varepsilon}{\varepsilon}$$

And finally we obtain

$$\nabla \times \nabla \times \boldsymbol{H} = \omega^2 \varepsilon \mu \boldsymbol{H} + \nabla \ln \varepsilon \times (\nabla \times \boldsymbol{H}) \tag{5.98}$$

On the other hand, if we apply the curl to equation (5.96), we will get

$$\nabla \cdot \boldsymbol{H} = 0 \tag{5.99}$$

And

$$\nabla \cdot (\varepsilon \boldsymbol{E}) = 0 \tag{5.100}$$

Expanding the left side of equation (5.100), we get

$$(\nabla \varepsilon) \cdot \boldsymbol{E} + \varepsilon \nabla \cdot \boldsymbol{E} = 0$$

Or

$$\nabla \cdot \boldsymbol{E} = -\boldsymbol{E} \cdot (\nabla \ln \varepsilon) \tag{5.100a}$$

With the help of another identity,

$$\nabla \times \nabla \times \boldsymbol{a} = -\nabla^2 \boldsymbol{a} + \nabla(\nabla \cdot \boldsymbol{a}) \tag{5.101}$$

We can rewrite equation (5.97) as follows

$$\nabla^2 \boldsymbol{E} \cdot \nabla(\nabla \cdot \boldsymbol{a}) + \omega^2 \varepsilon \mu \boldsymbol{E} = \boldsymbol{0}$$

by substituting into equation (5.100a), we get

$$\nabla^2 + \nabla(\boldsymbol{E} \cdot \nabla \ln \varepsilon) + \omega^2 \varepsilon \mu \boldsymbol{E} = 0 \tag{5.102}$$

Based on equations (5.101) and (5.99), equation (5.98) can be rewritten as

$$\nabla^2 H + (\nabla \ln \varepsilon) \times (\nabla \times \boldsymbol{H}) + \omega^2 \varepsilon \mu \boldsymbol{H} = 0 \tag{5.103}$$

If using t and z as the subscripts to express the transverse and longitudinal components of electromagnetic field, we can write the following equation

$$\nabla^2 E_t + \nabla\left(\boldsymbol{E}_t^{'} \cdot \nabla \ln \varepsilon\right) + (\omega^2 \varepsilon\mu - \beta^2)\boldsymbol{E}_t = 0$$

$$\nabla^2 H_t + (\nabla \ln \varepsilon) \times (\nabla \times \boldsymbol{H}_t) + (\omega^2 \varepsilon\mu - \beta^2)\boldsymbol{H}_t = 0 \tag{5.104}$$

And

$$j\beta E_z = \nabla \cdot \boldsymbol{E}_t + \boldsymbol{E}_t \cdot \nabla \ln \varepsilon$$

$$j\beta H_z = \nabla \cdot \boldsymbol{H}_t \tag{5.105}$$

Where β is the propagation constant.

5.5.3 Approximate scalar equation and the method of separation of variables

It is very difficult to solve the above vector wave equation. Similar to the planar waveguide, the following scalar equation is also a very good approximation for the channel waveguide.

$$\nabla^2 E_t + (n^2 k^2 - \beta^2) E_t = 0 \tag{5.106}$$

Unlike the former, E_t and n are functions of two variables with respect to x and y. In the general case, it is not an easy thing to solve equation (5.106) accurately. But if the square of refractive index can be expressed by

$$n^2(x, y) = n_0^2 + n_x^2(x) + n_y^2(y) \tag{5.107}$$

Then the two variables x and y of components of electromagnetic field can be separated. For example

$$E_t(x, y) = E_x(x) E_y(y) \tag{5.108}$$

Here $E_x(x)$ and $n_x(x)$ are functions of x, which are denoted as E_x and n_x in the following discussion; and $E_y(y)$ and $n_y(y)$ are functions of y, which are denoted as E_y and n_y, similar to equation (5.107), the square of propagation constant can be expressed by

$$\beta^2 = k^2 n_0^2 + \beta_x^2 + \beta_y^2 \tag{5.109}$$

Substituting equation (5.107) to (5.109) into (5.106), we get

$$E_y \frac{d^2}{dx^2} E_x + E_x \frac{d^2}{dy^2} E_y + \left[(k^2 n_x^2 - \beta_x^2) + \left(k^2 n_y^2 - \beta_y^2\right)\right] E_x E_y = 0$$

Or according to the separation of variables, we can write

$$\frac{d^2}{dx^2}E_x + \left(k^2n_x^2 - \beta_x^2\right)E_x = 0 \tag{5.110}$$

And

$$\frac{d^2}{dy^2}E_y + \left(k^2n_y^2 - \beta_y^2\right)E_y = 0 \tag{5.111}$$

So far, we has got two similar equations as that of planar waveguide, which can be solved by the methods mentioned in subsection 5.4, and need not be repeated here.

5.5.4 Other solutions of scalar equations

In practical applications, the refractive index of the waveguide medium satisfies the equation (5.107) in a very small number of cases, which makes the application of separation of variables introduced in the previous subsection be limited. To analyze various types of channel waveguides, people have proposed many methods, because of the limited space, we can neither discuss these methods one by one, nor can we analyze them deeply, in this subsection, we will intend to give a brief introduction of some of them, readers who wish to have a deeper understanding need to read other related literature.

In 1969, Marcatilu coming from Bell laboratory proposed a field shielding method,which utilizing the separation of variables to analyze the channel waveguide modes through a further approximation. In Fig. 5.19, the interface of the waveguide shown in Fig. 5.19(a) is perpendicular to the x-axis; the waveguide interface shown in Fig. 5.19(b) is perpendicular to the y-axis. Suppose the two waveguides are made of the same material, and the square of refraction index is $n_A^2 = \frac{n_1^2}{2}$, while the square of refraction indexes of substrate and covering layer are $n_B^2 = n_2^1 - n_1^2/2$, after they are

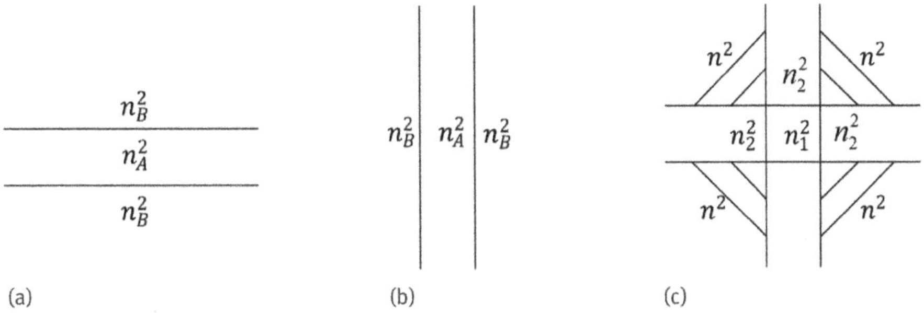

Fig. 5.19: Field shielding method (a) planar waveguide whose boundary is perpendicular to x-axis, (b) planar waveguide whose boundary is perpendicular to y-axis, (c) composite channel waveguide $n_A^2 = \frac{n_1^2}{2}$; $n_B^2 = n_A^2 - \frac{n_1^2}{2}$; $n^2 = 2n_2^2 - n_1^2$.

recombined by being overlapped according to x-axis and y-axis respectively, the refraction indexes are shown in Fig. 5.19(c), and the shadow is called as the field shielding area, in which the refraction index is $2n_2^2 - n_1^2$. If the optical field and refractive index of the field shielding area are not considered, we will obtain a typical buried channel waveguide, whose refractive index can be expresses by that of planar waveguide accordingly.

$$n^2 = n_x^2 + n_y^2$$

Under this condition, the optical field of channel waveguide can be separated relative to x and y,

$$E(x, y) = E_x E_y$$

while the propagation constant is determined by

$$\beta^2 = \beta_x^2 + \beta_y^2$$

where β_x and β_y are propagation constants of planar waveguide accordingly.

The field shielding method is similar in two senses: (1) the scalar wave equation is used instead of the vector wave equation; (2) the refractive index of the shield is neglected. In 1982, Akiba and his colleague, Haus, used the results of the screening method as a heuristic solution, and obtained the improved solution by the vector variational principle. Then, Kumar et al. proposed to use the perturbation method to correct the error caused by the difference between the refractive indexes of composite waveguide and buried waveguide, which can amend the propagation constant obtained by the scalar wave equation.

Another useful method for analyzing the channel waveguide is the effective refractive index method proposed by Knox et al, for a large number of practical waveguide structures, including ridge waveguides, buried waveguides and diffusion-type waveguides, using this method we can obtain the solutions which agree well with the more accurate digital calculation from computer or experimental results.

The basic idea of effective refractive index method is looking down the waveguide film on the plane of y-z from $-x$-direction. For homogeneous planar waveguides, their uniform effective refractive index N is independent on the coordinate of y, z; when the depth of waveguide d or refractive index n does a small change, the effective refractive index $N(y, z)$ will depend on the coordinate of y, z. Once we get the effective refractive index $N(y, z)$, we can determine the mode and propagation constant of channel waveguide by the method of planar waveguide.

Taking the ridge waveguide as an example, its cross section and vertical view are shown in Fig. 5.20, where n_1、n_2 and n_3 denote the refraction indexes of material respectively, which are the same as that of planar waveguide, and N_f and N_l denote the effective refraction index of the ridge and its both sides. Despite the refraction

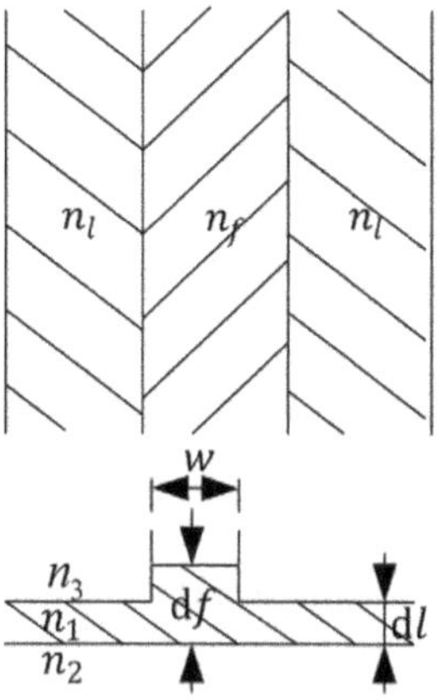

Fig. 5.20: The cross section and vertical view of the ridge waveguide. *df*—the depth of ridge; *w*—the width of ridge; *dl*—the depth of both sides of ridge.

indexes of this two material themselves being the same, N_f and N_l are different because the depths d_f and d_l are different.

It is divided into the following four steps to deal with the channel waveguide utilizing the effective index method

(1) First find out the regular depth of ridge and its both sides

$$D_f = kd_f\left(n_1^2 - n_2^2\right)^{1/2}$$

$$D_l = kd_l\left(n_1^2 - n_2^2\right)^{1/2} \tag{5.112}$$

(2) Determine the effective refraction index by the following equations

$$N_f^2 = n_2^2 + b_f\left(n_1^2 - n_2^2\right)$$

$$N_l^2 = n_2^2 + b_l\left(n_1^2 - n_2^2\right) \tag{5.113}$$

Under the condition of $n_1 - n_2 \ll n_2$, the effective refraction index is far less than

$$N = \left[n_2^2 + b\left(n_1^2 - n_2^2\right)\right]^{1/2} = n_2\left[1 + b\frac{n_1^2 - n_2^2}{n_2^2}\right]^{1/2}$$

$$\approx n_2 + b(n_1 - n_2)$$

So

$$N_f = n_2 + b_f(n_1 - n_2)$$

$$N_l = n_2 + b_l(n_1 - n_2)$$

(3) Calculate the regular depth of the equivalent waveguide

$$\Omega_{eq} = k\omega\left(N_f^2 - N_l^2\right)^{1/2} \tag{5.114}$$

Then determine the waveguide index b_{eq}.

(4) Finally get the effective index of channel waveguide

$$N^2 = N_l^2 + b_{eq}\left(N_f^2 - N_l^2\right) \tag{5.115}$$

We can analyze the optical field of channel waveguide utilizing the above results. For example, the propagation constant can be got by $\beta = kN$.

Example 5.3 suppose a ridge channel waveguide made of Ti:LiNbO$_3$, $n_1 = 2.234$: $n_2 = 2.214$, $n_3 = 1$, $d_f = 1.8\mu m$, $d_l = 1\mu m$, $\omega = 2\mu m$. Please find (1) the effective refraction N; (2) the propagation constant at $\lambda = 0.8\mu m$.

Solution:
(1) at first, we get $D_f = 4.2$, $D_l = 2.3$ by utilizing equation (5.112); second, $N_f = 2.227$, $N_l = 2.218$ can be obtained by equation (5.113); thirdly, we find $\Omega_{eq} = 3.14$, $b_{eq} = 0.64$; and finally, we get $N = 2.224$.
(2) $\beta = \frac{2\pi}{\lambda} N = 17.46\mu m$

5.6 Mode coupling theory in guided wave structures

Many phenomena that occur in physics or engineering fields can be seen as the processes of mode coupling. Typical examples include the diffraction of X-rays in crystals, the energy exchange between electron beams and slow waves in traveling wave tubes, and the scattering of light, sound wave and holographic gratings. Similarly, in integrated optics, the mode coupling theory is also a powerful tool to help us to understand and analyze a large number of important phenomena and devices. In this subsection we will give a brief introduction to some of the most basic questions on coupling theory.

5.6.1 Basic concepts of the directional coupling

In integrated optics, the most successful and widely used device is the directional coupler, which works by coupling together two modes travelling in the same direction, so it can be used as beam splitters, two-way switches, modulators, filters and polarizers.

Considering the lowest order modes in two symmetric planar waveguides, Fig. 5.21 shows the case that the two waveguides are far apart and thus no mode coupling occurs, and each symmetry mode is transmitted independently in its waveguide.

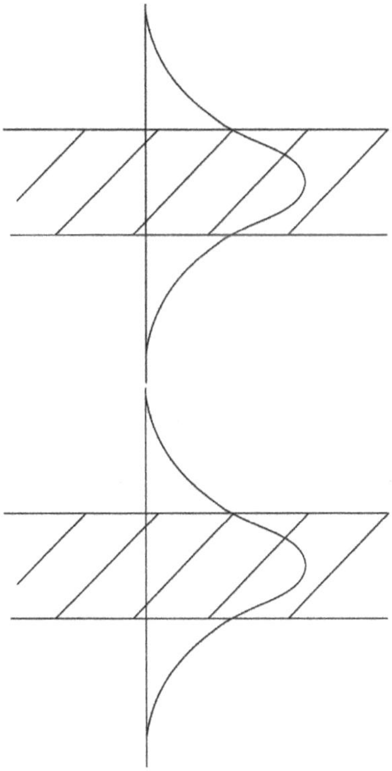

Fig. 5.21: Non-mode coupling in guided wave.

With the two waveguides being close to each other, the two modes begin to overlap partially, in the region of overlapping, it is hard to decide which waveguide any light in the overlap region actually belongs to. This means that there must be a coupling mechanism for light in one waveguide to be transferred to the other. If the two waveguides satisfy a specific set of conditions and are placed sufficiently close together (the interwaveguide gap required for this overlap is of the order of the waveguide width, i.e., a few of microns), this interchange of power is highly significant, reaching almost 100% in many cases. In this way, it is assumed that a finite input light only is transferred in one waveguide initially, and after being transferred a certain distance, part of the light will go completely into the other waveguide and thereafter continue to be transferred and gradually will start coupling back. So the power transfer process is periodic with distance.

Assuming the variation period is 2*L*, the transfer process can be shown in Fig. 5.22. Here the input is to Guide 1, at $z = 0$. And the complete coupling occurs at $z = L$ when all the light has been transferred to Guide 2. Then, at *z{\mequal}* 2*L*, the light will be completely coupled back to Guide 1, and so on. The above process is continuous.

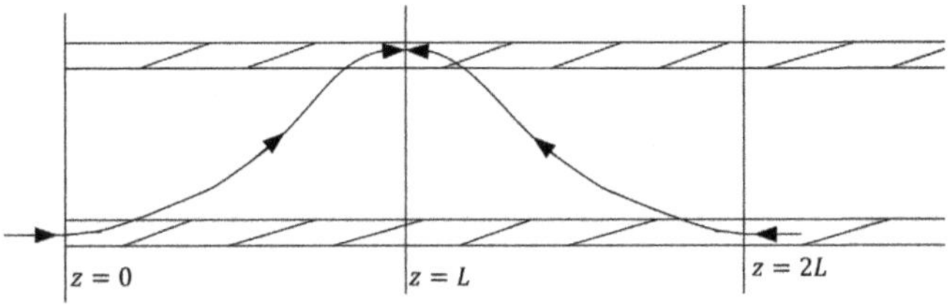

Fig. 5.22: The coupling process.

This 100% coupling requires that the two modes are fully synchronized (have the same propagation constant) so that the scattering contribution is added in phase. This usually requires that the two waveguides be identical.

If the two waveguides satisfy for 100% power transfer, they must be synchronous (i.e., have the same propagation constants) so that all the scattered contributions add up in-phase. This usually requires them to be identical.

5.6.2 Coupled mode equations

As we discussed in the previous subsection, the guided wave mode will be in the waveguide without radiation field originally due to the coupling effect, which means that there must be some kind of incentive mechanism. So, when we handle the problem of wave coupling, we need to deal with the incentive problem of waveguide mode first.

Considering the distribution of incentive source of various waveguide mode and expressing them by the complex amplitude of the corresponding induced polarization vector $P(x,y,z)$, then the coupled wave problem is to deal with the active Maxwell equations.

$$\nabla \times E = -j\omega\mu H$$

$$\nabla \times H = j\omega\varepsilon E + j\omega P \tag{5.116}$$

Setting two induced polarization vector (P_1 and P_2) as the source of field, and based on the identical relation of vector algebra, we can get

$$\nabla \cdot \left(E_1 \times H_2^* + E_2^* \times H_1\right) = (\nabla \times E_1) \cdot H_2^* - E_1 \cdot \left(\nabla \times H_2^*\right) + \left(\nabla \times E_2^*\right) \cdot H_1 - E_2^* \cdot (\nabla \times H_1) = -j\omega P_1 \cdot E_2^* + j\omega P_2^* \cdot E_1 \tag{5.117}$$

Or via the divergence theory, we can obtain

$$\iint_{-\infty}^{\infty} dxdy \frac{\partial}{\partial z}\left(E_1 \times H_2^* + E_2^* \times H_1\right)_z = -j\omega \iint_{-\infty}^{\infty} dxdy\left(P_1 \cdot E_2^* - P_2^* \cdot E_1\right) \tag{5.118}$$

Equation (5.117) and (5.118) are the differential form and integral form of the vector equation described for the coupling of two-waveguide respectively. If P_1 and P_2 are known, and a certain constitutive equation and boundary conditions are also considered, in principle, the required electromagnetic field components can be solved by any of the two above equations.

5.6.3 Scalar-coupled wave equations

Solving the vector coupled wave equation derived in the previous subsection is quite complicated, for simplicity, in this subsection, according to the above discussion, we will deduce the scalar forms and analyze them. And their solutions will be discussed in the next subsection.

We assume that the two waveguides are oriented in the z-direction, and are described in isolation by the refractive index distributions and $n_b(x,y)$ respectively. Assuming that the electric field is polarized in the *y*-direction, so the electric field $E_{ya}(x,y,z)$ and $E_{yb}(x,y,z)$ for both waveguides are got by

$$\nabla^2 E_{ya}(x,y,z) + n_a^2(x,y)k^2 E_{ya}(x,y,z) = 0 \tag{5.119a}$$

$$\nabla^2 E_{yb}(x,y,z) + n_b^2(x,y)k^2 E_{yb}(x,y,z) = 0 \tag{5.119b}$$

We assume that both waveguides are single mode, so the guided eigenmode solutions are

$$E_{ya}(x,y,z) = E_a(x,y)exp\left(-j\beta_a z\right) \tag{5.120a}$$

$$E_{yb}(x,y,z) = E_b(x,y)exp\left(-j\beta_b z\right) \tag{5.120b}$$

Where $E_{a,b}(x,y)$ is the transverse field of each waveguide mode, and $\beta_{a,b}$ is the corresponding propagation constant.

Substituting equation (5.120) into (5.119), and replacing ∇^2 with $\left(\nabla_t^2 + \frac{\partial^2}{\partial z^2}\right)$, we can get

$$\nabla_t^2 E_a(x,y) + \left[n_a^2(x,y)k^2 - \beta_a^2\right]E_a(x,y) = 0 \tag{5.121a}$$

$$\nabla_t^2 E_b(x,y) + \left[n_b^2(x,y)k^2 - \beta_b^2\right]E_b(x,y) = 0 \tag{5.121b}$$

Here the subscripts on the Laplacian indicate that the differential operation is performed in the transverse plane. These equations have the same forms with the scalar equation discussed in subsection 5.4, thus we can say that the field components (E_a and E_b) and the corresponding transfer functions (β_a and β_b) have been solved.

We assume that the complete coupler is described by a refractive index distribution *nT(x,y)*, which corresponds to both waveguides together. So the scalar wave equation for the complete coupler ism

$$\nabla^2 E_y(x,y,z) + n_T^2(x,y)k^2 E_y(x,y,z) = 0 \tag{5.122}$$

We further assume that the solution for $E_y(\mathrm{x},\mathrm{y},\mathrm{z})$ can be expressed as a linear combination of the eigenmodes in the two isolated guides, so we obtain

$$E_y(x,y,z) = A(z)E_a(x,y)exp(-j\beta_a z) + B(z)E_b(x,y)exp(-j\beta_b z) \tag{5.123}$$

Where *A(z)* and *B(z)* are amplitudes of the two eigenvalues, obviously, they vary with the propagation distance.

To act on both sides of equation (5.123) with operator $\nabla^2 = \nabla_t^2 + \frac{d^2}{dz^2}$, for simplicity, omitting the function argument, we can get

$$\begin{aligned}\nabla^2 E_y &= \nabla_t^2 E_a + \frac{d^2}{dz^2}\left[Aexp(-j\beta_a z)\right] + \nabla_t^2 E_b + \frac{d^2}{dz^2}\left[Bexp(-j\beta_b z)\right]\\ &= \left[A\nabla_t^2 E_a + \left(\frac{d^2A}{dz^2} - 2j\beta_a\frac{dA}{dz} - \beta_a^2 A\right)E_a\right]exp(-j\beta_a z) +\\ &\quad\left[B\nabla_t^2 E_b + \left(\frac{d^2B}{dz^2} - 2j\beta_b\frac{dB}{dz} - \beta_b^2\right)E_b\right]exp(-j\beta_b z)\end{aligned} \tag{5.124}$$

Substituting (5.123) and (5.124) into (5.122), we get

$$\begin{aligned}&\left[\left(\frac{d^2A}{dz^2} - 2j\beta_a\frac{dA}{dz}\right)E_a + n_T^2k^2AE_a + (\nabla_t^2E_a - \beta_a^2E_a)A\right]exp(-j\beta_a z) +\\ &\left[\left(\frac{d^2B}{dz^2} - 2j\beta_b\frac{dB}{dz}\right)E_b + n_T^2k^2BE_b + (\nabla_t^2E_b - \beta_b^2E_b)B\right]\exp(-j\beta_b z) = 0\end{aligned} \tag{5.125}$$

Via equation (5.121), we obtain

$$-\left(\nabla_t^2E_a - \beta_a^2E_a\right) = k^2n_a^2E_a$$

$$-\left(\nabla_t^2E_b - \beta_b^2E_b\right) = k^2n_b^2E_b$$

Substituting them into (5.125), we get

$$\left[\frac{d^2A}{dz^2}-2j\beta_a\frac{dA}{dz}+k^2\left(n_T^2-n_a^2\right)A\right]E_a exp\left(-j\beta_a z\right)+$$

$$\left[\frac{d^2B}{dz^2}-2j\beta_b\frac{dB}{dz}+k^2\left(n_T^2-n_b^2\right)B\right]E_b exp\left(-j\beta_b z\right)=0 \qquad (5.126)$$

In equation (5.126) there are two new functions, namely $n_T^2-n_a^2$ and $n_T^2-n_b^2$. Due to the square-law relationship between refractive index and relative dielectric, $n_T^2-n_a^2$ represents a perturbation in dielectric constant $\Delta\varepsilon_1$ to waveguide 1, caused by the neighboring waveguide 2, on the contrary, $n_T^2-n_b^2$. is a similar perturbation $\Delta\varepsilon_2$ caused to waveguide 2 by waveguide 1.This will be non-zero only in the case of the two waveguides being sufficient close.

Equation (5.126) can be further simplified, for example, we can ignore the second derivatives of the amplitudes ($\frac{d^2A}{dz^2}$ and $\frac{d^2B}{dz^2}$), because these represent modal envelopes, which will vary with distance slowly. In addition, we can divide both sides of equation (5.126) by $exp\left(-j\beta_a z\right)$ and obtain

$$\left[-2j\beta_a\frac{dA}{dz}+k^2\left(n_T^2-n_a^2\right)A\right]E_a+$$

$$\left[-2j\beta_b\frac{dB}{dz}+k^2\left(n_T^2-n_b^2\right)B\right]E_b exp\left(-j\Delta\beta z\right)=0 \qquad (5.127)$$

Where $\Delta\beta=\beta_b-\beta_a$ represents a mismatch in propagation constants between two waveguides.

Equation (5.127) is quite close to the relationship that we are looking for, but the latter should describe the change in wave amplitude A and B with the propagation distance, but equation (5.127) are three-dimensional equations, therefore, we must try to eliminate the dependence of variables $E_{a,b}$ and $n_T^2-n_{a,b}^2$ on coordinates x and y. Therefore, first, we multiply the whole of equation (5.127) by the complex conjugate of the transverse field in waveguide 1(i.e., E_a^*), and integrate over the cross-section of the waveguide, and get

$$-2j\beta_a\frac{dA}{dz}\iint dxdyE_a^*E_a+A\iint dxdy\left[k^2\left(n_T^2-n_a^2\right)E_a^*E_a\right]+$$

$$\left\{-2j\beta_b\frac{dB}{dz}\iint dxdyE_a^*E_b+B\iint dxdy\left[k^2\left(n_T^2-n_a^2\right)E_a^*E_b\right]\right\}exp\left(-j\Delta\beta z\right)=0 \qquad (5.128)$$

In equation (5.128), $\iint dxdyE_a^*E_b$ shown in the third term indicates the overlap of two modes, which can be ignored because the spatial overlap of the two modes is small. And $n_T^2-n_a^2$ shown in the second term is only non-zero in the evanescent region where the field function E_a is itself small. So the term can also be ignored. And then equation (5.128) becomes

$$-2j\beta_a \frac{dA}{dz} \iint dxdyE_a^*E_a + \left\{ B \iint dxdy\left[k^2\left(n_T^2 - n_a^2\right)E_a^*E_b\right] \right\} exp(-j\Delta\beta z) = 0 \tag{5.129}$$

Similarly, if we multiply the whole of equation (5.127) by E_b, then integrate over the cross-section of the waveguide, and do some approximate operation, we can get

$$-2j\beta_a \frac{dB}{dz} \iint dxdyE_b^*E_b + B \left\{ \iint dxdy\left[k^2\left(n_T^2 - n_a^2\right)E_b^*E_b\right] \right\} exp(+j\Delta\beta z) = 0 \tag{5.130}$$

Equation (5.129) and (5.130) can be further simplified. First of all, the propagation constant will not change sharply during the transfer process, in this way, as long as it is not on the index, we will assume $\beta_a \approx \beta_b \approx \beta$. Second, the symmetry of device will ensure that

$$\frac{\iint dxdy\left[k^2\left(n_T^2 - n_b^2\right)E_a^*E_b\right]}{\iint dxdyE_a^*E_a} \approx \frac{\iint dxdy\left[k^2\left(n_T^2 - n_a^2\right)E_b^*E_a\right]}{\iint dxdyE_b^*E_b}$$

Then, equations (5.129) and (5.130) can be simplified as

$$\frac{dA}{dz} + j\kappa B exp(-j\Delta\beta z) = 0 \tag{5.131}$$

$$\frac{dB}{dz} + j\kappa A exp(+j\Delta\beta z) = 0 \tag{5.132}$$

(5.131) and (5.132) are equations that eventually we are looking for, they describe the coupling in the amplitude of the mode between two waveguides, and thus are called as coupled mode equations of the waveguide. Where κ is decided by equation (5.133)

$$\kappa = \frac{k^2}{2\beta} \frac{\iint dxdy\left[\left(n_T^2 - n_b^2\right)E_a^*E_b\right]}{dxdyE_a^*E_a} \tag{5.133}$$

κ is called as the coupling coefficient, which is the most important parameter in waveguide coupling theory, Equation (5.133) looks complex, but it is not hard to find, in fact, $k^2/2\beta$ is a constant because it only depends on the wavelength of light; $\iint dxdyE_a^*E_a$ acting as the denominator is a normalization factor. So, in equation (5.133), the really important part is $\iint dxdy\left[\left(n_T^2 - n_b^2\right)E_a^*E_b\right]$.

In our planar waveguide, $n_T^2 - n_b^2$ is non-zero only within the core of waveguide a, thus this region will give the only contribution to the integral, that is to say, if you want κ to be large, you should make the evanescent tail of the field E_b penetrate waveguide a significantly. So κ is most strongly affected by the gap g between the two waveguides. Because the evanescent field roughly falls off exponentially, κ also depends exponentially on g approximately. Typically, g is of the order of the guide width, usually, it is a few microns.

In addition, the energy concentration of the mode also affects κ, for example, a poorly confined mode will have an evanescent field that extends a long distance from the core of the waveguide, which will result in a strong coupling. But on the other hand, the weak confinement will increase the loss, so the confinement need to be chosen optimally by a compromise. Finally, κ is also affected by the polarization characteristics of the field. Although we have discussed a scalar model here, the use of a birefringent electro-optic substrate will cause different coupling rates for TE and TM modes. This is obviously undesirable. Fortunately, careful choice of the fabrication parameters can make κ_{TE} equal to κ_{TM} in some cases.

5.6.4 Solutions of the scalar equations

To solve equation (1.131) and (1.132) more easily, we introduce two new variables, i.e., $a(z)$ and $b(z)$, and let

$$A = a\exp\left(-j\frac{\Delta\beta}{2}z\right), B = b\exp\left(+j\frac{\Delta\beta}{2}z\right)$$

Taking the derivative of the above two formulas with respect to z an obtain

$$\frac{dA}{dz} = \left(\frac{da}{dz} - ja\frac{\Delta\beta}{2}\right)\exp\left(-j\frac{\Delta\beta}{2}z\right)$$

$$\frac{dB}{dz} = \left(\frac{db}{dz} + jb\frac{\Delta\beta}{2}\right)\exp\left(+j\frac{\Delta\beta}{2}z\right)$$

Substituting these relationships into (5.131) and (5.132), we get

$$\frac{da}{dz} - ja\frac{\Delta\beta}{2} = -j\kappa b \tag{5.134}$$

$$\frac{db}{dz} + ja\frac{\Delta\beta}{2} = -j\kappa a \tag{5.135}$$

Equation (5.134) and (5.131), or equation (5.135) and (5.132) are completely equivalent respectively, but equation (5.134) and (5.135) have not the exponentials, thus they can be solved more easily, we assume that the boundary conditions are

$$a(0) = 1, b(0) = 0$$

Then we can get the solutions as follow

$$a(z) = \cos\left[\left[z\sqrt{\kappa^2 + (\Delta\beta/2)^2}\right] + j(\Delta\beta/2)\sin\left[z\sqrt{\kappa^2 + (\Delta\beta/2)^2}\right] / \sqrt{\kappa^2 + (\Delta\beta/2)^2}\right] \tag{5.136}$$

$$b(z) = -j\kappa\sin\left[z\sqrt{\kappa^2 + (\Delta\beta/2)^2}\right] / \sqrt{\kappa^2 + (\Delta\beta/2)^2} \tag{5.137}$$

Based on Poynting's Theorem, we can get the normalized power P_a and P_b which are considered more useful than the mode amplitudes in the waveguides, $\sqrt{\kappa^2 + (\Delta\beta/2)^2}$

$$P_a = cos^2\left[z\sqrt{\kappa^2 + \left(\frac{\Delta\beta}{2}\right)^2}\right] + \left(\frac{\Delta\beta}{2}\right)^2 \frac{sin^2\left[z\sqrt{\kappa^2 + (\Delta\beta/2)^2}\right]}{\kappa^2 + (\Delta\beta/2)^2} \tag{5.138}$$

$$P_b = \frac{\kappa^2 sin^2\left[z\sqrt{\kappa^2 + (\Delta\beta/2)^2}\right]}{\kappa^2 + (\Delta\beta/2)^2} \tag{5.139}$$

In particular, under the condition of synchronization, $\Delta\beta = 0$, equation (5.136) and (5.137) are simplified as follows

$$a(z) = cos(\kappa z) \tag{5.140}$$

$$b(z) = -j\sin(\kappa z) \tag{5.141}$$

Via equations (5.138) and (5.139), we can obtain

$$P_a = cos^2(\kappa z) \tag{5.142}$$

and

$$P_b = sin^2(\kappa z) \tag{5.143}$$

When $z=0$, $P_a = 1$, $P_b = 0$, the power emerges from waveguide a. After being transmitted by a distance of $z_1 = L = \pi/2\kappa$, then $P_a = 0$, $P_b = 1$, that is, all the power in waveguide a is coupled into the waveguide b. Again being transmitted by a distance of L, i.e., $z_2 = 2L$, then once again, $P_a = 1$, $P_b = 0$, the power returns back to the waveguide a. Figure 5.23 shows the periodic power transfer in two waveguides.

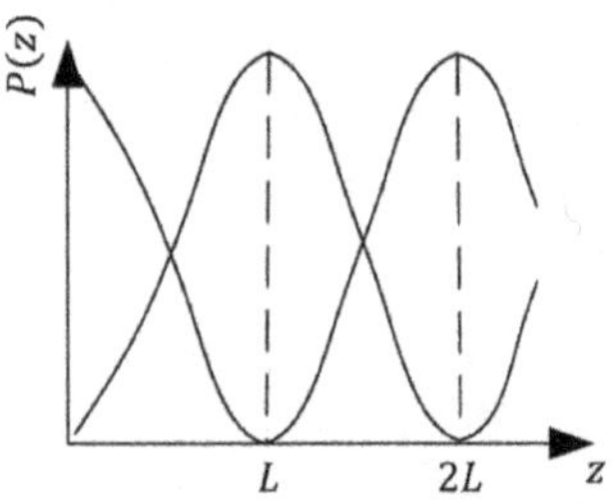

Fig. 5.23: The change of the power with distance Z in two waveguides.

According to equations (5.142) and (5.143), we obtain

$$P_a(z) + P_b(z) \equiv 1 \tag{5.144}$$

The total power is conserved during the process of transfer, because no loss in the whole process has been assumed, which is expected.

5.6.5 Periodic waveguide

Polarization $P(x, y, z)$ described in subsection 5.6.2 can be induced by various physical effects. For example, when there is a difference between the actual polarizability and the nominal one, i.e., $\Delta\varepsilon(x, y, z)$, the induced polarization is proportional to the field in the waveguide

$$P = \Delta\varepsilon E$$

In particular, if $\Delta\varepsilon(x, y, z)$ is a periodic function of z, the waveguide will also show the periodicity, this periodic waveguide has been applied in many optoelectronic devices, such as filters, grating couplers, DFB laser and MFA (mode field adaptor), etc. Being similar with the situation that the light is scattered by diffraction grating, the physical process taken place in the periodic waveguide is that the light will be scattered periodically, which can be regarded as a mode coupling process and studied by the mode coupling theory. If the two modes are approximately synchronous at a given optical frequency, we will directly use the solutions of coupling mode equations obtained in subsection 5.6.4 to study the periodic waveguide, and the rest of the study work is to find the coupling coefficient κ for practical issues.

As an example, in this subsection, we will study a corrugated planar waveguide, which is shown in Fig. 5.24. The depth of the waveguide is $d(z)$, which varies periodically with Z

$$d(z) = d_0 + \Delta d\cos\left(\frac{2\pi}{\Lambda}z\right)$$

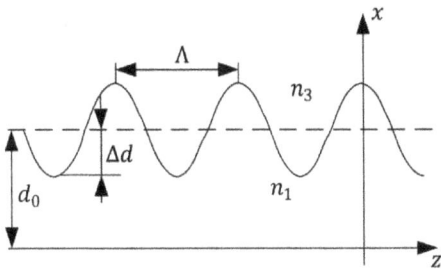

Fig. 5.24: The side view of corrugated planar waveguide. d_0—the average depth of waveguide; Δd—the amplitude of corrugated waveguide: Λ—the period of corrugated waveguide

where Λ is the period of corrugated waveguide.

d_0—the average depth of waveguide;

Δd—the amplitude of corrugated waveguide: Λ—the period of corrugated waveguide

We can use various methods to derive the coupling coefficient of corrugated waveguide, in this subsection we will employ the effective index method. We assume that the refractive index of waveguide is

$$N(z) = N + \Delta N cos(\kappa z)$$

According to these assumptions, corrugated waveguide looks like a cross-section of Bragg holographic grating. Via the grating theory, we can obtain the coupling coefficient described for Bragg diffraction intensity,

$$\kappa = \frac{\pi}{\lambda}\Delta N$$

For the corrugated waveguide, the coupling coefficient is

$$\kappa = \frac{\pi}{\lambda}\frac{\partial N}{\partial d}\Delta d \tag{5.145}$$

and $\frac{\partial N}{\partial d}$ can be deduced by the scattering relationship of the waveguide. For TE modes, it can be obtained by appropriate derivation

$$\frac{\partial N}{\partial d} = \frac{n_1^2 - N^2}{N d_{eff}} \tag{5.146}$$

where n_1 is still the nominal refractive index of the waveguide, and d_{eff} is the effective waveguide depth defined by equation (5.142).

Substituting equation (5.146) into (5.145), we can get the coupling coefficient of TE mode as follows

$$\kappa_{eff} = \frac{\pi}{\lambda}\frac{\Delta d}{d_{eff}}\frac{n_1^2 - N^2}{N} \tag{5.147}$$

The coupling coefficient of TM mode can be written as

$$\kappa_{TM} = \eta \kappa_{TE} \tag{5.148}$$

The coefficient η is given by

$$\eta = \frac{(N/n_1)^2 - (N/n_3)^2 + 1}{(N/n_1)^2 + (N/n_3)^2 - 1} \tag{5.149}$$

n_3 still denotes the refractive index of the cladding.

Via the analogy between corrugated planar waveguide and grating, we can get the relationship between the variation period of corrugated waveguide depth and grating vector K,

$$K = \frac{2\pi}{\Lambda} \tag{5.150}$$

Each groove is equivalent to grating fringe, and acts like a weak mirror, which would individually give only a very small reflection. However, if the optical path is a whole number of optical half-wavelengths (i.e., equation (5.151)), the components will overlap each other and result in a very strong combined reflection

$$n_{eff}\Lambda = m\frac{\lambda_0}{2} \tag{5.151}$$

where m is an integer, if $m = 1$, the equation can be simplified as

$$2n_{eff}\Lambda = \lambda_0$$

Or

$$\Lambda = \frac{\lambda_0}{2\pi}\frac{\pi}{n_{eff}} = \frac{\pi}{kn_{eff}} = \frac{\pi}{\beta} \tag{5.152}$$

Substituting into equation (5.150), we get

$$K = 2\beta \tag{5.153}$$

That is to say, when the first-order Bragg condition is satisfied, the magnitude of the grating vector corresponding to the periodic waveguide is exactly twice the propagation constant of waveguide. Then, we will take a simple example to let readers know the magnitude of Λ and K.

Example 5.4 If we assume that a periodical corrugated planar waveguide are made in Ti:LiNbO$_3$ ($n_s = 2.2$), and take the optical wavelength to be $\lambda_0 = 0.75\mu m$, evaluate Λ and K.

Solution: Because waveguides made by indiffusion are very weak, we may take $n_{eff} = n_s$, thus

$$\beta \approx \frac{2\pi n_s}{\lambda_0} = 1.84 \times 10^7, \Lambda = \frac{\lambda_0}{2n_s} = 0.17\mu m$$

Such grating is so dense that it is very difficult to be fabricated.

5.6.6 Waveguide mode transmission

Coupled waveguide mode can be obtained by mode coupling theory whose basis is the perturbation method. By means of this method, we can use the uncoupled mode to represent the nominal mode of coupled waveguide approximately.

Considering two waveguides a and b, the distance between them is D, when they are not coupled, they have the electric field distribution of ϕ_a and ϕ_b,and the corresponding propagation constant of β_a and β_b, and when they are close enough, that is, when D is small to some extent, they will be coupled and the coupling coefficient is

$$K = -i\kappa$$

Where $|\kappa|$ is the exponential function of D, and the coupled mode φ_i and φ_j may be expressed by a linear superposition of the uncoupled modes described above

$$\varphi_i = p\phi_a + q\phi_b, \varphi_j = -q\phi_a + p\phi_b \tag{5.154}$$

The ratio of the superposition coefficient is

$$f = \frac{q}{p} = -X + \left(X^2 + 1\right)^{1/2} \tag{5.155}$$

where $X = \frac{\Delta\beta}{2|\kappa|}$
is the parameter of coupled waveguide, and $\Delta\beta = \beta_a - \beta_b$ is the difference of propagation constant between the two waveguides. $f(X)$ varies with X, which is shown in Fig. 5.25.

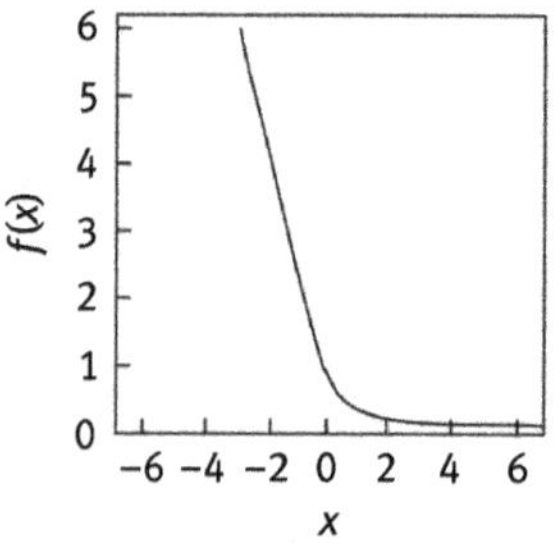

Fig. 5.25: $f = q/p$ varies with $X = \Delta\beta/(2|\kappa|)$.

The orthogonalization of uncoupled modes can be written as

$$\int_{-\infty}^{\infty} \phi_a \phi_a^* dx = \int_{-\infty}^{\infty} \phi_b \phi_b^* dx = 1$$

$$\int_{-\infty}^{\infty} \phi_a \phi_b^* dx = \int_{-\infty}^{\infty} \phi_a \phi_b^* dx = 0 \tag{5.156}$$

Via these relationships, the orthogonality of the coupled mode is obvious, in fact, via (5.154), we get

$$\int_{-\infty}^{\infty} \phi_i \phi_j^* dx = \int_{-\infty}^{\infty} \phi_i^* \phi_j dx = pq - pq = 0$$

While the normalization of the coupled mode is required to satisfy the following equation（5.157）

$$\int_{-\infty}^{\infty} \phi_i \phi_i^* dx = \int_{-\infty}^{\infty} \phi_j^* \phi_j dx = 1 \tag{5.157}$$

Substituting equation (5.154) into (5.157), via equation (5.156), we obtain

$$p^2 + q^2 = 1 \tag{5.158}$$

Via equation(5.155), we get

$$q^2 = f^2 p^2$$

The above formula and equation (5.158) represent two simultaneous equations, via them, we obtain

$$p^2 = \frac{1}{1+f^2} = \frac{1}{2}(X^2+1)^{-\frac{1}{2}}\left[(X^2+1)^{\frac{1}{2}} - X\right]^{-1}$$

Or

$$q^2 = \frac{f^2}{1+f^2} = \frac{1}{2}(X^2+1)^{-\frac{1}{2}}\left[(X^2+1)^{-\frac{1}{2}} - X\right] \tag{5.159}$$

Namely

$$p = \frac{1}{\sqrt{2}}(X^2+1)^{-\frac{1}{4}}\left[(X^2+1)^{\frac{1}{2}} - X\right]^{-\frac{1}{2}}$$

$$q = \frac{1}{\sqrt{2}}(X^2+1)^{-\frac{1}{4}}\left[(X^2+1)^{\frac{1}{2}} - X\right]^{\frac{1}{2}}$$

Particularly, under the symmetry condition, $\Delta\beta = 0$, so, X=0, $f(X) = 1$, via equation (5.159), we can get

$$\mathrm{p} = \mathrm{q} = \frac{1}{\sqrt{2}}$$

The coupled mode is simplified as

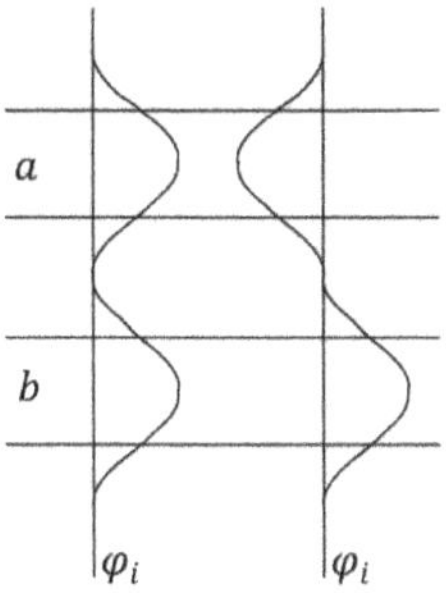

Fig. 5.26: Coupled waveguide modes: $\Delta\beta = 0$.

$$\varphi_i = \frac{1}{\sqrt{2}}(\phi_a + \phi_b), \varphi_j = \frac{1}{\sqrt{2}}(-\phi_a + \phi_b)$$

Figure 5.26 shows the coupled modes in the symmetry condition, it is easy to see, any mode of them has the same power.

Another extreme condition is that $\Delta\beta$ is quite big, which will result in a weak coupling, i.e., $|\kappa| \to 0$, $\mathrm{X} \to \pm\infty$, from Fig. 5.25, we can see, the value of $f(X)$ is close to zero or ∞, in this case the coupling mode power actually exists only in one of the two waveguides.

Although the accurate numerical solutions can be obtained for any waveguide system, the calculation process is rather tedious and needs to be performed separately for each waveguide. The advantage of mode coupling theory is to provide analytical expressions that depends directly on the waveguide parameters. It is worth noting that this method is the approximation of actual one, when we use it, the following requirements should be satisfied. First of all, there is sufficient spacing between the two waveguides; second, the two waveguides should have similar characteristics, for example, the difference of propagation constant ($\Delta\beta$) between them cannot be too large.

5.7 Semiconductor waveguide theory

The reason why semiconductor waveguide is particularly interesting is that it is possible to fabricate various components of integrated optics on the same substrate, including lasers, detectors, photoelectric switches, modulators, waveguide elements and electronic circuits. The theoretical basis of the semiconductor optical waveguide is discussed in this subsection. Some common devices will be introduced in the next subsection.

Due to the fact that the waveguide made of any material is required to control the variation of the refractive index in the plane normal to the direction of the power flow, several methods to control the refractive index of semiconductor are introduced at the beginning of this subsection, using these methods, the refractive index in certain region can be higher than that in the surrounding region; after that, the conditions for the planar waveguide and single mode operation are discussed, and channel waveguide and the coupling effect of modes are briefly introduced.

5.7.1 Methods for altering the refractive index of semiconductor

To generate a waveguide, it is necessary to construct a region in the semiconductor which has a higher refractive index than the surrounding region. Several common methods will be introduced in the following discussion of this subsection.

5.7.1.1 Free carrier effect

Research results show that the refractive index of semiconductor is closely related to the concentration of free carriers in it. In general, the higher the carrier concentration is, the lower the refractive index is. In this way, for the two same semiconductor material, if one with the higher free carrier concentration is used as the substrate, and another with the lower carrier concentration is used as the waveguide layer (shown as Fig. 5.27a), the light can propagate along the waveguide layer when the depth and refractive index of the layer reaches certain critical values (shown as Fig. 5.27b).

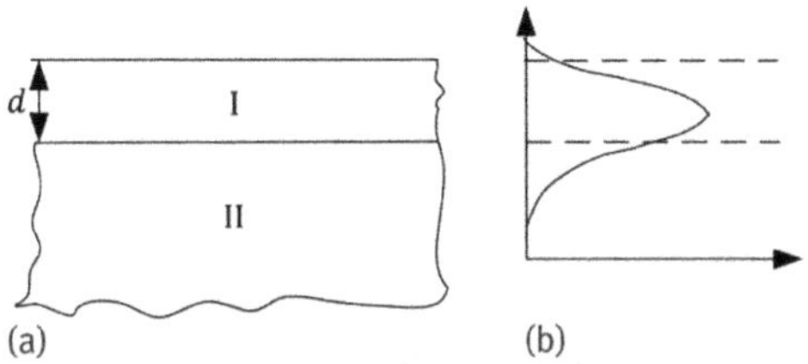

Fig. 5.27: GaAs homojunction planar waveguide: (a) waveguide structure; (b) relative intensity in each layer. I: low carrier concentration region; II: high carrier concentration region.

It is easy to understand that the refractive index drops with the rising of the free carrier concentration. Because the free carriers contribute to the generation of the conductance, namely, the higher the free carrier concentration is, the smaller the dielectric constant is, and the smaller the refraction index which is proportional to the square root of the dielectric constant is.

We suppose that the dielectric constant of the material without free carrier is ε, so the real part of the dielectric constant of the material with free carriers of N_{fc} in a cube of unit volume is

$$\varepsilon' = \varepsilon - \varepsilon_{fc}$$

Where

$$\varepsilon_{fc} = \frac{N_{fc} e^2 \lambda_0^2}{4\pi^2 c^2 m^*} \tag{5.160}$$

where e is the electron charge, m^* is the equivalent mass of carriers in semiconductors. The refractive index of the material is

$$n_2 = \left(\frac{\varepsilon - \varepsilon_{fc}}{\varepsilon_0}\right)^{\frac{1}{2}} = \left(\frac{\varepsilon}{\varepsilon_0}\right)^{\frac{1}{2}}\left(1 - \frac{\varepsilon_{fc}}{\varepsilon}\right)^{\frac{1}{2}} = n_1\left(1 - \frac{\varepsilon_{fc}}{\varepsilon}\right)^{\frac{1}{2}} \tag{5.161}$$

where $n_1 = \left|\frac{\varepsilon}{\varepsilon_{fc}}\right|$,denotes the refractive index of material without carrier.

In general, ε_{fc} is much smaller than ε, via equation (5.161), we can get

$$n_2 \approx n_1\left(1 - \frac{1}{2}\frac{\varepsilon_{fc}}{\varepsilon}\right) = n_1\left(1 - \frac{1}{2}\frac{\varepsilon_{fc}}{n_1^2 \varepsilon_0}\right)$$

$$= n_1 - \Delta n \tag{5.162}$$

where

$$\Delta n = \frac{N_{fc} e^2 \lambda_0^2}{8\pi^2 \varepsilon_0 n_1 c^2 m^*} \tag{5.163}$$

For GaAs, $n_1 = 3.5$, $\mathrm{m} = 0.067 m_e$, $m_e = 9.1 \times 10^{-31} kg$, which is the mass of free electron, $e = -1.602 \times 10^{-19}$. Substituting them into equation (5.163),we obtain

$$\Delta n = 1.8 \times 10^{-21} N_{fc} \lambda_0^2$$

where, the unit of N_{fc} is cm^{-3}, and the unit of λ_0 is μm. For $\lambda_0 = 1\mu m$ and $N_{fc} = 2 \times 10^{18} cm^{-3}$, we can get

$$\Delta n = 0.0036 \tag{5.164}$$

If this material is used as the substrate, the same material without free carrier is used as the waveguide layer, then the waveguide can be generated when the depth of the latter is enough.

5.7.1.2 The effect of changing composition

As we all know, the refractive index of semiconductor material is a function of the impurity component, so the doping of semiconductor can change the refractive index. Generally speaking, the change of the refraction index caused by altering the composition ratio of semiconductor impurities is more obvious than that caused by altering the free carrier concentration. So the epitaxial growth technique is very useful for fabricating semiconductor waveguide.

Figure 5.28 shows a GaAs/AlGaAs heterojunction planar waveguide (a) and the relative intensity distribution in each layer of the waveguide mode (b). An Al-doped AlGaAs substrate with low concentration of free carriers is grown on a GaAs layer with higher concentration of free carriers, and the former is covered by GaAs waveguide films with low concentration of free carriers. With the increase of the concentration of Al, the refractive index of AlGaAs drops. The drop degree of the refractive index is related to the surrounding temperature and the working wavelength. Figure 5.29 shows the change of the refractive index with the dopant concentration at several different wavelengths, and the temperature is 295 K. As can be seen from the figure, for $\lambda > 1.0\mu m$, the refractive index drops linearly with x (the percentage of Al concentration). Relative to undoped GaAs,

$$\Delta n \approx 0.45x \tag{5.165}$$

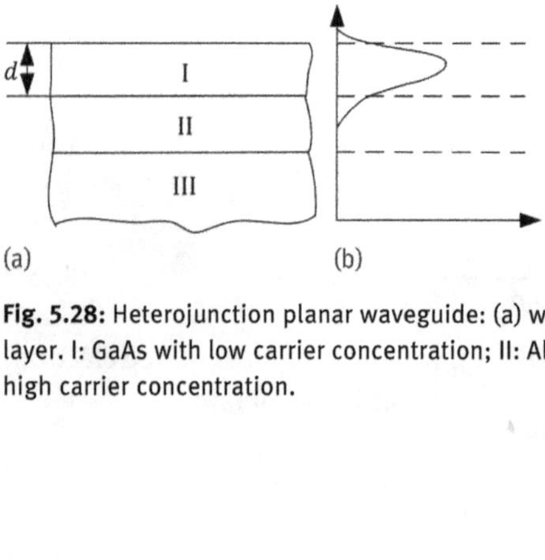

Fig. 5.28: Heterojunction planar waveguide: (a) waveguide structure; (b) relative intensity of each layer. I: GaAs with low carrier concentration; II: AlCaAs with low carrier concentration; III: GaAs with high carrier concentration.

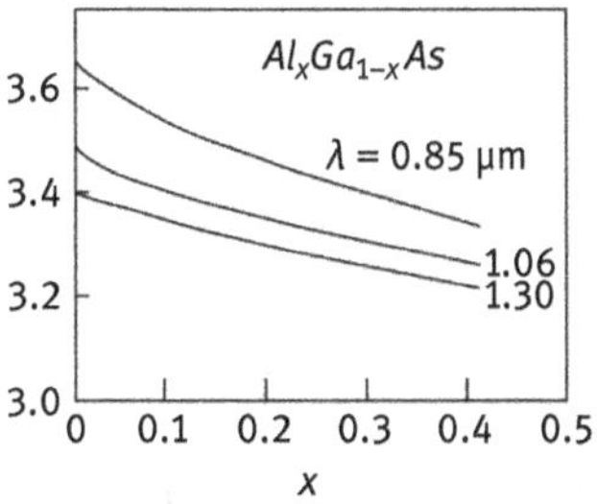

Fig. 5.29: When T=295 K, the change of the refractive index with x (the percentage of Al-doped concentration) at several different wavelengths.

In this way, for $x = 0.2$, we can get $\Delta n \approx 0.09$. It is really much larger than the refractive index change caused by the free carrier effect.

With the development of molecular beam epitaxy (MBE) and chemical vapor deposition (CVD), some new devices have been developed, such as quantum well and superlattice, which are not discussed in detail due to the limited space.

5.7.1.3 Electric field effect

For the material with linear electro-optic effect, the refractive index will change with the external electric field. III-V semiconductor with zincblende (43 m point group symmetry) belongs to this class.

For the electro-optic tensor, the components which are not equal to zero are only $\gamma_{41} = \gamma_{52} = \gamma_{63}$. They vary with the wavelength. For GaAs, at $\lambda = 1.3\mu m$, the electro-optic coefficients are approximately $-1.4 \times 10^{-10} cm/V$.

5.7.1.4 Deformation effect

When the semiconductor material is deformed, the lattice constant and the energy band gap vary accordingly, which will induce the change of refractive index. For simple compression and tensile deformation, the change of refractive index is similar to that caused by the change of temperature and pressure. Because III-V semiconductor has large photoelastic coefficient, the small deformation can induce significant changes in refractive index.

The electric field effect and deformation effect in semiconductor waveguide are often ignored, which will have a certain influence on the study of waveguide. We should pay attention to them when using the waveguide.

5.7.2 Semiconductor planar waveguide

Examples of homojunction- and heterojunction- semiconductor planar waveguides are shown in Figs. 5.27 and 5.28, respectively. In an asymmetric three-layer waveguide structure with only one high refractive index film lying above a low refractive index substrate, the refractive indexes of the three layers satisfy the relationship $n_1 > n_2 > n_3$ and $n_3 \approx 1$ generally.

As discussed earlier in this chapter, the TE and TM mode which can propagate in planar waveguide depend on the depth of the waveguide layer (a) and the difference of the refractive index between the waveguide layer and the confinement layer. For the mth mode, the relationship between the cut-off depth of the waveguide layer (d_m) and the refractive indexes of the three layer can be expressed as follows

$$\frac{2\pi}{\lambda} d_m \left(n_1^2 - n_2^2\right)^{1/2} = tg^{-1}\left[r_0 \left(\frac{n_2^2 - n_3^2}{n_1^2 - n_2^2}\right)^{\frac{1}{2}}\right] + (m-1)\zeta \tag{5.166}$$

Where, $r_0 = \begin{cases} 1, & \text{for TE mode} \\ \frac{n_1}{n_2}, & \text{for TM mode} \end{cases}$

ζ is a constant, and for GaAs, its value is 5-15051.

When $n_3 = n_2$ and $m = 1$, it is obvious that $d_1 = 0$ via equation (5.166). That is to say, there is no cut-off depth for the lowest order mode of the symmetric planar waveguide. However, when the depth is very small, most of the energy will escape from the high refractive index waveguide layer.

We take the homojunction planar waveguide mentioned in the above subsection as an example, $n_1 = 3.5$, the concentration of free carriers in the substrate is $N_c = 2 \times 10^8 cm^{-3}$. For $\lambda = 1\mu m$, $\Delta n = 0.0036$, $\left(n_1^2 - n_2^2\right)^{1/2} \approx (2n_1\Delta n)^{\frac{1}{2}} \approx 0.159$. We let $n_3 = 1$, and substitute these data into equation (5.166), then we can get the cut-off depth of the mth mode as follows.
For TE mode,

$$d_m = 1.52779 + 3.15051(m-1)$$

For TM mode

$$d_m = 1.57138 + 3.15051(m-1) \tag{5.167}$$

It can be seen that these above two are quite close, especially when m is a larger one. Figure 5.30 shows the variation of the cut off depth (d_m) with the carrier concentration in the substrate at $m = 1$ and $m = 2$. Because the cut-off depth of the same order mode is very close for TE mode and TM mode, only the situation of TE mode is shown in Fig. 5.30.

The cut-off depth of the heterojunction waveguide mode is mainly determined by the dopant concentration, for example, in $GaAs/AL_xGa_{1-x}As$ waveguide, if x=0.3, then $\Delta n = 0.135$, when $\lambda = 1\mu m$, the cut off depth of the mth mode is
For TE mode

$$d_m = 0.21156 + 0.51938(m-1)$$

For TM mode

$$d_m = 0.25565 + 0.51938(m-1) \tag{5.168}$$

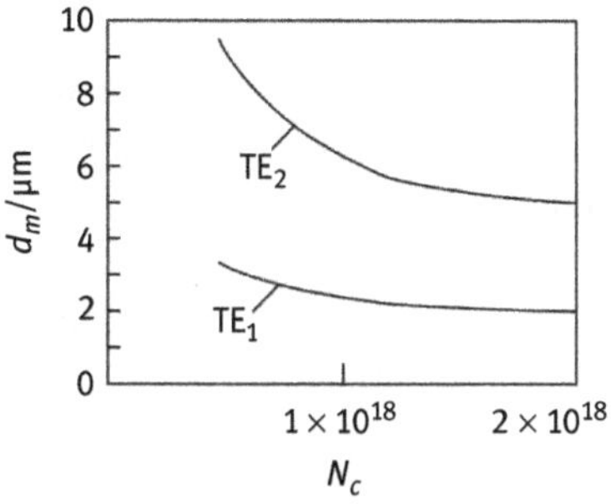

Fig. 5.30: Variations of the cut-off depth of the first two TE modes ($d_{1,2}$) with the carrier concentration in a homojunction planar waveguide.

Comparing equation (5.167) with (5.168), it is easy to find that the cut off depth of the heterojunction mode is much smaller than that of the homojunction mode. For a GaAs/AlGaAs heterostructure planar waveguide, Fig. 5.31 shows the variation of the cut-off depth with the percentage concentration of Al in the substrate in the first two order TE modes. As can be seen, from x>0.1, with the increasing of the concentration of Al, the dropping rate of the cut-off depth is very low.

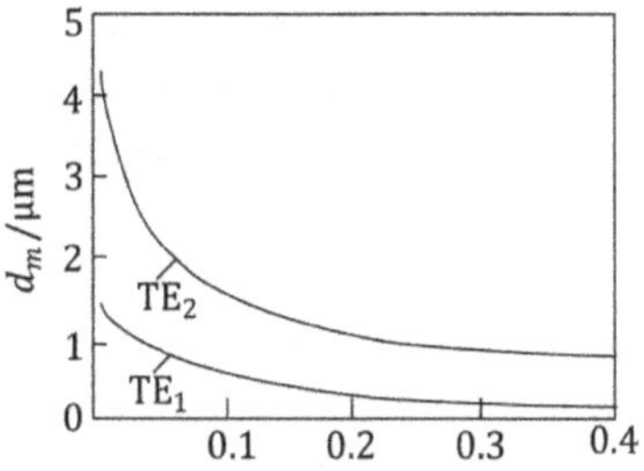

Fig. 5.31: Variations of the cut-off depth with the percentage concentration of Al in the substrate in the first two order TE modes in a GaAs/AlGaAs heterostructure planar waveguide.

5.7.3 Channel waveguide

The planar waveguide is only confined to one-dimensional coordinate normal to the direction of light propagation, and the wave will not be refracted to other directions, so it is relatively simple to deal with. However, from a practical point of view, the waves in the integrated optical path often need to be confined to the two-dimensional coordinate normal to the propagation direction. The condition of obtaining a channel waveguide is that the refractive index in x- and y-direction (We assume the wave propagates along the z direction) vary according to the requirements. As discussed in the previous subsection, it is possible to satisfy the condition through changing the concentration of free carriers or dopants in semiconductor.

The structure of the channel waveguide is basically the same as the ones mentioned previously, but its substrate and waveguide layer are made in materials with appropriate dopants or free carriers. For example, Fig. 5.32 (a) is a kind of strip-loaded waveguide, in which the substrate and the waveguide layer are both made in GaAs, but the former has the higher concentration of free carriers, while the latter has the lower one. Figure 5.32 (b) is a kind of buried channel waveguide, the waveguide layer is made in GaAs material purely, but the substrate is doped with Al.

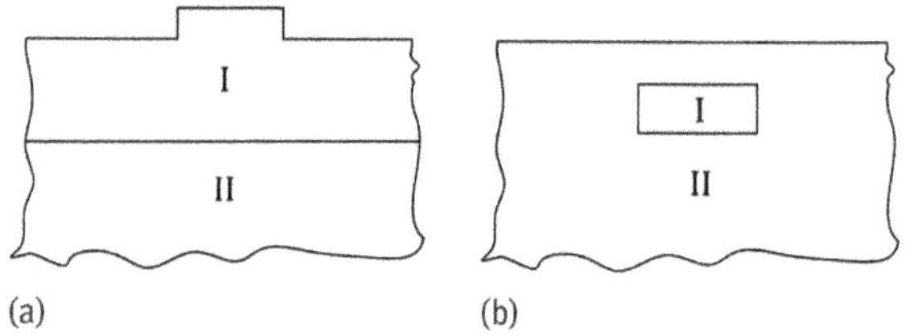

Fig. 5.32: Semiconductor channel wave guides: (a) strip-loaded waveguide; I: GaAs with low carrier concentration; II: GaAs with high carrier concentration. (b) Buried waveguide; I: pure GaAs; II: $Al_xGa_{1-x}As$.

The channel waveguide may also be composed of the semiconductor substrate and the above metal or metallic oxide strip. The channel waveguide shown in Fig. 5.33 is formed by preparation of metal strips on GaAs substrates, and the deformation caused by the stretching of the metal bars in the semiconductor can lead to the variation of the refractive index of the substrate which is adjacent to the metal bar.

In general, it is difficult to obtain the analytical expression of the channel waveguide modes. In addition, the mode of channel waveguide is neither pure-TE mode nor pure-TM mode. If the transverse electric field is mainly in the *x*-direction, the mode is usually represented by E^x_{mn} ; if the transverse electric field is mainly in the *y*-direction, the mode is usually represented by E^y_{mn}, in which *m* and *n* denote the order of the *x*- and *y*-direction mode respectively. $m=n=1$ is the lowest order mode. It should be noted that, although these modes are not pure-TE modes or pure-TM modes, but E^x_{mn} and E^y_{mn} can be considered as quasi- TE and{\mminus}TM modes in most cases.

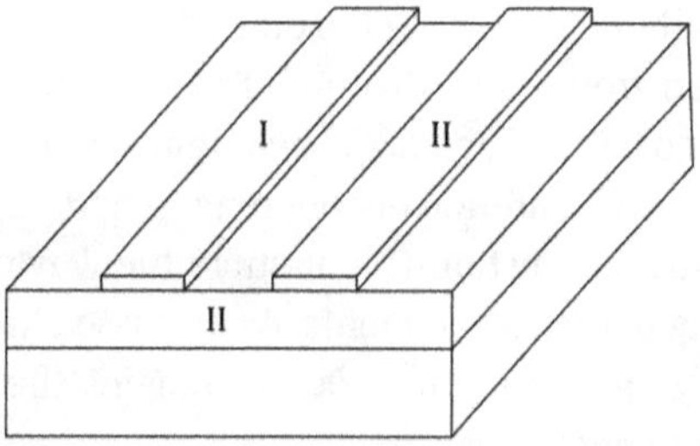

Fig. 5.33: Metal strip-loaded channel waveguide.

Under a certain condition, the approximate solution of channel waveguide can be obtained by the effective refractive index method. The strip-loaded waveguide (shown in Fig. 5.34) is divided into two regions: A and B. Each region is regarded as a three-layer planar waveguide which is infinitely extended in y-direction, and the substrate is GaAs with high concentration of free carriers. The waveguide layer is obtained by epitaxial growth, and has a low concentration of free carriers. The depth of the two regions are d_a and d_b respectively. The cladding is air, so $n_3 \approx 1$. The effective refractive index of waveguide layer can be calculated by the propagation constant, $N = \frac{2\pi}{\lambda}\beta$.

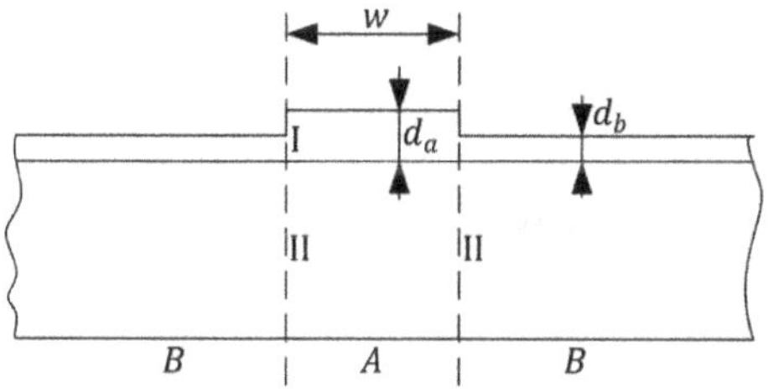

Fig. 5.34: Analysis of strip-loaded waveguide with effective refractive index method. I:low carrier concentration epitaxial layer; II: high carrier concentration GaAs.

If only one mode is allowed for each of the two planar waveguide regions, the only effective refractive index can be determined for each region. While for the strip-loaded waveguide, there is only one peak in x-direction, thus m=1. If more than one mode is allowed in the planar waveguide region, the problem is more complicated, and the analyses show that the modulus of the strip-loaded waveguide depends on D_2/D_1 and D_2/W, where D_1, D_2 or W is called as the effective length of the waveguide, which is equal to d_1,d_2,w, plus the attenuation distance through which the electric field in the corresponding region attenuates to $\frac{1}{e}$ times of the initial value respectively.

The results obtained by the effective refractive index method is very accurate when the strip-loaded waveguide and the two planar waveguides forming the strip-loaded waveguide have only one mode and the channel groove is shallow. With the deepening of the groove, the transverse confinement is enhanced, and there will be more modes. The accuracy of this method will gradually become worse. In general, the propagation constant obtained by the effective index method is higher. But if the refractive index difference between the substrate and the waveguide layer is not too large, the effective index of the waveguide may be less than the refractive index of substrate for the high order mode, which can be transmitted to the evanescent layer. In this way, the strip-loaded waveguide may vary from single-mode to multi-mode, and finally turn back to single mode.

The exact solutions of other channel waveguides can also be obtained by the effective index method. Moreover, the larger the difference of the channel waveguide length between the two directions is, the more accurate the result is.

5.7.4 Coupling effect

Coupling phenomena have been discussed in subsection 5.6, but as mentioned before, one of the important advantages of semiconductor waveguide is that it is possible to fabricate different devices on the same substrate, which makes the various coupling effects more obvious in the integrated circuit with the semiconductor optical waveguide. Therefore, in this subsection we will further discuss some main coupling effects, including modes being coupled from one waveguide to another, the forward traveling wave being coupled to the backward traveling wave, and the coupling between the two modes such as TE mode and TM mode. All of these couplings are inherently dependent on the wavelength and, in some cases, it is possible to enable the coupler to be effective only in a narrow band near the selected wavelength.

In any coupling phenomenon, to obtain a high power transmission percentage, it is usually necessary to synchronize the output wave with the input wave. Whether the coupling between the non-waveguide devices and the waveguide devices or the coupling between the waveguides, the waveguide or mode all should have the same propagation constant. Here to see some specific coupling effects.

5.7.4.1 Coupling between the waveguides

As mentioned in subsection 5.6, the coupling between the waveguides can be studied either by solving the actual nominal mode of the coupled system or by the mode coupling theory.

The modes of the coupled waveguide system can be obtained by the method similar as the single waveguide, which is solving the wave equation under certain boundary conditions. When the coupled waveguide system consists of a small number of planar waveguides, it is straightforward. However, with the increase of the number of waveguides, the problem becomes complicated. For example, it is almost impossible to obtain the exact analytical expression for the channel waveguide with two-dimensional optical confinement, but only some approximate analytical solutions are present. In particular, if the isolated waveguides constituting the coupled mode system are all single-mode, the number of the system mode is generally equal to the number of waveguides. The propagation constant of the coupled mode can be obtained easily via the propagation constants of the isolated waveguides.

Figure 5.35 (a) indicates a twin-waveguide coupler consisting of the two identical waveguides. There are two modes, one is the symmetric mode, with the propagation constant of β_A, the other is the asymmetric mode, with the propagation constant of β_B and $\beta_A > \beta_B$. If the phases of mode A and B at $z=0$ are shown in Fig. 5.35(b), the constructively superposition of the two modes occurs in the left waveguide, while the destructively superposition occurs in the right one, namely the light is mainly confined to the left one.

(a)

(b)

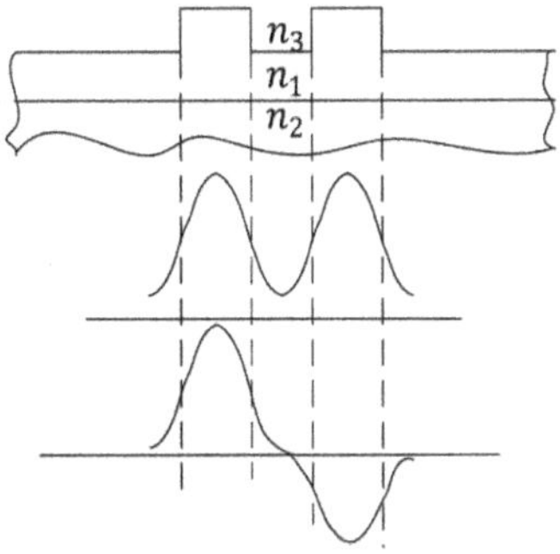

Fig. 5.35: Coupling between the two same waveguides at $z = 0$.

Due to the difference of speed between the two modes, after the waves propagate a certain distance along z-axis, the field distributions of the two modes are shown in Fig. 5.36, then there is a certain phase difference in the left waveguide, which can be expressed as follows:

$$\mathbf{\Phi}_L = (\beta_A - \beta_B)z \tag{5.169}$$

(a)

(b)

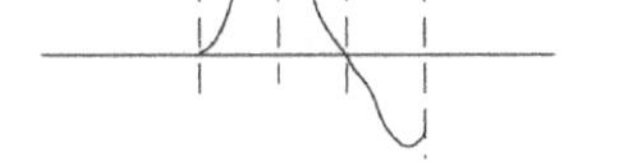

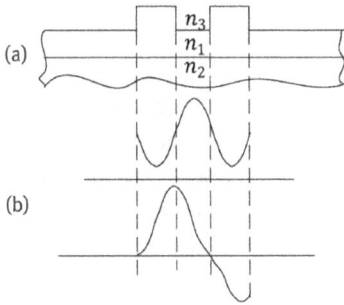

Fig. 5.36: Coupling between the two same waveguides at z.

The phase difference in the right one is reduced from π to

$$\mathbf{\Phi}_R = \pi - (\beta_A - \beta_B) \tag{5.170}$$

In particular, when the propagation distance of z reaches

$$L_c = \frac{\pi}{\beta_A - \beta_B} \tag{5.171}$$

Via equations (5.169) and (5.170), we obtain $\mathbf{\Phi}_L = \pi$, $\mathbf{\Phi}_R = 0$. This means the two modes are combined completely destructively in the left waveguide; while they are combined constructively in the right one (as shown in Fig. 5.37), here, the light intensity is mainly confined to the right waveguide. For a twin-waveguide coupler composed of

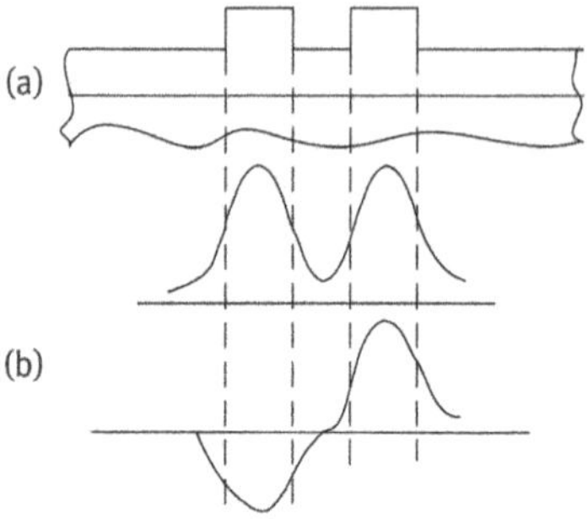

Fig. 5.37: Coupling between the two same waveguides at $z = \frac{\pi}{\beta_A - \beta_B}$.

the two same waveguides, the modes at $z = 0$, and $z = L_c$ are symmetrical with respect to the center of the two modes.

The actual power which transfers from one waveguide to another is dependent on the input-and output optical structure, and the coupler itself. If the waveguide is weakly coupled, the modes A and B are the linear superposition of the isolated waveguide approximately. If the input and output have the same structure as the isolated waveguide of the system, the power losses of input and output are small. For the twin-waveguide coupler with the same waveguides, it is possible to make the power of one waveguide be coupled into another completely. If the two waveguides are different, the transmission of the two modes in the input waveguide will not be completely offset after they propagate a distance of $z = L_c$, so there will be no "complete" power transmission. In addition, for the coupler with the two non-identical waveguides, the high power transmission can be obtained only in the case of the two non-identical waveguides with the same propagation constant.

5.7.4.2 Coupling between the modes induced by perturbation

A small perturbation of the waveguide size or refractive index may induce the coupling between the guided modes. This coupling is usually caused by the inevitable tolerance and inaccuracy during the process of the waveguide fabricating, so it is not desirable and is difficult to eliminate. However, there are some man-made periodic perturbations which are useful for a guided wave mode being coupled to another. For example, the periodic grating, can be used in DFB laser (distributed feedback laser) and DBR laser (distributed Bragg reflection laser) to make the forward traveling wave be coupled to the backward traveling wave, it also can be used in the frequency selective waveguide filter.

The periodic grating waveguide can be analyzed with the mode coupling theory or the eigenmode method. If the mode coupling theory is used, the coupling coefficient of forward- and backward- mode is calculated by the perturbation of the effective refractive index, then the coupled mode equations obtained are similar to the equations used to determine the coupling between waveguides. In the perturbation region, the forward traveling mode drops exponentially with the propagation distance, at the same time, the power is coupled to the backward mode, so the power of the latter rises exponentially with the propagation distance, which is shown in Fig. 5.38, and the amount of transmitted light and reflected light is determined by the range of the perturbation region.

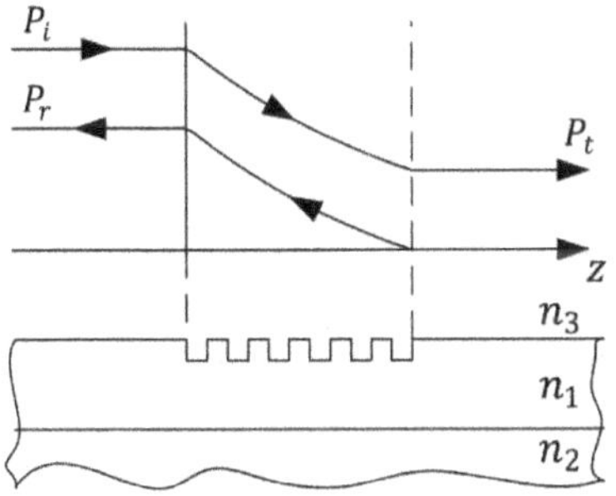

Fig. 5.38: Periodic perturbation effects in optical waveguides: P_i, incident power; P_t, transmission power; P_r, reflected power.

If the grating has a vertical side wall, it can be regarded as the waveguide chain formed by two-section waveguides which appear interchangeable. The two waveguides have the width of w_1 and w_2 respectively, and the similar propagation constant of β_1 and β_2. The light is reflected on each interface. If the optical path difference caused by the difference between w and β is a whole number of multiples of $\pi/2$, namely,

$$\beta_1 w_1 - \beta_2 w_2 = (2m-1)\frac{\pi}{2}, \quad \mathrm{m} = 1, 2, 3, \ldots \tag{5.172}$$

The light reflected from different interfaces are combined constructively, which will generate a strong backward traveling wave. For small perturbations, the grating period of $W = w_1 + w_2$ is related to the average propagation constant of $= \frac{1}{2}(\beta_1 + \beta_2)$, which can be expressed as follows

$$W = \frac{m\pi}{\beta} = \frac{m\lambda}{2n} \tag{5.173}$$

For smaller m, the grating period given by equation (5.173) is generally small. For example, the typical refractive index of III-V semiconductor is n=3.5, when $\lambda = 1.3\mu m$, $\frac{\pi}{\beta}$ is only 200 nm. Therefore, for DEB and DBR in GaAs/AlGaAs, the grating period is usually the number of times of π/β.

Besides the two above kinds of coupling phenomenon, if the refractive index is anisotropic, then, when the two principal axis of refractive index ellipsoid is beyond the planar waveguide plane there will be another kind of coupling between modes, which is the coupling between TE mode and TM mode. Normally, the typical refraction index of III-V semiconductor is isotropic, but it can become anisotropy by applying the external electric field or internal stress, so as to realize the coupling between TE mode and TM mode. As space is limited, no further discussion will be discussed.

5.7.5 Losses in semiconductor waveguides

When light is transmitted in the semiconductor waveguide, there will be a certain amount of energy losses. The main reasons of the losses are the scattering of the transverse evanescent mode and the light absorption in materials. Two loss mechanisms and the loss induced by waveguide junction discontinuities are briefly introduced in this subsection.

5.7.5.1 Absorption loss

Various types of light absorption will occur in each layer of the semiconductor composed of the waveguide. In n^{+}- and p^{+}- type semiconductor, the light absorption by free carrier is an important loss mechanism, and the absorption coefficient can be approximately expressed as

$$\alpha = \frac{ge^3\lambda_0^2 N_c}{4\pi^2 m^{*2}\mu C^3 n\varepsilon_0} \tag{5.174}$$

Where, μ is the mobility of carriers, and the factor g depends on the relationship between the time of the carrier scattering and the energy, and for the phonon scattering, g is slightly larger than 1, while for the ion impurity scattering, g is approximately 3.

Via equation (5.174), we can know that the absorption coefficient is directly proportional to the concentration of the free carrier (N_c) and the square of the wavelength in vacuum (λ_0^2), and is inversely proportional to the carrier mobility (μ), but the relationship between the absorption coefficient and λ_0^2 is approximate, which is related to the dependence of g on the wavelength.

Via equation (5.174), the absorption loss of each layer is obtained, then, the total absorption coefficient of the waveguide can be expressed as

$$\alpha = \sum_{i=1}^{3} \alpha_i P_i / P_T \tag{5.175}$$

α_1 and P_i is the absorption coefficient and the optical power of the ith layer, respectively; P_T is the total power of the guided mode.

5.7.5.2. Scattering loss

The scattering loss mainly occurs at the turning or bending section of the waveguide. Due to the characteristics of light confinement, when the guided wave propagates in the bend waveguide, the fields propagate along the top and bottom surface of the waveguide layer with the different propagation distances or different propagation speeds, which will induce the guided wave mode be coupled to the evanescent wave. This problem can be solved by the mode coupling theory. Then the energy loss per radian is expressed as a function of waveguide parameters

$$\alpha = \frac{8.686\times 2b(1-b)\Delta nR}{w\sqrt{2n\Delta nb+\lambda}} \times exp\left\{-\frac{8\pi\Delta n}{3\lambda}\sqrt{\frac{\Delta n}{n}}R\left[1-(1-b)\left(1+\frac{wn}{4\Delta n(1-b)R}\right)^2\right]_{3/2}\right\} \tag{5.176}$$

where Δn is the difference of the refractive index between waveguide layer and substrate; R is the radius of curvature of the bending waveguide; b is the numerical factor. Figure 5.39 shows the results for GaAs single mode waveguide based on equation (5.176). The radius of curvature of the waveguide is 1 mm, and the wavelength is 1.3 μm.

As can be seen from Fig. 5.39, the larger Δn is, the less the loss is. However, to ensure single-mode operation, for a waveguide with a certain width, there is a maximum Δn. For example, when the waveguide width is 1 μm, as long as $\Delta n \le 0.35$, the single-mode operation can be maintained, and the scattering loss can be reduced to $1\times 10^{-3} dB\cdot rad^{-1}$. Under this condition, $\Delta n = 0.0285$ should be taken for obtaining the loss of $0.5 dB\cdot rad^{-1}$. When the waveguide width is increased to 1.5 μm, $\Delta n \le 0.0275$ is necessary to maintain the single-mode operation, and the scattering

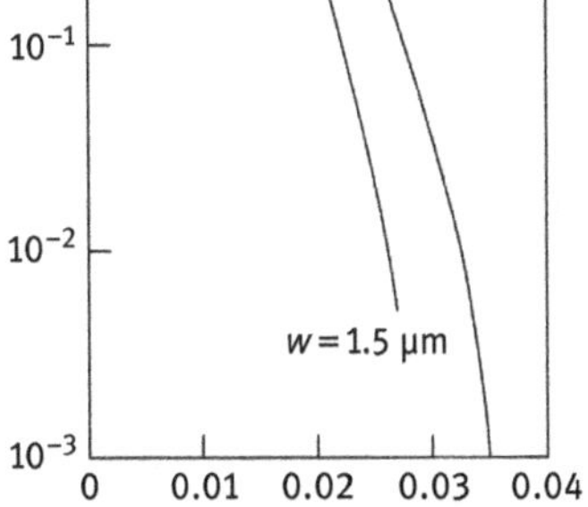

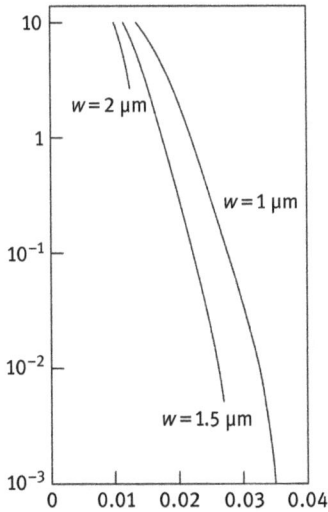

Fig. 5.39: Variation of scattering loss coefficient of GaAs single mode waveguide with Δn, w is a parameter of wave guide width.

loss is approximately $7.5\times10^{-3}dB\cdot rad^{-1}$. To get the loss of $0.5dB\cdot rad^{-1}$, $\Delta n=0.0225$. When the waveguide width is increased to $2\mu m$, the condition of $\Delta n 0.0155$ is required, and the minimum scattering loss is $3dB\cdot rad^{-1}$. At this time, if we still want to get low loss transmission, we can only increase the radius of curvature of the waveguide.

5.7.5.3 Power loss induced by wave guide junction discontinuities

Imperfect lithography process will lead to the junction discontinuities of the wave guide, and it is necessary to match the refractive index between the ending faces of waveguides by filling the silica gel. In these cases, the guided wave modes are coupled in the two wave guides which are well collimated but have a small gap between the ending faces, which will also result in a certain power loss. When there is a gap, the power can be written as

$$P=P_0e^{-al} \tag{5.177}$$

Where P_0 is the power when we assume that the gap does not exist; l is the length of the gap, α is the power attenuation coefficient caused by the gap, which depends on the geometric structure of the wave guide, the wave guide material and the working wavelength.

Figure 5.40 shows an example, in which the waveguide is divided into three regions. The waveguide in region I acts as a radiation source to emit radiation to the separation area II, and the distribution of the radiation light intensity is a power of the cosine function of the axial angle, that is

$$I(\theta)=I_0cos^m(\theta) \tag{5.178}$$

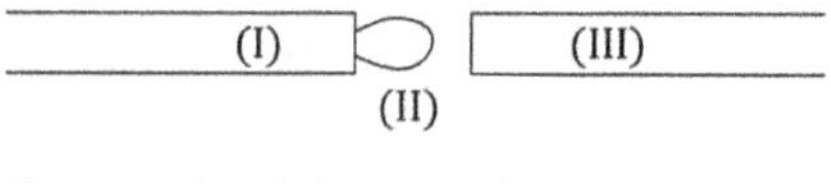

Fig. 5.40: Coupled modes of the waveguides with longitudinal misalignment.

Where I_0 is the light intensity along the central axis, and I is the light intensity along the direction with an angle of θ to the central axis.

The power coupled to the input-end of the output waveguide (region III) is obtained by calculating the overlap between the radiation type in region II and the waveguide cross-section in region III. When the gap is small, via equation (5.177) we can get

$$ln\frac{P_0}{P}=al$$

That is, the logarithm of the relative power varies linearly with l, and the orders of magnitude of the proportional coefficient α is approximately $10^{-2}\mu m$.

In the theoretical analysis of the optical waveguide, when the above condition is satisfied, the propagation constant of the mode can be written as

$$\beta = \beta_r + j\beta_i = \beta_r + j\frac{\alpha}{2}$$

Where β_r is the real part of the propagation constant, which does not vary with the waveguide gap; β_i is the imaginary part of the propagation constant, which is equal to half of the attenuation coefficient. In this way, once α is calculated or measured, β_i can be given immediately.

5.8 The new progress of waveguide theory

In recent years the new progress of waveguide theory is mainly reflected in the improvement of the coupled mode theory; using the nonlinear materials such as switch, second harmonic generation to manufacture the waveguide to obtain a desired function; analyzing the dispersion characteristics of waveguide and other relevant issues precisely. This subsection will give a brief introduction to the first two aspects.

5.8.1 Second harmonic generation in a nonlinear waveguide

In this subsection, take Ti: $LiNbO_3$ as the example, the second harmonic generation in a waveguide and waveguide-resonator is mainly introduced.

5.8.1.1 Second harmonic generation in channel waveguide

For the second harmonic generation, one of the most important parameters is the conversion efficiency of the fundamental mode (pumping) power P_w converting to the second-harmonic power P_{210}.

$$\eta = \frac{P_{2w}}{P_w}$$

The second harmonic generation can also be studied by the mode coupling theory, and the results show that the conversion efficiency significantly higher than that in the same bulk materials devices can be got from the waveguide with proper design. Figure 5.41 is a typical contrast example, the material is Ti: $LiNbO_3$, and the fundamental mode power is 1 mW, the effective area of interaction between the fundamental mode and second harmonic is 260 μm^2. The abscissa indicates the length of the waveguide l. Curve (a) shows the harmonic conversion efficiency of lossless waveguide, and curve (b) and (c) show the conversion efficiency of lossy waveguide and bulk materials respectively, for the fundamental mode, the waveguide loss

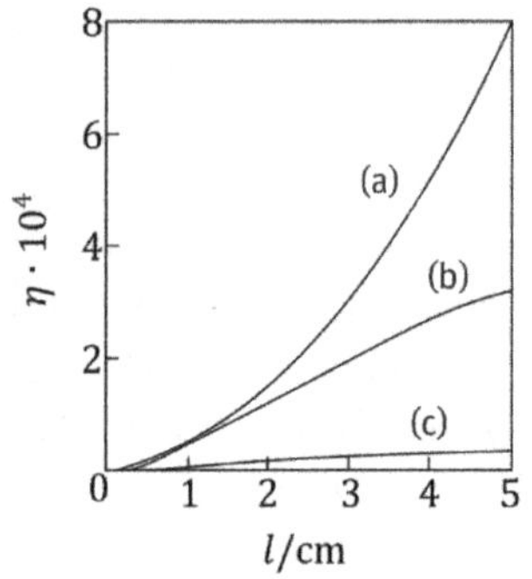

Fig. 5.41: Second harmonic conversion efficiency as the function of the length of waveguide: (a) lossless waveguide, (b) lossy waveguide, (c) bulk materials.

coefficient is $\alpha_w = 0.3dB \cdot cm^{-1}$, for the second harmonic, $\alpha_{2w} = 1.0dB \cdot cm^{-1}$ we can see that, to give full play to the advantages of the waveguide devices in the second harmonic conversion, it is necessary to make the waveguide with low loss as far as possible.

5.8.1.2 Second harmonic generation in waveguide-resonator

Although the nonlinear waveguide can significantly improve the efficiency of second harmonic generation, but for the low power laser source (e.g., mW), we also want to have higher conversion efficiency, a waveguide resonator is such the device that has the higher conversion efficiency.

Figure 5.42 shows a typical waveguide-resonator, the reflected mirror which the pumping light P_w incident into is called the front reflected mirror, and expressed by M_f; the other mirror is called the rear reflected mirror and expressed by M_r. The reflectivities of the above two mirrors are R_f and R_r respectively. The things we are interested in are symmetric resonator,

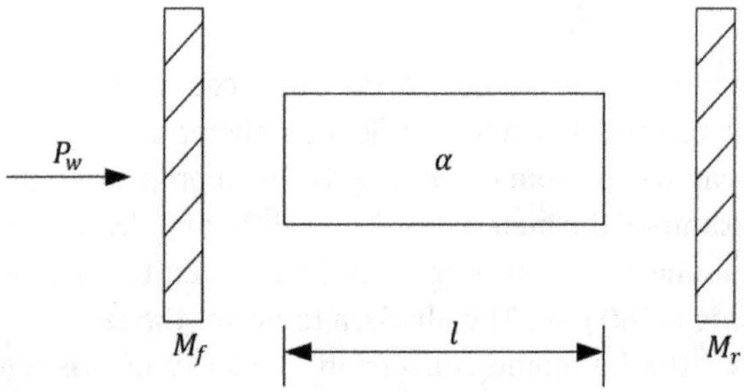

Fig. 5.42: A typical waveguide resonator.

$$R_f = R_r = R$$

And matching resonator

$$R_f = R_r \exp(-2\alpha_w l)$$

Figure 5.43 shows the changes of conversion efficiency of second harmonic generation with the interaction length between the fundamental wave and harmonic in channel waveguide without cavity, symmetric waveguide resonator and matching waveguide resonator. For the above three cases, we suppose the fundamental mode ($\lambda = 1.09\mu m$) power is 1 mW, the phase matching conditions are well satisfied, and the mode attenuation coefficients are

$$\alpha_w = 0.023cm^{-1} \approx 0.1 \times dB \cdot cm^{-1}, \alpha_{2w} = 2\alpha_w$$

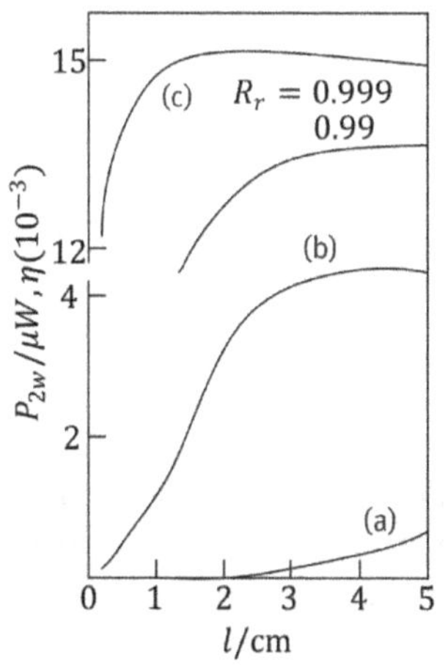

Fig. 5.43: The changes of η and P_{2w} with l in channel waveguide without cavity, symmetric waveguide resonator and matching waveguide resonator: (a) channel waveguide without cavity; (b) symmetric waveguide resonator: R=0.9; (c) matching waveguide resonator: $R_r = 0.99$, 0.999.

From Fig. 5.43, we can see that matching resonant is by far the best device to produce the effective second harmonic, whose harmonic conversion efficiency is higher than that of symmetrical resonator, and, the rear mirror reflectivity is close to 1, the higher is the conversion efficiency, the smaller is the dependence on the interaction length. Making use of the latter feature, the shorter material can be used to achieve the higher conversion efficiency of second harmonic in principle, which is attractive to the user.

The dependence of the conversion efficiency on the interaction length being weakened can be understood as follows: with the increase of the length of the nonlinear interaction, the growth of the fundamental mode power drops according to the same rule, and the above two effects exactly cancel out each other.

On the other hand, the conversion efficiency of the second harmonic in the waveguide cavity is also strongly dependent on the loss of the waveguide. Figure 5.44 shows the results of matching waveguide resonator, the interaction length of the fundamental and harmonic is 4 cm, and the other parameters are the same as the former. It can be concluded that as long as the loss is low, the matching resonator can be used as the second harmonic generator, which can achieve more than 10% conversion efficiency for the fundamental mode with mW power.

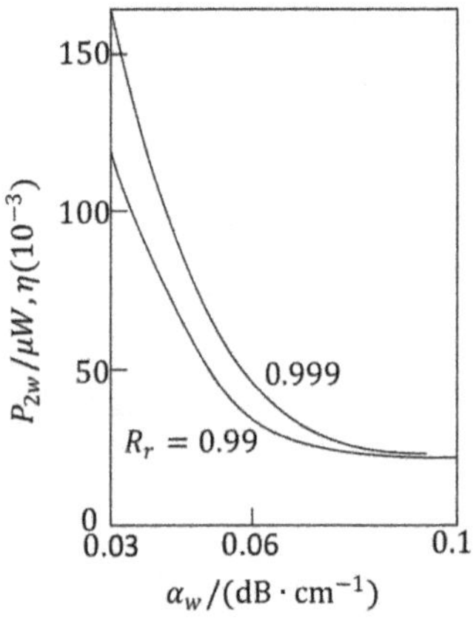

Fig. 5.44: The change of harmonic power of matching resonator with α_w.

In the past, the study of the waveguide is mainly limited to the dielectric material or semiconductor structure, or using different materials on each layer alternatively, or making stripes on the surface of some material with epitaxial growth method. In recent years some people have proposed to utilize the metal-semiconductor-metal (MSM) structure to achieve the waveguide coupling and optical switch based on the nonlinear interactions between the metal and the semiconductor. According to the report, people can obtain high contrast switchs by using such devices, combined with the existing standard silicon optoelectronic technology, high contrast switchs effectively play a role in the bias control and other applications. Readers who wish to know them in depth can read the relevant literature.

5.8.2 Non-orthogonal coupled mode theory of waveguide

The conventional coupled mode formulas are built based on each waveguide modes being orthogonal, which is a simple and intuitive way; while more rigorous coupled mode theory is based on the superposition of each waveguide and appeared power cross terms because of the non-orthogonality between the waveguide modes. To make the theories of different waveguide coupled modes be self-consistent, the cross power terms are necessary. Although the non-orthogonal coupled mode theory is more complex, it can obtain more accurate results than the conventional orthogonal coupled mode theory in most practical applications. However, the conventional orthogonal coupled mode theory is quite accurate for describing the weak coupling between two nearly identical waveguides. For different waveguides, a self-consistent coupled mode theory can be derived by redefining the coupling coefficient and the coupling length and power exchange between the two distant waveguides can be predicted accurately.

There are many different coupled mode formulas in the literature, and the choice of the specific formula depends on the waveguide structure and the desired accuracy. Generally speaking, it is still a problem that the coupled mode theory is used to analyze the range and accuracy of optical waveguide.

Figure 5.45 shows the simplest form of the coupled waveguide system, which is a directional coupler consisting of two waveguides which are parallel and close to each other. It is assumed that the waveguide is isotropic, linear and non-loss, which is called the uniform directional coupler. The above choice of modes and media is not suitable for some cases, however, the main properties of the coupled mode theory are similar.

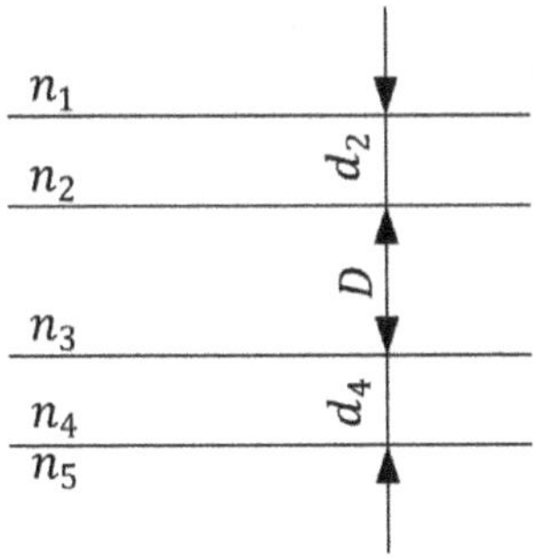

Fig. 5.45: Directional coupler of waveguide.

It is difficult to find the exact analytical solution for the practical directional couplers which are made up of channel waveguides. Although we can calculate its rigorous numerical solution by the computing technology, but it is lack of intuitive and physical understanding. Coupled mode theory is useful for understanding the physical processes in directional coupling devices. In this subsection we do not intend to discuss this issue in depth, but simply introduce the results obtained by using the conventional orthogonal coupled mode theory and improved non- orthogonal theory to deal with uniform directional couplers.

Being different from the orthogonal coupled mode theory described in subsection 5.6, the solution of the coupling equation must be expressed as a linear superposition of different waveguide modes, that is

$$E = a_1(z)\boldsymbol{e} + a_1(z)\boldsymbol{e}_2,\ \boldsymbol{H} = a_1(z)\boldsymbol{h}_1 + a_2(z)\boldsymbol{h}_2$$

Where $\mathbf{e}_i$ and $\boldsymbol{h}_i$ are vector waveguide modes, a_1 and a_2 are amplitudes of the according modes respectively. The amplitude of the $\boldsymbol{i}$th waveguide mode is

$$b_i(z) = a_i(z) + \boldsymbol{X}a_j(z)$$

Where X is the coefficient depending on the non-orthogonality of mode, while the power is

$$P_i(z) = |b_i(z)|^2 = |a_i(z) + \boldsymbol{X}a_j(z)|^2$$

Therefore we can get the coupling length of the power in two waveguides. The parameters of the waveguide structure shown in Fig. 5.45 are

$$n_1 = n_3 = n_5 = 3.200, n_2 = 3.250, n_4 = 3.300$$

$$d_4 = d_2 = 1.0\mu m, \lambda = 1.5\mu m$$

The results obtained by the orthogonal and non-orthogonal coupled mode theory are given and compared with the numerical results, which are shown in Fig. 5.46.

From Fig. 5.46, we can see when D is smaller and the coupling is strong, the results obtained by the above two coupled mode theory significantly deviate from the exact solutions, with the increase of D, the two results are closer to the exact solutions, but no matter what the situation is, results obtained by non-orthogonal coupled mode theory are more close to the exact solutions.

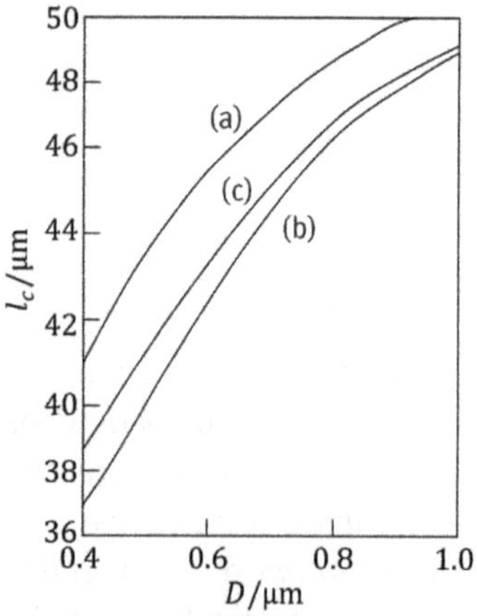

Fig. 5.46: Coupling length obtained by different methods: uniform directional coupler: (a) orthogonal coupled mode theory; (b) non-orthogonal coupled mode theory; (c) rigorous numerical solution.

5.9 Waveguide devices in insulating crystals

The most versatile waveguide devices with simplest structure are switches and modulators, for example, they can be used as the servo or protective devices, bypass switches in the field of optical networks, or used as the programmable delay line in the field of signal processing and so on. Besides, the high speed optical switches used for time division multiplex communication system allow several low bitrate channels to share the broadband of a single-mode optical fiber, or the most basic optical switches used as external modulators for signal encoding usually have two or more than two alternative output ends, and for symmetric system, the switches require for a corresponding number of the input ends. The significant parameters of the switches include the required switch driving voltage, the crosstalk between open state and conduction state, and the optical insertion loss. For the time division multiplex communication system and signal encoding, the speed of switches is important as well. Owing to space constraints, this subsection only introduces the directional coupler, balanced bridge interferometer, cross-coupled waveguide switches and some of their variants.

5.9.1 Directional couplers

Directional couplers are versatile devices. Different directional couplers have different abilities such as wave-front filter, polarization selection and so on.

Directional couplers consist of a couple of same strip-loaded waveguides which are placed sufficiently close together (shown as Fig. 5.47), in which the light from one waveguide would be coupled into the second one by evanescent coupling. The coupling coefficient of per unit length κ depends on waveguide parameters, wavelength λ and the gap g between the two waveguides. Besides, characteristics of directional couplers can also be expressed by the difference of propagation constants $\Delta\beta$ between the two waveguides and interaction length L, here:

$$\Delta\beta = \frac{2\pi}{\lambda}(N_2 - N_1)$$

N_1 and N_2 are the effective refractive index of the two waveguides respectively.

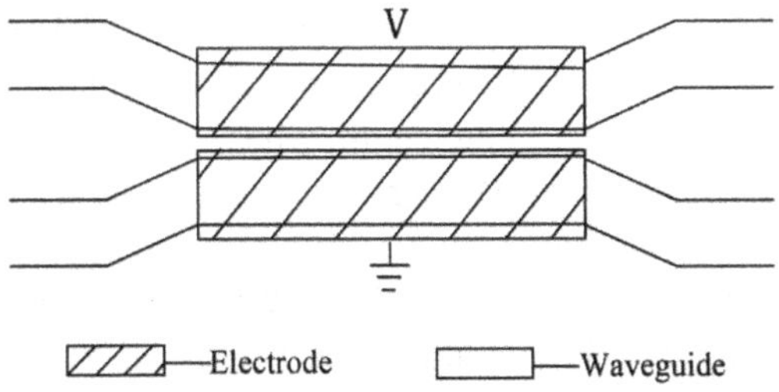

Fig. 5.47: Directional coupler switch.

As shown in Fig. 5.47, one of the waveguides takes positive power supply, with the driving voltage of V, and another leads to ground. The purpose of applying voltage on waveguide is to adjust possible phase mismatch through changing the refractive index of waveguide induced by electro-optic effect. For example, if the condition of electrode is shown as Fig. 5.47, *Z*-cut $LiNbO_3$ will show a strong electro-optic effect.

The working characteristics of directional couplers and some other important Ti: $LiNbO_3$ waveguides could be analyzed by mode coupling theory. Assuming the intensity of the incident light which is to the first waveguide is I_0, so the intensity of the light coupled into the second one is:

$$I = \eta I_0 \tag{5.179}$$

$$\eta = \frac{1}{1+(\beta/2\kappa)^2} sin^2 \kappa L \left[1+(\beta/2\kappa)^2\right]^{1/2} \tag{5.180}$$

Where η is the light intensity (or power) coupling efficiency.

If the two waveguides composing the coupler are identical, in the condition of no voltage being applied, $\Delta\beta = 0$, via equation (5.180), we get

$$\eta = sin^2 \kappa L \tag{5.181}$$

Especially when

$$\kappa L = (2m-1)\frac{\pi}{2} \quad m = 1, 2, \cdots \tag{5.182}$$

Then $\eta = 1$, which is called the complete crossing between the two waveguides, while when $m = 1$, the interaction length is called the coupling length, which is still expressed by L_c, and shown as follows:

$$L_c = \frac{\pi}{2\kappa} \tag{5.183}$$

The typical value of coupling length of Ti:$LiNbO_3$ waveguide is 0.2~10 mm. When $\Delta\beta \neq 0$, the first term shown on the right of equation (5.180) will less than 1 forever, and the second term will never more than 1, thus $\eta < 1$, which means the complete coupling would not happen whatever the value of κL is.

Standard directional couplers have two important disadvantages. First, to realize the complete coupling under the condition of no voltage being applied, the interaction length L should be a whole number of the coupling length, which makes their fabrication technology be more difficult (however this problem can be solved easily if the coupling coefficient κ could be effectively changed by electro-optic method). Secondly, if L is several times more than L_c, then $\Delta\beta L$ is required to be a bigger value, and for this, the higher driving voltage is required too. For the switches requiring small crosstalk between the conduction state and open state, the first problem is more serious, because that the modulator has a higher loss under the conduction state when $L \neq nL_c$. While for the high-speed modulation under low voltage, the second one is more serious. We can decrease the gap between electrodes to lower the required driving voltage.

The above two problems could be solved by changing $\Delta\beta$ spatially. The most important way is to reverse voltage to change the sign of $\Delta\beta$ (shown in Fig. 5.48). Sometimes this kind of device has multi-joint structures, if the number of joints is, then the total coupling efficiency is:

$$\eta = sin^2 \kappa_{eff} L \tag{5.184}$$

Where $\kappa_{eff} = \frac{M}{L} sin^{-1}\sqrt{\eta_m}$

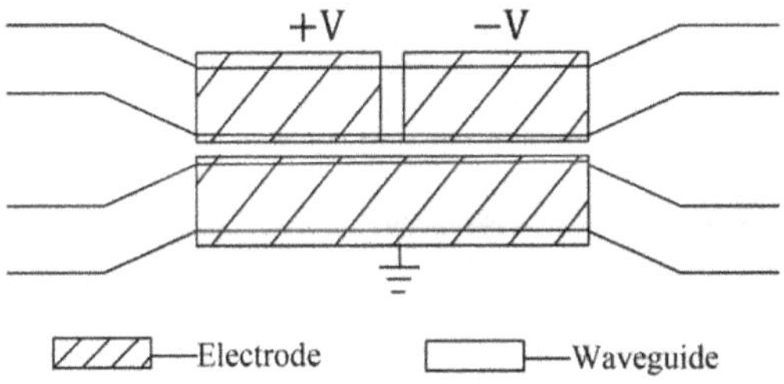

Fig. 5.48: The directional coupler with reversed $\Delta\beta$ control.

η_m is the coupling efficiency of waveguide whose length is L/M. With this method, the crossing state and the blocking state could be realized. In fact, people have measured the extreme crosstalk in this kind of waveguide.

5.9.2 Balanced bridge interferometers and cross-coupled waveguides

As shown in Fig. 5.49, the balanced bridge interferometer-type optical switch has two input ends and two output ends, which is made by the principle of conventional Mach–Zehnder interferometer. In the center section of this interference switch, two waveguides could not be coupled because they are sufficiently far apart, the relative phase between the two waveguides could be adjusted by the electro-optical effect. While the two waveguides are close together on the input end and output end, in which the directional couplers are formed respectively. The incident light entering into any of the two waveguides is split by the directional couplers on the input end,

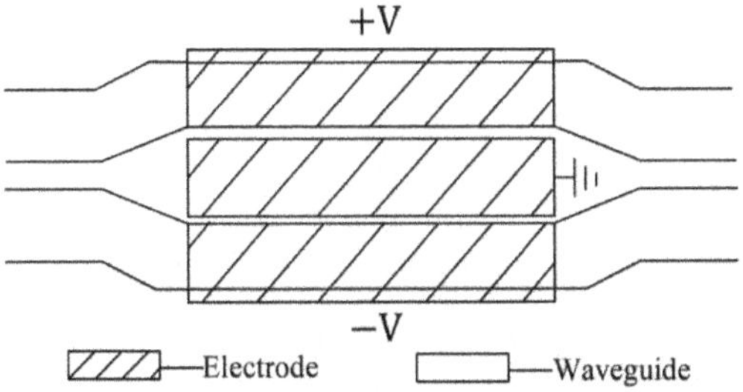

Fig. 5.49: The balance bridge interference switch.

and coupled at the directional coupling region on the output end after being transmitted a distance without coupling. Via mode coupling equations, the coupling efficiency of power could be calculated, and the result is

$$\eta = cos^2 \frac{\Delta\beta L}{2} \tag{5.185}$$

Where L is the length of electrode, which is used to adjust the phase shifting.

Equation (5.185) indicates that the interference switch has the periodic response to $\Delta\beta$, which is different from the directional coupling switches but similar to multi-joint inversed $\Delta\beta$ modulators. Besides, the phase shifting from the coupling state to the blocking state is $\Delta\beta L = \pi$, and both these two states could be realized by electrical methods.

To get the ideal switch state, i.e., low crosstalk state, both splitter and compositor are 3 dB ideal couplers (i.e., both splitting ratio and compositing ratio are 50% to 50%), and the loss in each waveguide is equal. if there exists little coupling error, $\eta = 1$ would not be realized. The way to get 3 dB couplers is to let κL be equal to $\pi/4$ and $\Delta\beta$ be equal to zero (i.e., the waveguides are the same), which means. When $\kappa L > \pi/4$, we can adjust $\Delta\beta$ to get 3 dB couplers by utilizing the electro-optic method.

The working principle of Y-junction interferometer shown in Fig. 5.50 is the same as that of the balanced bridge interferometer introduced above. It has two 3 dB couplers at its both ends. Differently, Y-junction interferometer has only one input end and one output end. The incident light is split into two components by splitters, and each component propagates along one waveguide of the interferometer, and in the center region, the two waveguides are sufficiently far apart so that the evanescent coupling and crosstalk are avoided. Typically, the optical paths of the two waveguides are the same, therefore the two components will add up in-phase at the output couplers and continue propagate along the output waveguide if no phase shifting is applied to the two waveguides.

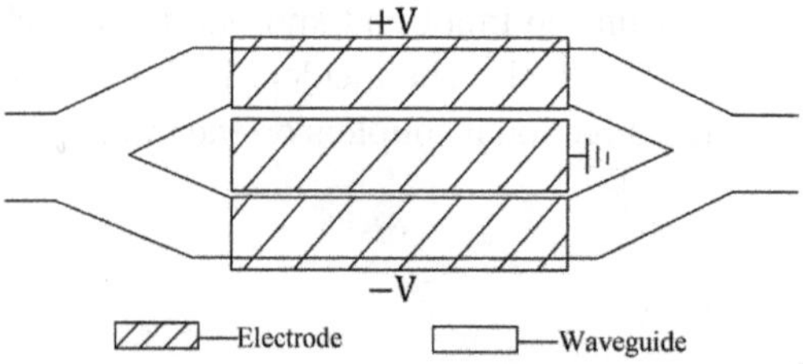

Fig. 5.50: Y-junction interferometer.

This switches have simple structure and can be used widely. A variant of this kind of switches is adopting the asymmetric structure at the output of interferometer and supplying two available output ends, which is a one-to-two switch.

Another type is composed of cross-coupled waveguides which have many different forms, one of them is shown in Fig. 5.51. This kind of waveguide made in Ti: $LiNbO_3$ has been widely used in many photoelectric devices. The early devices adopt multimode waveguides and the work principle is based on total reflection. Later, this kind of device adopt the single mode waveguides, which could also be seen as zero-gap directional couplers or interferometer. It is convenient to study symmetrical directional couplers by considering the whole structure of two eigenmodes. The eigenmode are not coupled and propagates with the different speed with regard to the symmetrical structure. Under the ideal condition, both crossing and blocking states with low crosstalk could be realized

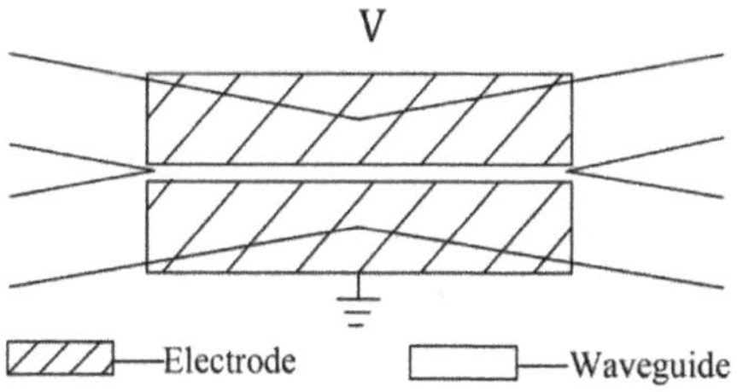

Fig. 5.51: Cross-coupled waveguide switch.

Being an important element for multiplex communication system, single-mode waveguide filters have a long history. The key parameters of filters include central wavelength, the width of passband, the peak of filtering efficiency, sidelobe level, electrical tunability and so on. The desirable passband width depends on the specific application. For instance, the broad band filter is adopted to minimize tolerance, on the contrary, for multiplex communication system with good controlled source, the narrow-band filter is required. Thus for system designers, it is important to design a filter with a wide range of bandwidth. As similar as switches and modulators, the filter can also be divided into coupling mode and interference type, which will be introduced later in this subsection.

5.9.3 Interference filters

Figure 5.52 shows an example of interference filter. It is actually a Y-junction interferometer that is composed of two waveguides with different lengths. Both the splitter and the compositor have the same function to the corresponding devices in interference modulators. However, unlike the interferometer with two same waveguides, the optical phase difference in compositor strongly depends on the wavelength because there is a physical path difference of ΔL between the two waveguides, that is

$$\phi = \frac{2\pi}{\lambda} N\Delta L$$

where N is the effective refractive index. Correspondingly, the transmission of light intensity is periodic, and could be expressed as:

$$I_{out} = I_{in} cos^2\left(\frac{\pi N\Delta L}{\lambda}\right) \tag{5.186}$$

where I_{in} and I_{out} denote the input light intensity and output light intensity respectively. The null to null bandwidth of filter is:

$$\Delta\lambda = \frac{\lambda^2}{N\Delta L} \tag{5.187}$$

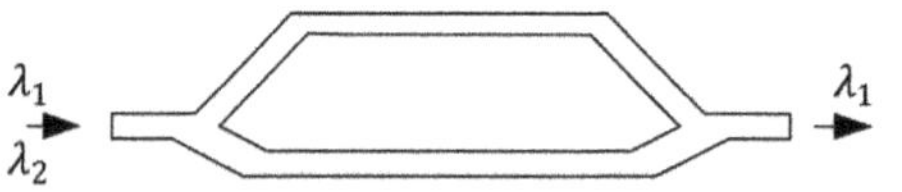

Fig. 5.52: Interference filter that is composed of two waveguides with different lengths.

Another example of interference filter is show in Fig. 5.53, in which the two waveguides has the same physical lengths but different optical paths. We assume that this filter has a 3 dB splitter and compositor, and in the center section, the modes are not coupled. The light enters into the dual-mode waveguide from single-mode waveguide and is split into two single-mode waveguides after propagating a distance of L.

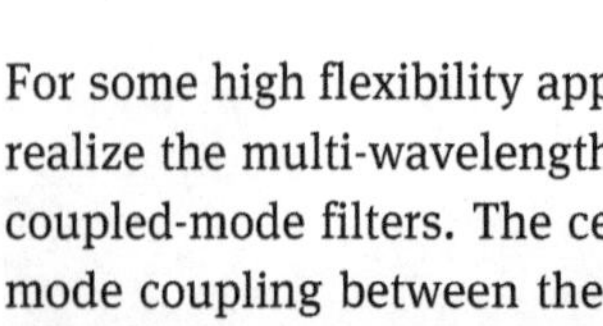

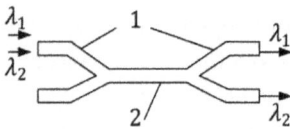

Fig. 5.53: Multi-wavelength interference filter. 1—Single-mode waveguide; 2—Multi-mode waveguide.

5.9.4 Coupled-mode filters

For some high flexibility applications, we use the filter with a passband response to realize the multi-wavelength division. The passband response could be realized by coupled-mode filters. The central wavelength of this kind of filters depends on the mode coupling between the two waveguides with different propagation constants. This subsection will introduce two general technologies. The first is to realize the phase matching for desirable wavelength by utilizing the periodic coupling, which could realize the narrowband filtering. The second is to select a parameter which is considered as a function of wavelength, and different for the two waveguides, and overlapped at the central wavelength of the desirable filter, using this method we could get a relatively wide passband.

5.9.4.1 Tunable filters with mode conversion

For the birefringent materials such as LiNbO3 substrate, the conversion between electro-optic TE mode and TM mode depends on the wavelength. The effective coupling between nonsynchronous TE mode and TM mode could be realized by periodical coupling with the electrode period of Λ, and Λ satisfies a certain phase matching condition. However the phase matching condition apparently depends on the wavelength and can be strictly satisfied only at the central wavelength λ_0. That is to say, TE mode and TM mode would be effectively coupled at the central wavelength. When $\lambda \neq \lambda_0$, the value of phase mismatching is

$$\Delta\beta = \frac{2\pi}{\Lambda}\left(\frac{\lambda_0}{\lambda} - 1\right) \tag{5.188}$$

The conversion efficiency or filter response with respect to the wavelength λ should be given by the standard coupling function i.e., equation (5.187), and the bandwidth of this filter is

$$\frac{\Delta\lambda}{\lambda} = \frac{\Lambda}{L}, \quad \Delta\lambda = \lambda - \lambda_0 \tag{5.189}$$

Due to the large birefringence, the filter using $LiNbO_3$ waveguide has very narrow bandwidth. To realize the signal separation, a polarization beam splitter is needed to select the conversion component. By utilizing the interdigital electrode, the mode conversion between TE mode and TM mode whose phases are matched hasbeen realized in x-cut or z-cut $LiNbO_3$. No matter which one is used, the x-component of electric field could be coupled to the strong non-diagonal coefficient $r_{51}(= 28 \times 10^{-10} cm \cdot V^{-1})$. If the electrode with the length of 0.5~6 mm is applied, the bandwidth of the filter in visible light can reach 0.5~5 mm.

5.9.4.2 Tunable directional coupling filters

As shown in Fig. 5.54, it is a specially designed directional coupler with the function of wavelength choice, which could realize the broadband filtering.

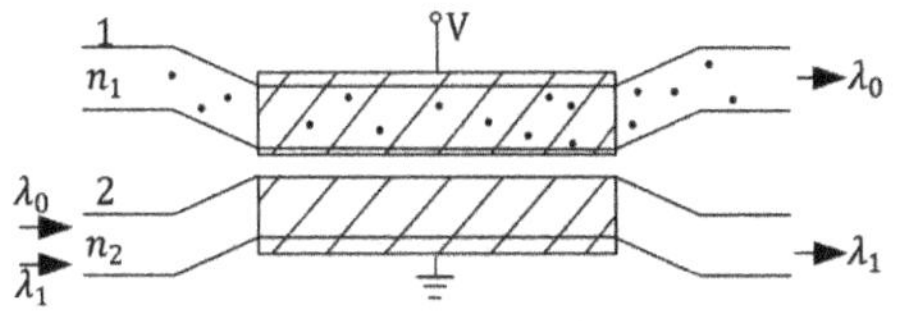

Fig. 5.54: Tunable directional coupling filter.

Couplers consist of two waveguides with different widths and refractive indexes, in which the narrow one has a higher refractive index. Because of their different size and refractive index, the two waveguides have different dispersion characteristic of $n(\lambda)$. However, we can make the two waveguides have the same effective refractive index at the central wavelength λ_0 by proper design (shown in Fig. 5.55a). The phase matching at this wavelength makes it possible that the complete crossing between the two waveguides can be realized. However the crossing could hardly occur at the wavelength of λ that is much different from the central wavelength of λ_0, as shown in Fig. 5.55(b). Thus the signal we need can be separated from other components of input wavelength spatially. The response of filter is still given by equation (5.187). For the component that satisfies $\lambda \neq \lambda_0$, the value of mismatching is

$$\Delta\beta = \frac{2\pi}{\Lambda_{eff}}\left(\frac{\lambda_0}{\lambda} - 1\right) \tag{5.190}$$

And the bandwidth is

$$\frac{\Delta\lambda}{\lambda} = \frac{\Lambda_{eff}}{L} \tag{5.191}$$

Where

$$\Lambda_{eff} = \left[\frac{d}{d\lambda}(n_2 - n_1)\right]_{\lambda=\lambda_0} \tag{5.192}$$

Λ_{eff} is the efficient period of the filter.

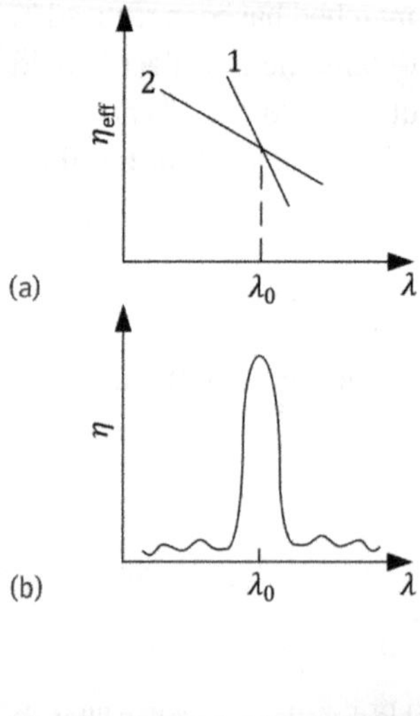

Fig. 5.55: The response of directional coupling filter: (a) the variation of effective refractive index with the wavelength; (b) the variation of coupling efficiency with time.

The discussions above have been proved by the tunable filters which adopt Ti:$LiNbO_3$ waveguides. For example, the interaction length of the two waveguides is 1.5 cm, when λ_0 =0.6 μm, the measured 3 dB bandwidth of the filter is 20 nm, and when λ_0 =1.5 μm, it reaches 70 nm. This kind of devices have shown very particular functions in wavelength division multiplex switch system.

5.9.5 The polarization selection devices

In optical system, especially the coherent light system, we need to modulate and control the polarization of light or divide the light with different polarization. It has been proved that the appropriately designed Ti:$LiNbO_3$ waveguides have above functions.

There are various methods to define the polarization of the light. It is convenient to define it by the polarization angle of θ and the phase of ϕ. Normalized complex amplitudes of TE mode and TM mode could be expressed as:

$$\begin{pmatrix} A_{TE} \\ A_{TM} \end{pmatrix} = \begin{pmatrix} cos\theta \\ sin\theta e^{j\phi} \end{pmatrix} \tag{5.193}$$

Where ϕ is the phase difference between TE component and TM component, and the relative amplitude between them can be got by ϕ. There are some exceptions, when $\phi = 0$, the light is linearly polarized along the direction of θ; when $\theta = 0$, only the TE mode is polarized and when $\theta = \frac{\pi}{2}$, only the TM mode is polarized. When $\theta = \frac{\pi}{4}$ or $\phi = \frac{\pi}{2}$, we will get the right-handed circularly polarized light. For passive Ti:$LiNbO_3$ waveguides, the light linearly polarized along the main axis has the characteristics of polarization maintaining. For example, when the light enters into the $z-$ cut $LiNbO_3$, the mode would be kept during the propagation process. Furthermore, for the elliptical polarized incident light, the relative amplitudes of TE component and TM component of the output wave do not alter. However, the relative phase of these two components will alter if the waveguide has birefringence characteristics.

The polarization selection devices can be classified into two kinds: one is the polarizer, and the other has the function of splitting light according to different polarization states. The simplest polarizer is the waveguide device with metal coating, in which the induced current may cause the loss of $10dB \cdot cm^{-1}$ with respect to a TM polarization component but unchanged loss with respect to TE mode. The loss of TM mode will increase if the thin dielectric layer is added between the waveguide and the electrode, because the coupling resonance between TM mode and metal coating is intensified. Under these conditions, the difference of loss coefficients between TE mode and TM mode reaches $35dB \cdot cm^{-1}$.

Another way of polarizing the light is based on the dependence of the refractive index of waveguide on the polarization, which derives from the fabrication process of the waveguide. For example, during the process of fabricating $LiNbO_3$ waveguides by proton-exchange method, the refractive index of materials with respect to the extraordinary light (i.e., n_e) increases while the one with respect to the ordinary light (i.e., n_0) decreases. Thus, the waveguides in z-cut $LiNbO_3$, which are fabricated by the proton-exchange method can only transmit the TM mode. The polarizing region with high extinction ratio could be fabricated on the substrates by the combination of proton exchange method and Ti diffusion method. With the help of this technology, the polarizer which has the extinction ratio of 40 dB and low residual loss has been fabricated successfully.

Compared with the polarizer, the linear polarization beam splitter, also is called the analyzer, is a more versatile device, it can split the TE mode and TM mode spatially, which can be realized by specially designed Y-junction beam splitter, cross-coupled waveguide or directional coupler. Considering the directional coupler shown in Fig. 5.56, as long as we make x or $\Delta\beta$ strongly depend on the polarization state, the polaroid then can be achieved.

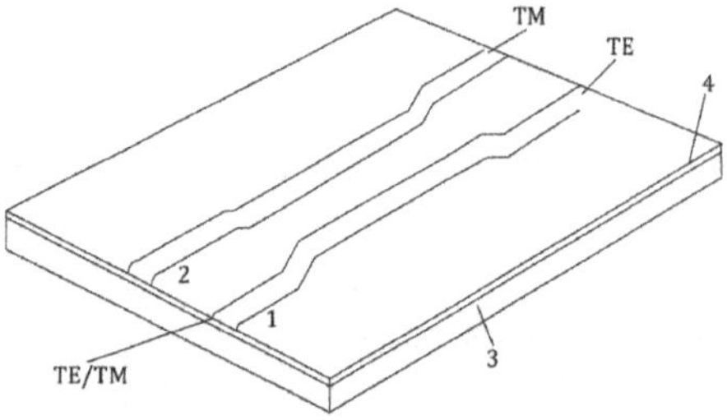

Fig. 5.56: Directional coupler used as a polaroid direction coupler. 1、 2—Ti diffusion waveguide; 3—$LiNbO_3$ substrate; 4—SiO_2.

For Ti:$LiNbO_3$ waveguides, the differences of refractive index between the substrate and the waveguide (i.e., Δn) with respect to the TE mode and TM mode are not equal generally, so κ depends on the polarization too. The difference of Δn between the two polarization components can be increased or decreased by selecting the different diffusion parameter, therefore we can choose the diffusion parameter and the interaction length of L to get the equations as follows

$$\kappa_{TE}L = \pi, \kappa_{TM}L = \frac{\pi}{2} \tag{5.194}$$

Typically, the length of the device is approximately 1 cm. When the condition of equation (5.193) is satisfied, the TE component of the light from waveguide 1 will stay in itself, while the TM component is coupled to waveguide 2, so that the different components can be separated spatially. In principle, being limited by the manufacturing tolerance, this kind of polarization beam splitter is a passive device which does not need the electrode.

We can generate an analyzer by utilizing the dependence of $\Delta\beta$ on the polarization. Taking the device shown in Fig. 5.56 as an example, for one of the two waveguides, we directly add the metal electrode on it, and for the other one, we add a crash pad before adding the electrode on it, and then, the dependence of $\Delta\beta$ on polarization is realized. The metal coating can carry TM mode and change its propagation constant apparently but has a week influence on the propagation constant of TE mode. By selecting the length of L, TE component will be coupled to the second waveguide and the TM component will remain in the input waveguide, which can be used to the separate the TE mode and TM mode spatially. In fact, it is difficult to separate these two modes strictly, however, the analyzer fabricated with $Ti:LiNbO_3$ waveguide could realize quite low polarization crosstalk. If the dependence of κ on the polarization is weak, and the optoelectronic inducing of $\Delta\beta$ strongly depends on the polarization, the crosstalk between the two modes will be further decreased.

5.9.6 Transmission gratings

Figure 5.57 is the plan view of waveguide transmission grating. The light enters into the left side of the grating with the Bragg angle, then two transmitted waves emerge from on the right side of the grating, which correspond to the 0th-order diffraction wave and the 1th-order diffraction wave respectively. The above result indicates that we can disturb the refractive index of the waveguide slightly to make the incident beam be deflected with a large angle. The reason is that the working foundation of this kind of devices is the diffraction effect instead of the refraction effect. Due to the minor change of refractive index in guided-wave optics is typical, in this sense, the grating is an ideal waveguide element.

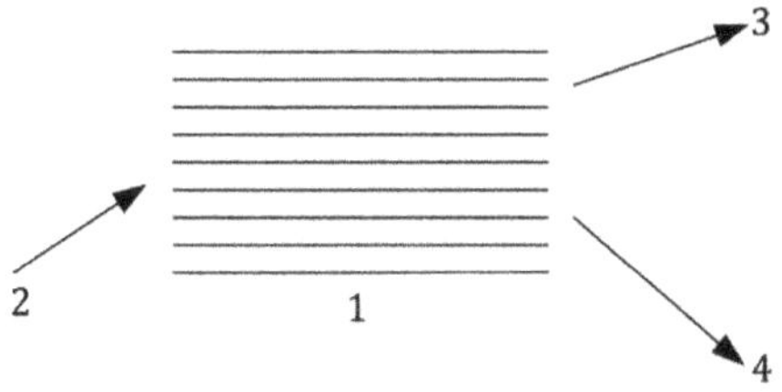

Fig. 5.57: The plan of transmission grating: 1—grating; 2—incident wave; 3—transmitted wave (0th order diffraction wave); 4—1th order diffracted wave.

Because larger beam deflection allows a higher numerical aperture, the diffraction devices could be used as the more effective waveguide lens. Figure 5.58 shows an example of the grating being used as the linearly tuned waveguide lenses. Here a slow variation in the grating period produces a corresponding change in the deflection angle over the lens aperture, which results in an approximate focussing of incident beam. A linear chirp, i.e., the linear variation in period, is especially suitable for focussing the off-axis input wave shown in Fig. 5.58. In more general cases, the grating fringes must be curved, and the local fringe orientation and spacing is found by local application of the Bragg condition.

There are two main disadvantages for grating lens: First, because of their inherent chromatic dispersion, the focus spot moves with the variation of wavelength, and can be seriously aberrated. Secondly, the selectivity of angle limits the field-of-view seriously. However, in some applications these disadvantages may be unimportant, for example, in a telescope beam expander, if a fixed angle and wavelength are used.

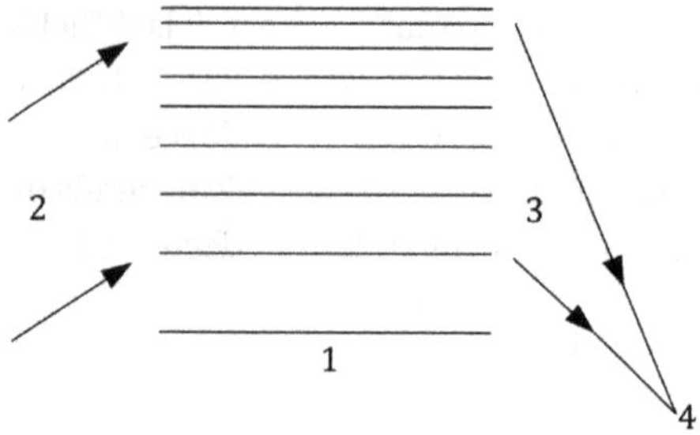

Fig. 5.58: Linear chirped grating lens: 1—grating; 2—off-axis input wave; 3—diffracted wave; 4—the focus of diffracted wave.

5.9.7 Reflection gratings

Figure 5.59 shows a reflecting grating whose structure is different from the transmission grating mentioned above. The gating fringes are oriented parallel to the input boundary. The input wave is the same as before, but the two diffraction orders emerge from different sides of the grating. The 0th order is transmitted, but the first-order is reflected. In general, the reflection grating require much smaller grating periods.

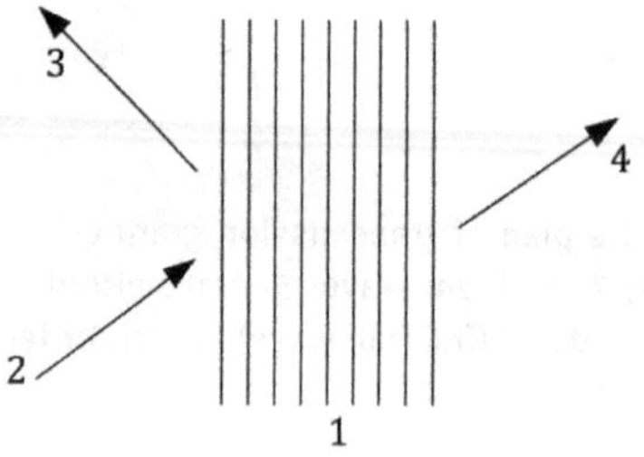

Fig. 5.59: The plan view of reflection grating: 1—gratingincident wave; 3—transmitted wave (0th order diffraction wave); 4—reflection wave.

Figure 5.60 shows a grating whose fringes are oriented at 45° to the edges of a boundary, which has the advantage of making a relatively narrow input beam be diffracted as a much wider output beam. Therefore the device can act as a compact beam expander; which could replace a common one. However, the grating strength must be kept consistent along its length to ensure a uniform amplitude distribution in the output beam.

5.9.8 Electro- and acousto-optic gratings

The schematic diagram of electro-optic grating is shown in Fig. 5.61, in which a periodic metal structure, known as an interdigital electrode, is placed on the surface of the waveguide fabricated in an electro-optic material (e.g., $LiNbO_3$ or $LiTaO_3$).

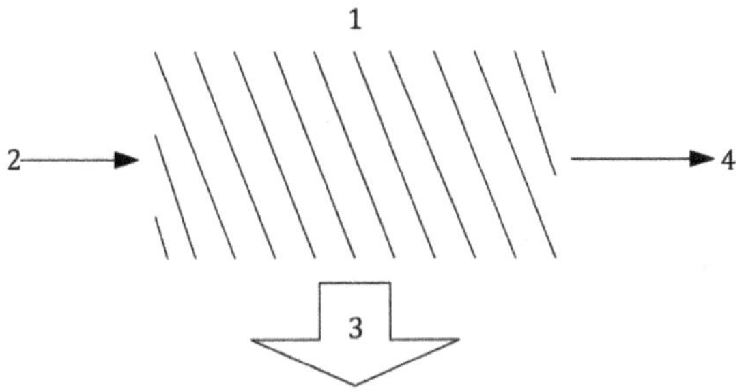

Fig. 5.60: The plan view of beam expander grating: 1—grating; 2—incident wave; 3—diffracted wave; 4—transmitted wave.

When a static voltage is applied, the voltages of the interdigital electrode alternate in sign, this will result in a periodic variation in electric field beneath the electrodes, which induces a corresponding variation of the waveguide refractive index through the electro-optic effect, so the grating is produced. If no voltage applied to the electrodes, there is no grating. Therefore the grating may be switched on and off optionally. The linewidth of electrodes is limited to approximately 0.5~1 μm because of the lithographic process. If the electrode pitch is equal to the linewidth, the grating period is approximately 2~4 μm. Under these above conditions, the beam deflection angle is rather small, despite this, the electrodes may be so sufficiently long that the device acts as a volume grating, so the diffraction efficiency can be high. 98% efficiency has been achieved with Ti-indiffused waveguides on $LiTaO_3$ substrates. Therefore the electro-optic grating may act as an effective modulator, whose speed is limited to approximately 1 GHz.

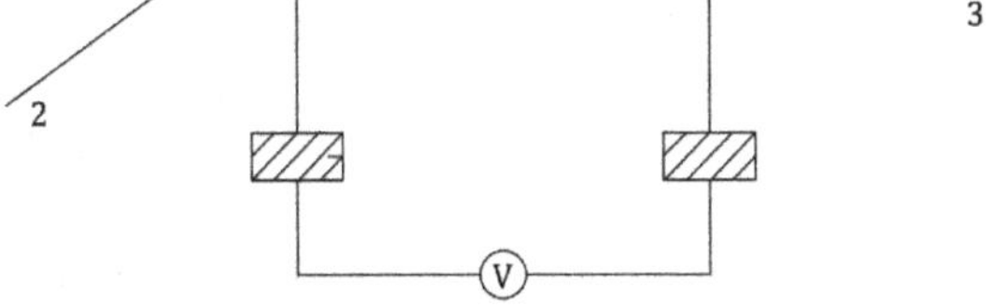

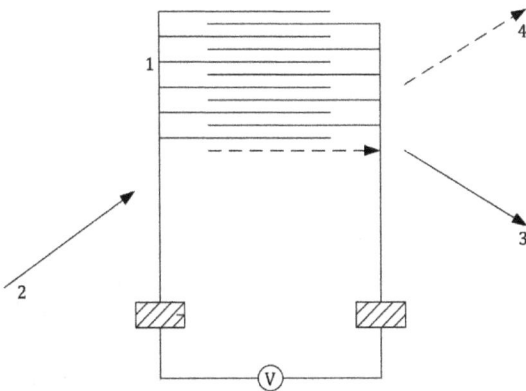

Fig. 5.61: The schematic diagram of electro-optic waveguide grating: V —direct voltage; 1—interdigital electrode; 2—incident beam; 3—the diffraction beam when the grating is switched on(the electric field is applied); 4—the diffraction beam when the grating is switched off.

The acousto-optic grating is another switchable gating whose schematic diagram is shown in Fig. 5.62. It is a more versatile device, which can steer a beam as well as modulating it. Like the electro-optic grating, its periodic electrode structure is placed on the surface of the waveguide fabricated in a piezo-electric or acousto- optic material. However, in some cases, non-piezo electric materials (e.g., Si) can also be used if a layer of piezo-electric material (typically ZnO) is deposited between the electrodes and the substrate.

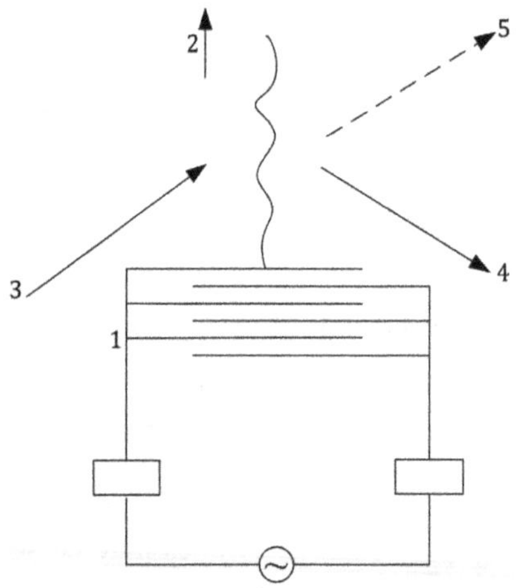

Fig. 5.62: The schematic diagram of acousto-optic grating—alternating voltage: 1—interdigital transducer; 2—traveling acoustic wavel; 3—incident wave; 4—the diffracted wave when the grating is switched on (the electric field is applied); 5—the transmitted wave when the grating is switched off (the electric field is not applied).

The electrode structure known as an interdigital transducer does not itself act as the grating. However, when it is excited by an AC radio-frequency voltage, it will create a time-varying, spatially-periodic electric field in the material beneath the interdigital transducer, which can induce a corresponding stress distribution through the piezo-electric effect. This kind of stress distribution corresponds to the standing acoustic wave, which can be decomposed into the sum of two travelling waves. Therefore, the final result is creating two acoustic waves which emerge from beneath the transducer and travel in opposite directions. Because the waves propagate near the surface of the material, they are called as surface acoustic waves. In general, only one wave is required, and the other is removed by using a surface absorber. The remaining wave can induce a variation in effective index through the acousto-optic effect, which acts as the grating. Similar to the electro-optic grating, the acousto-optic grating can be switched on and off, which also can act as a modulator. Its speed is approximately 1 GHz which is mainly limited by the acoustic propagation losses. Moreover, the grating period may be varied in a certain range by changing the RF frequency, and the variation range depends on the transducer bandwidth. While the variation of grating period can also alters the diffraction angle, so the acousto-optic grating can be used as the beam deflector. Compared with electro-optic grating, one more important characteristic of acousto-optic grating is that the diffraction orders are all frequency-shifted via the Doppler effect. So the acousto-optic grating may also be used as a frequency modulator.

5.9.9 Grating couplers

Another function of grating is realizing the phase matching of waveguide modes and radiation modes, forming a similar coupled device as the prism coupler, this device is called as the grating coupler.

Although the function of grating coupler is similar to that of prism coupler, the former have many important advantages. First, the grating coupler is small, flat, solid, and fully integrated with the waveguide. Secondly, it is possible to design grating couplers with a varying periodicity, which can have multiple functions, for example, they can couple a guided beam into free space, and focus it at the same time. Figure 5.63 shows the working principle of the chirped grating coupler, which contains a grating with linearly varying period. A guided mode entering into the coupler will be diffracted into a free-space beam, whose direction of propagation depends on the local grating period. In this case, the output will be a cylindrical wave which converges on a focus line above the waveguide. However, by chirping the grating period simultaneously or curving the grating fringes, an output which converges at the focus can be produced which is important in some applications.

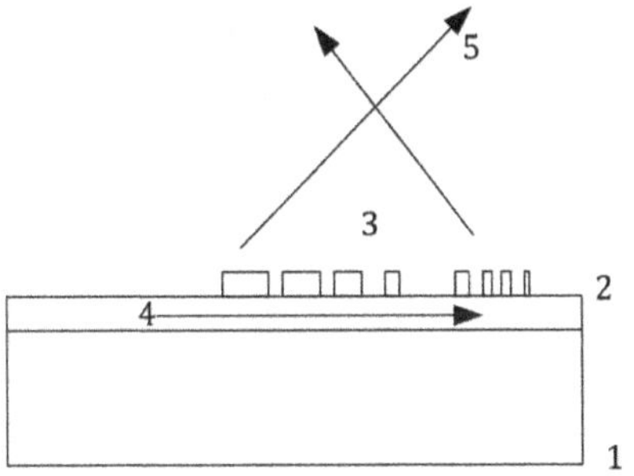

Fig. 5.63: Linear chirped grating coupler: 1—substrate; 2—waveguide; 3—grating; 4—waveguide mode; 5—diffraction focused beam.

5.10 Semiconductor waveguide device

Although the contents discussed in earlier subsections are also applicable to the semiconductor waveguide, but semiconductor waveguide has its own characteristics in both manufacturing process and device characteristics, so in this subsection we intends to focus on it. First, the semiconductor passive waveguide device is described, and then the semiconductor electro-optic conductor modulator is introduced. Finally, a brief introduction is given to the semiconductor integrated circuit.

5.10.1 Semiconductor passive waveguide

In this subsection, we will introduce the semiconductor channel waveguide devices, semiconductor coupled-waveguide devices, and, the bending and bifurcation of semiconductor waveguides.

5.10.1.1 Channel waveguide

Some of the early GaAs channel waveguides are manufactured by the ion bombardment method, if the ion bombardment method is combined with photolithography, we can also produce the ridge GaAs waveguide, CdTe waveguide and ZnTe waveguide. However, the ion bombardment method will cause serious optical loss while decreasing the carrier concentration and increasing the refractive index of the material. Although this loss can be reduced by properly annealing the material after bombardment, it is still difficult to make the channel waveguide meet the requirements in most modern applications.

To solve the above problems, by combining epitaxial growth and photolithography, the homogenous rib channel waveguide manufacturing technology based on the material of III-V semiconductor has been widely studied. The basic principle of this method is shown in Fig. 5.64, first, the epitaxial layer with low carrier concentration is grown epitaxially on the substrate, and then the excess material is removed by using photoetching method to get ridge channel waveguide. The epitaxial growth is realized by LPE or VPE method, according to early reports, for the light of $1.06 \mu m$ wavelength, the loss of GaAs single mode homojunction ridge waveguide manufactured by VPE technology is no more than $1 cm^{-1} (4 dB \cdot cm^{-1})$; similarly, the loss range of GaAs waveguide depending on LPE technology and InP waveguide utilizing VPE technology is 6~8 $dB \cdot cm^{-1}$. Later, the loss can be further reduced by using more advanced metal organic chemical vapor deposition technology (MOCVD), MBE method and heterojunction structure. It is said that currently the lowest optical loss of GaAs/GaAlAs waveguide can be produced by this method.

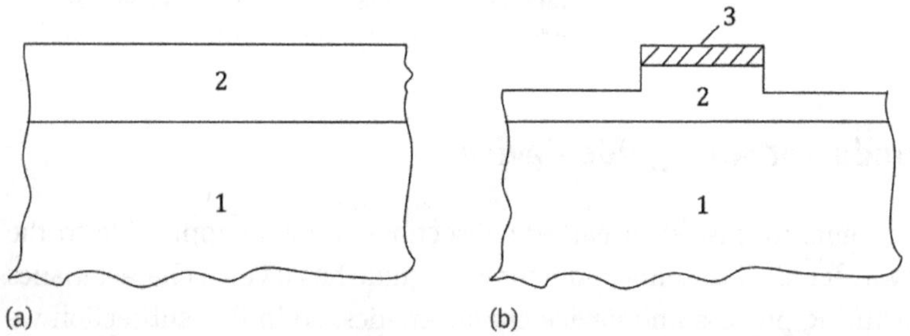

Fig. 5.64: The fabrication of semiconductor ridge waveguide: GaAs (a) epitaxial growth of low carrier concentration layer (b) ridge etched by proper mask. 1–High concentration carrier region; 2–low concentration carrier region; 3–mask.

During this period, the deep-etched single-mode homojunction ridge waveguide can be made in the materials of GaAs and InP. The reproducibility of such kind waveguide is quite poor because of its dependence on the substrate carrier concentration, the size and shape of ridge, especially the width of the bottom of the ridge. In addition, its loss is higher than that of shallow-etched ridge waveguide, which may be induced by the scattering from the side of ridge.

To apply the electric field to the ridge waveguide, we can adopt Schottky barrier contactor, metal-oxide-semiconductor (MOS) type or p-n junction type contactor. However, the Schottky barrier contactor will produce interference and increase the loss of waveguide if it is placed directly on the waveguide layer, and the MOS type contactor placed on GaAs can cause the DC drift, which is only suitable for a transient application; by contrast, the p-n junction type contactor is suitable for GaAs waveguide and InP waveguide.

For the homojunction ridge waveguide, the lower limit of optical loss depends on the loss of free carriers in the substrate. For example, for GaAs waveguides, if the wavelength is 1.3 μm, the typical value of this loss is 2~4 $dB \cdot cm^{-1}$, which can be reduced by replacing the substrate.

One approach that has been successful is to add a SiO_2 layer, as shown in Fig. 5.65, the undoped GaAs layer is usually grown on a SiO_2 layer by a lateral epitaxial growth technique, and the growth source is the GaAs below. The orientation of the SiO_2 strip must make the lateral growth rate be much higher than the vertical growth rate, which is generally determined by the experimental direction. For example, in a (100)-oriented GaAs substrate, the angle between SiO_2 strip and the cleavage plane of material must be 10° to achieve a good lateral growth, which limits the orientation of the waveguide to a certain extent. To provide lateral constraints, the rib can be etched on the growth layer. For this kind of waveguide, the loss of light with the wavelength of 1.06 μm is 2 $dB \cdot cm^{-1}$, which is nearly half of that of the waveguide directly grown on the substrate.

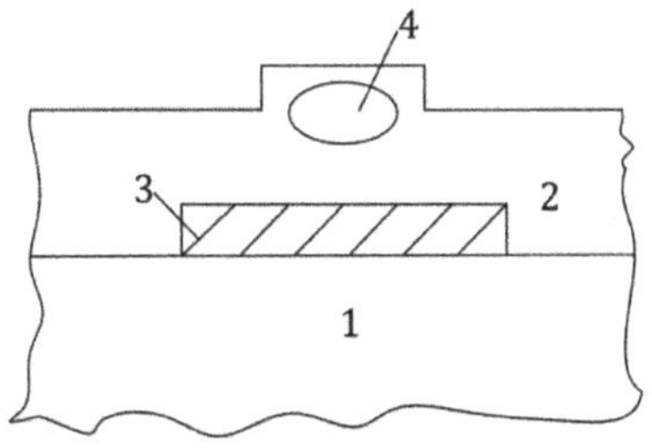

Fig. 5.65: GaAs ridge waveguide grown on SiO_2: 1–high concentration carrier region; 2–low concentration carrier region; 3–SiO2 layer; 4–light.

Based on the advantages of MBE and MOCVD growth methods, the method to further reduce the waveguide optical loss is to use a AlGaAs constrained layer with a lower loss coefficient (as shown in Fig. 5.66). The depth of the AlGaAs layer is usually larger than the attenuation length of the optical mode in the region, so the boundary wave

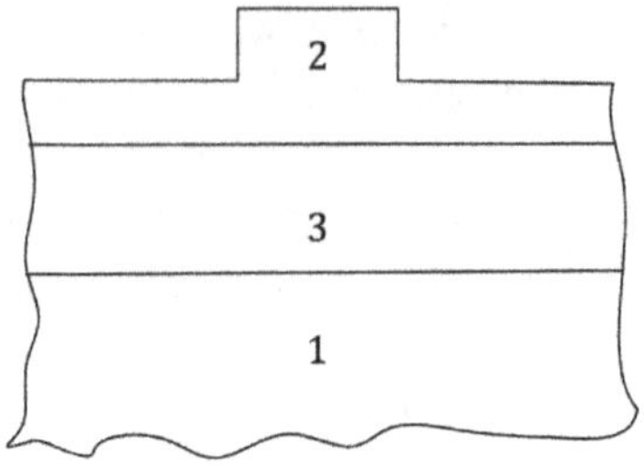

Fig. 5.66: GaAs/AlGaAs heterojunction ridge waveguide: 1–high concentration carrier region; 2–low concentration carrier region; 3–AlGaAs.

between the layer and the GaAs substrate can be neglected, otherwise the guided wave mode will be coupled to the GaAs substrate. The kind of heterojunction waveguide with less than 1 $dB\ cm^{-1}$ loss has been fabricated based on MOCVD method.

Another example is the InGaAsP/InP ridge waveguide with heterojunction structure, the loss rate of single-mode InGaAsP/InP waveguides fabricated by LPE is approximately 7.3 $dB\ cm^{-1}$; and for a light with the wavelengths of 1.15 μm, the loss rate of the same waveguide grown by MOCVD method is 11.5 $dB\ cm^{-1}$.

The most promising waveguide for integrated optical circuits may be embedded heterostructure, which has a lower loss and excellent optical confinement performance, and is compatible with heterojunction semiconductor lasers and detectors. As an example, Fig. 5.67 shows the InGaAsP/InP embedded heterostructure waveguide, the undoped InP buffer layer, InGaAsP waveguide layer and InP layer with high p-type carrier concentration are grown on the InP layer with high n-type carrier concentration successively. The InP thin layer above can be considered as the mask when the InGaAsP waveguide layer is being etched. By controlling the composition and size of the four layers, the lateral confinement characteristics can be controlled, and a single mode waveguide can be fabricated.

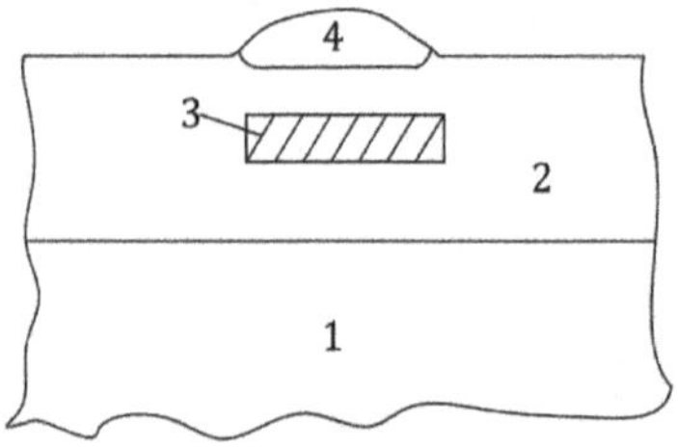

Fig. 5.67: GaInAsP/InP embedded heterostructure waveguide: 1–high n-type carrier concentration region; 2– undoped InP; 3–GaInAsP; 4 InP with high P-type carrier concentration.

Similar waveguide structures based on AlGaAS/GaAs are also fabricated successfully. If AlGaAs is exposed to air, there may be some problems with the epitaxial layers growth; however, if the entire manufacturing process is carried out in a high-vacuum environment, there will be no difficulties.

In III-V semiconductor waveguide, it is noteworthy that residual deformation produced during the process of fabricating a heterojunction layer may lead to residual birefringence, which is caused by the photoelastic effect of material. $(N_{TE} - N_{TM})$ is of the typical order of 10^{-4}. Due to the influence of internal electric field, the birefringence phenomenon may be more obvious in the modulator structure, which will be introduced in the next subsection.

Other waveguides based on semiconductors include the structures fabricated by multilayer insulating materials being grown on Si or other semiconductor substrates. An example is shown in Fig. 5.68, first the SiO_2 transition layer with low refractive index is grown on a Si substrate, and then the Si_3N_4 waveguide layer with high refractive index is grown on the SiO_2 transition layer. The loss of the device shown in Fig. 5.68 can be as low as 0.1 $dB \cdot cm^{-1}$.

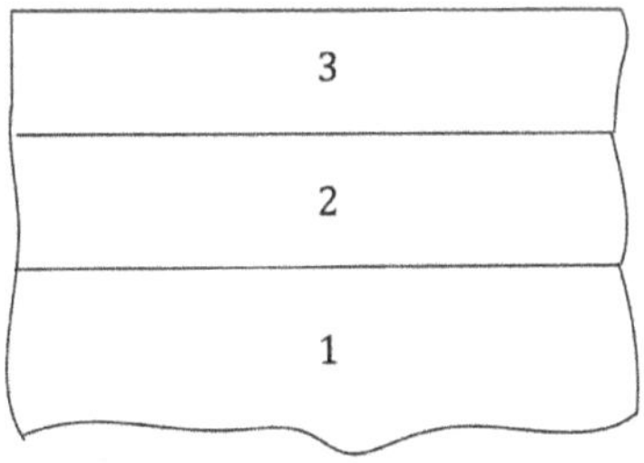

Fig. 5.68: Bilayer dielectric waveguide on Si substrate: 1–Si; 2–SiO_2; 3–Si_3N_4 2. Semiconductor waveguide coupler

Optocoupler is a very important component for large scale integrated optical circuit. Various types of dual waveguide and multi waveguide couplers have been extensively studied and are recommended to be used for the improvement of sampling and filtering techniques, power division and recombination, and the integration of inputs and outputs of interferometer. In addition, the coupled laser array, which is composed of many parallel stripe lasers, has attracted great interest.

The optocoupler consists of two or more waveguides whose fields overlap because of being so close to each other. At present, there are plenty of waveguides being composed of two single-mode ridge waveguides. Figure 5.69 shows an example of a buried heterostructure waveguide coupler, the waveguides constituting the coupler can be placed side by side as shown in figure (a) and can be placed up and down as shown in figure (b).

Figure 5.70 shows the GaAs $P^+ - n^- - n^+$ channel waveguide directional coupler, here P^+/n^+ denotes the high p-type and n-type carrier concentration region respectively. n^- denotes the low n-type carrier concentration region. The width of waveguide is 8 μm, the distance is 4 μm, and the thickness of P^+ is 2 μm, for the different depth of the epitaxial layer t, the change of coupling length with the wavelength L_C are given in Fig. 5.71. In all cases, the coupling length decreases with the increase of wavelength, which can be also observed in the ridge waveguide coupler.

Once the coupling length is known, for the desired wavelength, it is possible to fabricate a directional coupler with any desired power coupling rate in principle.

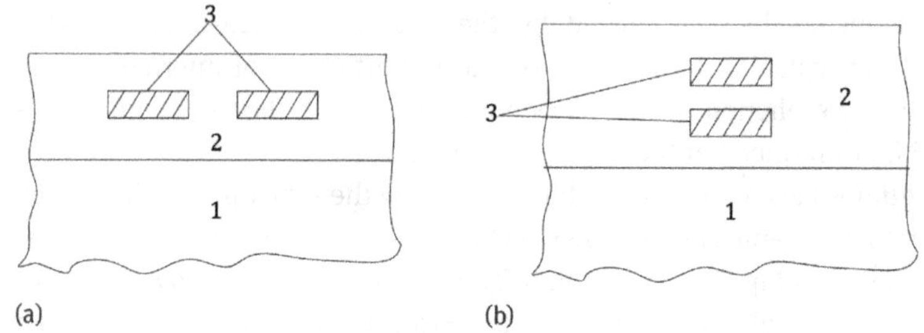

Fig. 5.69: Buried waveguide coupler: GAlnAsP/InP (a) two waveguides placed side by side; (b) two waveguides placed up and down. 1–High carrier concentration region; 2–low carrier concentration region; 3–GalnAsP

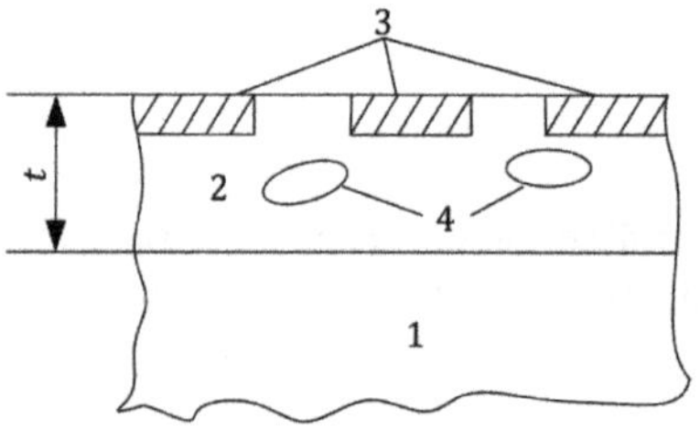

Fig. 5.70: GaAs channel waveguide coupler: 1–high carrier concentration region; 2–low carrier concentration region; 3–p-type carrier region; 4–light

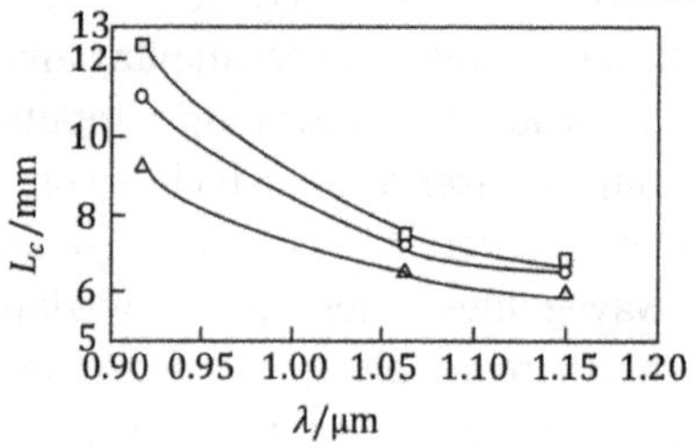

Fig. 5.71: The change of coupling length with the wavelength: _t = 5.8μm; o_t = 4.8μm; Δ_t = 4.5μm.

However, there will be no perfect coupling between the input and output waveguides, the biggest extinction ratio is usually less than 20 dB. In addition, it is difficult to make the waveguide length equal to the coupling length, so it is often necessary to provide an appropriate combination of $\Delta\beta$ to adjust the effective coupling length to equal to the actual interaction length.

5.10.1.2 The turning, bending, and bifurcation of waveguide

In many applications, it is desirable to change the direction of light propagation, and therefore, bending and bifurcation waveguides are very important devices.

Curved waveguides either have smooth shape shown in Fig. 5.72(a), or have the shape with a series of turning sections shown in Fig. 5.720(b). Just as all optical

(a)

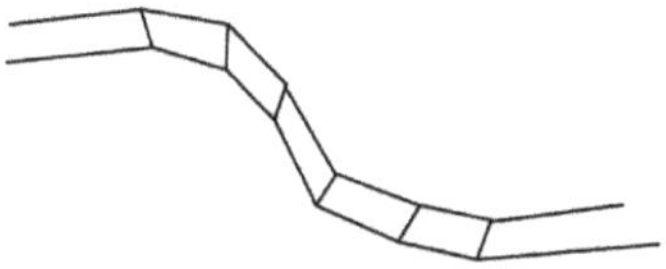
(b)

Fig. 5.72: The shape of curved waveguide. (a) Smooth curved waveguide; (b) curved waveguide with multi turning sections.

waveguide propagation, there is a certain amount of power loss into the radiation mode at the turning position. Of course there will be reflection of some power light, which can usually be ignored unless the effective refractive index at the turning position has obvious sudden change. In some cases, if the length and angle of each segment are chosen appropriately, it is also possible that the radiation mode is coupled into the waveguide in a backward coherent manner.

To achieve the resonance turning angle and the radius of curvature of the waveguide under the condition that the loss is minimal, the transverse confinement of the waveguide must be carefully controlled. For most applications, the transverse confinement of ridge waveguide is generally poor, however, the deep etched ridge waveguide is an exception, whose transverse confinement is better. So, a lot of turning and Y-bifurcation structure are fabricated by this kind of waveguide. In the Y-bifurcation waveguide, the loss does not decrease significantly with the decrease of the bifurcation angle, which indicates that there is an inherent mode matching loss at the top of the Y in the Y-bifurcation deep etched waveguide.

In the semiconductor integrated optics circuit, a more practical method for fabricating the turning and bifurcation device is using the embedded heterogeneous junction waveguide. In this case, the waveguide parameters can be more flexible and the scattering loss is also expected to be lower. Figure 5.73(a) shows a GaInAsP/InP embedded heterojunction Y-bifurcation waveguide structure, and Fig. 5.73(b) shows the change of the transmittance T with the bifurcation angle θ. Where T denotes the ratio of the output power of the two-bifurcation to the input power of the straight-section. The depth of $Ga_{0.17}In_{0.83}As_{0.4}P_{0.6}$ waveguide region is approximately 0.2 μm and the width is approximately 2 μm. As can be seen from Fig. 5.73, it is really possible to obtain a transmittance greater than 90% (i.e., the loss is approximately 1 dB) by using the embedded heterostructure waveguide.

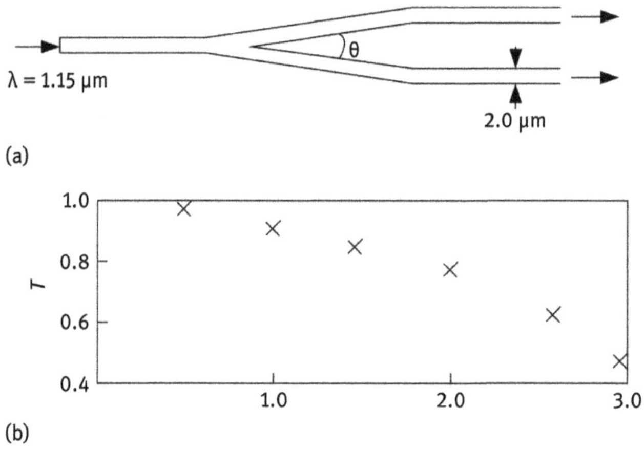

Fig. 5.73: GaInAsP/InP Y-bifurcation waveguide. (a) Waveguide structure; (b) the variation of T with θ.

5.10.2 Electro-optic waveguide modulator

5.10.2.1 Phase modulator

The first reported (1963) waveguide modulator is used to reverse the bias voltage of the GaP diode, and its modulation mechanism is the linear electro-optic effect. The phase modulation based on the linear electro-optic effect can be up to $18^{\circ} V^{-1}.mm^{-1}$ in GaAs/GaAlAs planar waveguide.

At zero bias, the not completely empty heterojunction phase modulator with the depth of submicron has a very good phase modulation performance ($56^{\circ} V^{-1}.mm^{-1}$), and is suitable for coupling chip and integrated laser device, whose bandwidth is from a few GHz to dozens of GHz.

The GaAs planar waveguide mainly used for modulating the traveling waves has been reported, which can be used to realize the frequency shift of 10.6 μm laser. Its power conversion efficiency is up to 2%, and the optical power is approximately 10 W, furthermore, the conversion efficiency can be further improved if a thinner waveguide layer with a high resistance cladding is employed.

5.10.2.2 Directional coupler switch

As previously mentioned, directional coupler can be composed of two closely-spaced waveguide, if we modulate the propagation constant of one to another to destroy their synchronization, we will obtain the coupler switch.

The early coupler switch is composed of a pair of coupled planar waveguides, and the first reported GaAs channel waveguide switch with the extinction ratio of 13 dB used a metal gap structure and required a 35 V driving voltage. Another types of switches include a rib structure, or photoelastic confinement device etc. The first high extinction ratio device can get the extinction ratio of 25 dB, which has a metal gap structure and uses $\Delta\beta$ to reverse each electrode having independent bias voltage.

5.10.2.3 Interference modulator

The change of propagation constant required from the interferometric modulator used as a switch is $\sqrt{3}$ times smaller than that required from the directional coupler switch mentioned above. Moreover, the change of output light intensity with the external electric field is periodic, these above characteristics make the interferometer be attractive in many applications. A GaAs interferometric modulator fabricated by the Y-bifurcation coupler or three-core waveguide coupler has been used for power splitting, which has an extinction ratio of approximately 20 dB and a flat bandwidth response.

5.10.2.4 Electro-absorption modulator

One kind of the main electro-absorption modulator is the short waveguide homojunction or heterojunction device with Schottky electrode. The extinction ratio and the bandwidth of GaAsInP modulator is approximately 20 dB and approximately 1.6 GHz, respectively, while the GaAlAs/GaAs heterostructure has an extinction ratio of 30 dB and a bandwidth of 3 GHz. The driving voltages of above devices are less than 10 V, and the insertion losses of them are approximately 10 dB.

It should be pointed out that the electric-absorption can also enhance the quadratic electro-optical effect, which has been confirmed in the material of GaInAsP. In addition, the electric-absorption of GaInAsP has the obvious bidirectional feature, that is, the attenuation of TM mode activated by a given electric field exceeds that of TE mode, this effect has also been observed in GaAs device and can be attributed to the bidirectional feature of enhanced tunneling field effect induced by the rise of the degeneracy of valence band activated by electric field. A further discussion of this issue is beyond the scope of this book.

The electro-absorption array has broad application prospects in signal processing. For example, the CCD array fabricated on a planar waveguide can realize the spatial modulation of light, when we combine it with waveguide lens and modulated laser, this kind of structure can realize the functions of Fourier transform and correlator.

In addition to the waveguide modulators described above, the main waveguide modulators also include multiple quantum well structures and devices based on nonlinear optics. The discussion of their working principles are beyond the scope of this book, readers who are interested in them need to read other related literature.

5.10.3 Optoelectronic integrated circuit

Integrated optoelectronics is very close to the semiconductor integrated optics, which utilizes the electronics technology to integrate a large number of optoelectronic components on the same substrate. For the study of integrated optoelectronics, the long-term goal is to realize the integrated circuit with complex structure and comprehensive functions; and the recent study direction is to fabricate the integrated transmitting devices, receiving devices, conversion devices, addressable LED, laser arrays, detector arrays, laser/waveguide and detector/waveguide structures by using GaAs and InP.

The basic problem of optoelectronic integration is how to solve the contradiction between optical components and electronic components on the aspects of material composition, film thickness, and heat emission, which makes the high resolution lithography become quite difficult, and may require developing the selective epitaxial growth techniques and fabricating the waveguide devices on the lithography groove of the substrate.

In addition, another problem of the fabrication of integrated optoelectronic receivers is that the PIN field effect transistor (FET) has the highest sensitivity, even if it has the minimum equivalent noise dark current. On the one hand, we hope that the mutual conductance of FET is maximal; on the other hand, we also require that the sum of capacitance of all components including photodiode, FET and interconnection lines is minimal, which is a difficult task.

An alternative is to use a photoconductive detector and preamplifier, which has the inherent advantage of being more compatible with the fabrication technology of FET. However, due to the lifetime of minority carrier, the response speed of the device is generally low, and its sensitivity is poor due to the increase of dark current. Nevertheless, it is attractive for some applications that require less sensitivity because it is easy to be fabricated.

For the fabrication of waveguide lasers, one of the difficulties is the reflection mirror. The original device used a separate structure, the length of the chip is equal to the length of the laser cavity. Later, a more complex loop with the non-separable mirrors was developed, in which the two mirrors were not placed at the edge of the chip and can be placed in any position on the surface of the chip. The GaInAsP laser array using this technology has been fabricated successfully.

As early as the mid 80s, an integrated laser device using an advanced quantum well structure with low threshold and multi transistor drive circuits was reported. The different layers of the chip grown by MBE method are used for the electronic part and the optical part respectively. The laser device has a reflector fabricated by dry-etching and a threshold current of 40 mA.

An important problem of the development of the laser waveguide circuit is to keep it having the low dislocation density, so that we can prolong the lifetime of the lasers. For example, GaAs lasers with long life require that the dislocation of substrate is fewer than $10^3 cm^{-2}$. However, the higher dislocation density is more favorable for the uniformity of the transistor threshold. In recent years, with the development of optoelectronic technology, people have got the high quality electronic circuits based on the substrate with high-impedance and low dislocation density, so the photoelectric integrated system can be optimized.

5.11 Application examples of optical waveguide

The previous subsections have discussed some typical optical waveguide devices. In this subsection, we will briefly introduce these devices and some application examples of optical waveguide systems composed of them.

5.11.1 The planar integrated optic RF spectrum analyzer

The planar integrated optic radio frequency (RF) spectrum analyzer is a kind of single waveguide system. The function of this system is to perform a real-time, parallel, spectral analysis of RF signals. Such a system, which can be used to monitor enemy radio or radar signals, might form a key component in an aircraft-based electronic countermeasures system. This operation needs a Fourier transformation of the received radio signal to be carried out. Although electronics can be used to actualize Fourier transformation, lens has the ability to perform a Fourier transformation on a spatial distribution of light. So it is more straightforward to use the optical method. When the optical method is selected, the advantages of integrated optics are lightweight, small size and rugged package. Once the system is aligned before package, it won't be misaligned with the influence of shocks and vibration (e.g., during takeoff and landing).

Figure 5.74 shows an example of the mentioned system, which consists of a planar integrated optic chip integrated with prism coupler, grating beam expander, acoustic-optic modulator, Fresnel lens array, etc. The substrate of the chip is fabricated with some kinds of piezoelectric, acoustic-optic materials such as Ti:$LiNbO_3$ waveguide.

The system mentioned in Fig. 5.74 works as follows. First, the light of an external laser is guided into the waveguide by the prism coupler. At this moment, the beam is relatively narrow. Second, the beam is passed to a corrugated Bragg grating, which has a rectangular boundary and whose fringes are oriented at 45° to the input beam direction. The grating is used to expand the beam cross-section and to deflect it through a right-angle. The expanded beam then has an interaction with a travelling surface acoustic wave that is excited when the RF signals pass through the interdigital transducer on the piezoelectric material. As a result, there are two waves propagating on the surface of the substrate: the guided optical wave and the surface

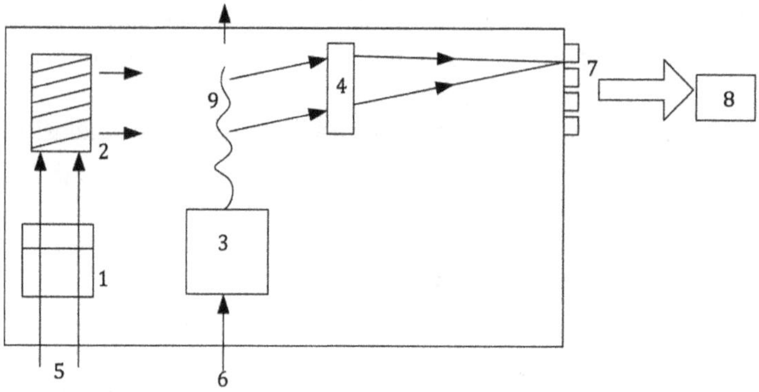

Fig. 5.74: The planar integrated optic RF spectrum analyzer.

acoustic wave (SAW). The SAW acts as the travelling phase grating to diffract the beam. The wavelength of the grating which is created by the SAW is rather long, so its diffraction takes place in the Raman regime and has many diffraction orders which is not what we want. However, the deflection angle of the first diffraction order, which is seen as the most important order, is approximately linear with the RF frequency. For simplicity, we choose to ignore all the higher diffraction orders in the discussion, and consider the first diffraction order as the diffracted beam. Under the agreement, we can say the diffracted beam is linear with the RF frequency.
An overlay Fresnel lens is used to focus the diffracted beam to a point somewhere on the focal plane. Because the lens can actualize the Fourier transformation of the optical amplitude, the distance between the focus and the optical axis is also linear with the deflection angle. An external, linearly-spaced array of photodiodes is placed at the focal plane, so when the RF drive is tuned to the correct frequency, each diode can receive a signal. Thus, the output of the detector can represent the power spectrum of the RF input, which is then passed to the electronics for further processing. Nowadays, it is reported that the typical performances of the integrated optic RF spectrum analyzer are 200–500 MHz bandwidth, 4–8 MHz resolution, 2 μs response time and dynamic ranges greater than 20 dB.

5.11.2 The waveguide chip connector

This is an example of guide wave optics used in very large scale integrated (VLSI) electronics. As the working speed and complexity of the circuitry rise further, the connections, among the integrated circuits and between the chips will encounter serious problems. They are mainly caused by two factors. The first factor is the limitation on the signal bandwidth caused by the RC time constants in the circuit. This limitation makes the propagation delays become more significant than the gate delays. The second factor is the interference of the stray capacitances, which becomes more significant with the improvement of the integrated degree.

Because of the above reasons, we suggest to use the optical interconnects to take the place of the electronic links. Optical interconnects involve the connections between free spaces as well as optical waveguides. Figure 5.75 shows a sketch of the connector which was reported in the late 1980s and is based on channel waveguide technology. It shows that four LSI chips on the same substrate are connected by the waveguide. This connector system is a hybrid and its optical part consists of semiconductor laser diodes (LD) as which the InGaAsP devices with the operation wavelength of 1.3 μm are extensively used recently, a multimode ridge waveguide circuit and photodiodes (PD). The waveguide circuit is composed of four guides and every guide has the function of emitting as well as receiving signals. Partial feedback signals from the mixer are tapped by the angled mirrors and then are passed to the PD receptor. Waveguides are fabricated with the high refractive index materials such as $SiO_2{\cdot}TiO_2$ and a thick SiO_2 buffer layer is used to separate the guides from the Si substrate. In this example, the transmission capacity tested is 1 Gbit/s. Although it is an early result, it has clearly showed the potential of these devices.

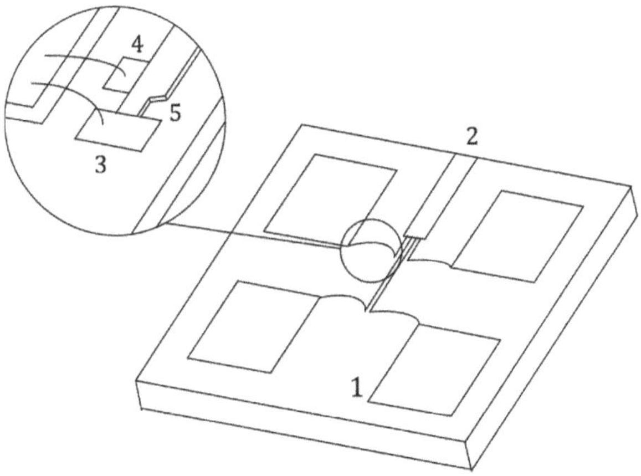

Fig. 5.75: The optical waveguide chip-to-chip connector circuit.

5.11.3 The channel waveguide A/D converter

The operation speed of most universal algorithms is fairly slow, so the quick analog-to-digital (A/D) conversion is a more difficult mission in electronics. In electronics, the operation speed of the parallel A/D converters is faster, however, they are so expensive that it is commonly consider that optical A/D converters have a better prospect.

Optical A/D converter is based on Mach–Zehnder interferometer with the feature of periodic response. We have introduced this kind of interferometer in the above section that it is composed of two Y-branch waveguides, linked by straight waveguides. The incident light is split into two components at the first Y-branch and then they pass along the straight guides. When arriving at the second Y-branch, these two components are recombined. Either or both arms of the interferometer have surface phase-shifting electrodes, which allow the relative phase of the recombining components to be adjusted as required. If the two components are in-phase, the output is high, and on the contrary, the output is low. More generally, as shown in Fig. 5.76(a), the output R_m changes periodically with the voltage V of the electrodes. However, if an appropriate static phase shifter is used to bias the interferometer, the sinusoidal response will be shifted by a quarter of a period. In this situation, if the optical output R_S is detected, and then use an electronic comparator whose threshold value T is set at a level of half the peak output, the comparator output will vary as a square-wave with voltage, as shown in Fig. 5.76(b). The latter corresponds to a 1-bit binary representation of the applied voltage which realizes the A/D conversion.

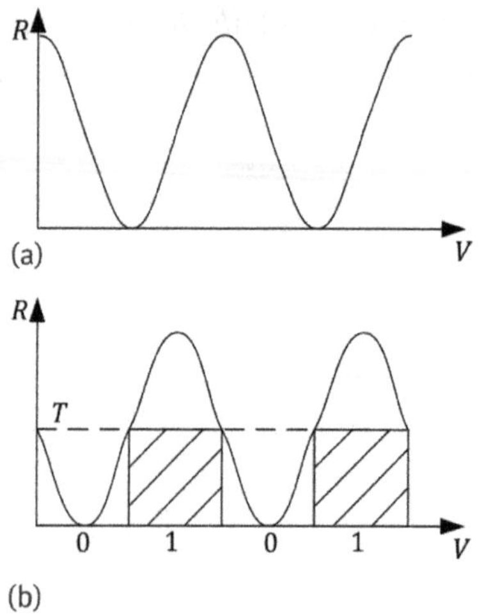

Fig. 5.76: The output of the Mach–Zehnder interferometer variation with electrode voltage. (a) Basic type (b) Biased type.

Moreover, the output of the Mach–Zehnder interferometer depends on electrode length. For example, the sensitivity of the device with short electrode length is lower than that with long electrode length. Figure 5.77 shows the output characteristics of two similar interferometers with the electrode lengths of L and $L/2$. Clearly, both of the responses are periodic with voltage, but the sensitivity of the interferometer whose electrode length is L is twice as the other one. If the additional processing described above, static phase-shifting, detection and thresholding, is applied to each interferometer, the output of the comparator will all vary as square waves, but the period of the interferometer with electrode length of L is half as the other one. Taken together, they yield a two-bit binary representation of the drive voltage.

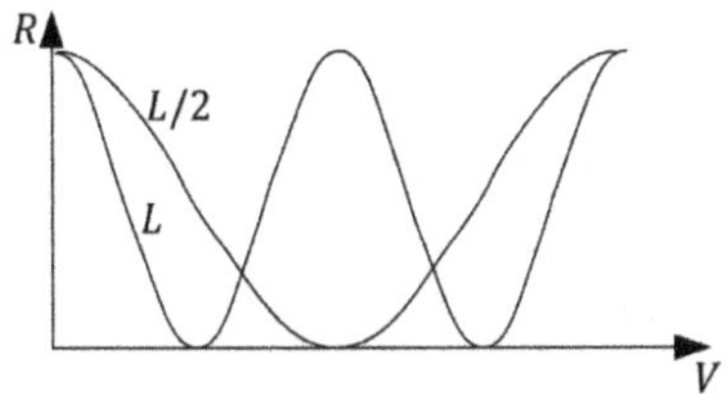

Fig. 5.77: Responses of two Mach–Zehnder interferometers with the electrode lengths of L and $L/2$.

5.11.4 Guided-wave optical communication

A main application of guided-wave optics is high speed optical communications. Among all the optical communication technologies, the point to point communication, which has already reached a very high level, can be used to transmit information at very high bit rates over extremely long distance. However, the single users cannot occupy all the bandwidth of the system and as a result, point to point communication is a waste of the guided wave optical transmission circuit. In fact, to avoid the waste, the system is usually shared by lots of users, or multiplexed. The two major schemes used are time division multiplexing (TDM) and wavelength division multiplexing (WDM). In the TDM scheme, every user can only occupy a set fraction of the available time, and in the other scheme, user is assigned with a specific wavelength interval to communicate. Next, we will simply introduce TDM and WDM.

5.11.4.1 Time division multiplexing

The operational principle of TDM is shown in Fig. 5.78. Here, the N users all communicate on the same carrier wavelength. However, they cannot use the shared channel at will; instead, they are allocated with specific time slots. In Fig. 5.78, channel 1 is assigned to communicate at the time slot centered on t_1, and channel 2 is assigned to that on t_2, etc. Information from all the N users is combined together in chronological order by the multiplexer (MUX), and then is transmitted through the shared channel. On the other side of the channel, the signals are again separated by the de-multiplexer (DEMUX).

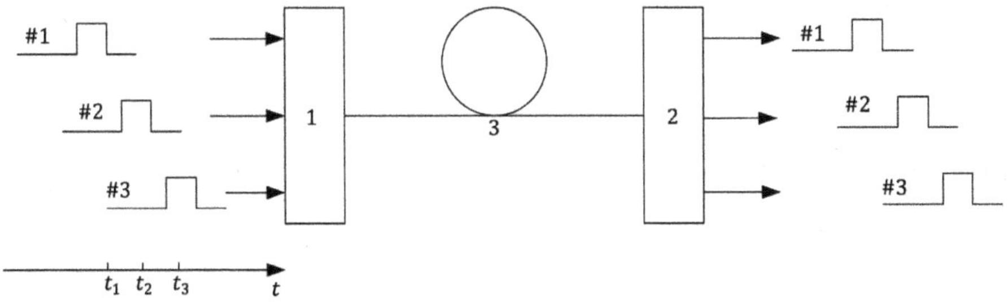

Fig. 5.78: Operational principle of optical TDM communication system.

5.11.4.2 Wavelength division multiplexing

Figure 5.79 shows the operational principle of another kind of multiplex communication system which is known as the WDM communication system. First, a set of N users are assigned with wavelengths λ_1, λ_2 ... λ_N to communicate. These signals may be emitted by a set of lasers with different wavelengths. Every user can independently modulate his source so that the light with wavelength λ_n becomes the carrier of information for channel n. Information from all of N channels is combined together by the MUX, and then passed to a shared channel for transmission. On the other side of the channel, the signals are separated by the DEMUX and passed to their final destinations individually.

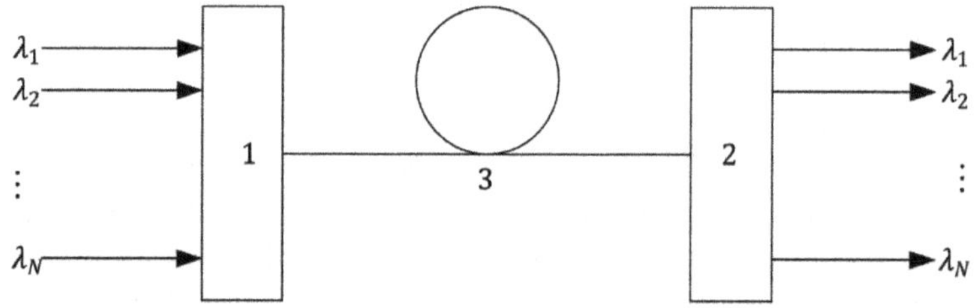

Fig. 5.79: Operational principle of optical WDM communication system.

A set of discrete optical filters can be arranged in series to form the MUX. For example, a set of asymmetric directional coupler filters might be used as the basic combiner element or we can use mode-conversion filters as well as Bragg gratings. Different kinds of filters have different response peak wavelengths but every filter couples its maximum power of response peak wavelength to shared channel. By this way, all the N channels can be combined into a single output without loss of power. The DEMUX is a similar network as MUX, operating backwards.

The advantage of WDM communication system is that different users can use the shared channel meanwhile so the total bandwidth of WDM is N times of the discrete channel bandwidth. As a result, WDM system has higher communication efficiency. However, the main disadvantage of WDM is that when there are too many channels and their central wavelengths are close to each other, it will be a very difficult mission to fabricate the MUX and the DEMUX.

5.12 Introduction of MOEMS

5.12.1 Introduction

In the last ten years or so, micro-optical elements, such as diffractive microlens and refractive microlens, have received great attentions in the development of optical systems. It has been proved that these elements have many functions such as focusing, improving optical efficiency, developing digital image technology, separating color, reshaping and transforming beams, etc. In the first-generation miniaturized optical system, micro-optical elements are combined with electrical circuits and sometimes with motor elements (such as scanning mirrors and piezoelectric actuators). The development of micromachine and microelectric motor technologies promotes the progress of microelectronics technology and as a result, miniaturized optical system is developed to the extent of nearly monolithic integration. The combination of micro-optics, microelectronics and micromachine forms a kind of brand new and wide-ranging micro-opto-electro-mechanical system (MOMES) such as laser scanner, adaptive mirror, digital micromirror device (DMD), fiber distributed data interface (FDDI), three-dimensional adjustable Fabry–Perot etalon, optical shutter, MEM switch, optical interconnection, data storage device, MEM corner cube mirror and etc.

In this section, we will first introduce diffractive microlens and refractive microlens, and then, MOEM system will be briefly introduced. Finally, we will show optical scan device as a typical example of this kind of system.

5.12.2 The diffractive microlens

The first breakthrough of micro-optics in optical processing field is the development of the diffractive micro lens. Diffractive microlens is one kind of micro-optical element whose diameter ranges from tens microns to hundreds microns and thickness is only at wavelength magnitude. If the diffractive micro lens is made of materials with high refractive index such as silicon and GaAs, its speed can reach $f/0.3$ (f is the focal distance). Even if it is made of quartz or glass matrix, its speed can also reach $f/1$.

Diffractive microlens is the approximation of Kiron lens and can be fabricated with binary optical method by multilayered lithography. The lithography depth of the mth layer is

$$d(m) = \frac{\lambda}{(n-1)2^m} \tag{5.195}$$

where λ is the wavelength of free space light; n is the refractive index of lens material. When the incident light energy is diffracted, the efficiency of the first level is

$$\eta = \left(\frac{\sin(\pi/2^m)}{(\pi/2^m)}\right)^2 \tag{5.196}$$

Equation (5.196) shows that the diffractive efficiency increases with the increase of m and when m is large enough, then $\eta \rightarrow 1$. For example, when $m = 2$,

$$\eta = \left(\frac{\sqrt{2}/2}{(\pi/4)}\right)^2 \approx 0.81$$

When $m = 3$, then $\eta \approx 0.95$ and when $m = 4$, then $\eta \approx 0.99$. However, η is smaller than the theoretical value above because of the processing factors. For example, when $m = 3$, the measured value of η is only approximately 0.9. This kind of film microlenses with high diffractive efficiency can be used to design many complicated optical systems which cannot be achieved by lumpy optical elements. Recently, high-quality microlens array fabricated with this technology is sold all over the word and they have great attractions in the applications of many passive or active optical sensors.

In the design of binary optical microlens, there are three important optical fabrication parameters: wavelength λ, *f*-number (F) and the minimum feature size L. In the typical devices which are fabricated with three layer lithography, the relationship between these three parameters is

$$L = \frac{\lambda F}{4} \tag{5.197}$$

where

$$F = f/d$$

f is the focal distance; d is the diameter of the microlens.

In most labs, L ranges from 0.5 to 1 μm. As a result, for the light with the wavelength of 0.632 μm or 0.850 μm, the speed of the devices is limited to arbitrary value between $f/6$ and $f/3$. Nevertheless, for the light with mid infrared wavelength, we can get high speed (low F value) microlens array.

5.12.3 The refractive microlens

To some extent, refractive microlens is complementary to diffractive microlens. For example, in optical system which requires high speed response for short wavelength, diffractive optical elements are limited by processing size L according to the discussion in last section. Under these circumstances, refractive microlens is an attractive low-cost alternative.

Rockwell Corporation uses active ion etching technology on fused silica, bulk silicon, CdTe, GaAs, InP, and GaP or deposits Ge film on fused silica and Al_2O_3 to fabricate refractive microlens. The diameter of the lens ranges from 300 μm to 500 μm and the *f*-number ranges from $f/0.76$ to $f/6$. The magnification of the lens can be expressed as follows

$$M = \frac{1}{2(a/r)\left\{n\left[1-(a/r)^2\right] - \left[1-(na/r)^2\right]\right\}} \tag{5.198}$$

where n is the refractive index of material; a is the radius of the lens; r is the curvature radius of the lens. Equation (5.198) is fairly accurate, especially when it is used for high speed (low F value) microlens.

5.12.4 MOEM system

Since the end of 20th century, microfabrication technology in the field of integrated circuit and microchip has developed rapidly. The combination of tiny mechanical motion and circuit develops a new technology called micro-electro-mechanical system (MEMS). With the development of MEM filed, a lot of new subsystems are invented such as micromotor and switch, both of whose size is only millimeter range.

The structure of MEM can be fabricated with bulk materials as well as film materials. In the former case, chemical method or dry etching is used directly to carve out three-dimensional graph on the materials. In the other case, MEM structure can be formed on the surface of the substrate without carving the substrate itself.

The application of MEM technology is mostly concentrated on the design and development of the microsensor and microactuator. Microsensor and microactuator based on MEM technology have inherent capability of overall integration. They are important elements of small intelligent mechanical system and their size ranges from micron to millimeter, which makes the size of the device decrease by orders of magnitude. Moreover, because of that, the performance of the device is improved and the prize drops sharply.

MEM and micro-optical technology both have an important common characteristic that is their compatibility with the integrated circuit. This characteristic ensures that the final products can be mass-produced at a low cost. The combination of three major technical fields, namely, electronics, mechanics, and optics, requires the successful integrated optical processing. Using this new technology, micro-optical elements can be fabricated on a single silicon chip in an integrated manner. Using computer-assisted design and microfabrication technology, microdevices such as microlens array, mirror, beam splitter and grating have been fabricated on silicon substrate. It has already been proved in experiment that three-dimensional Fresnel microlens can reduce the divergence angle of beam, which is from single mode fiber and whose wavelength is 1.3 μm, from 5.0° to 0.33°.

6 Light detection and detector

Light is a signal with a very high frequency, for example, the magnitude of frequency of visible light is 10^{14} Hz. The frequency of infrared radiations with wavelength of approximately 10 μm is as high as 3×10^{13} Hz. So far, no instrument exists with such a high response speed. Thus, it is necessary to convert light to other physical quantities that can be easily detected. Because electrical signal measurement is very mature and the relationship between light and electricity is very close, it is the most common method is to convert optical signal into electrical signal. This conversion takes place in photoelectric detectors, and the mechanism is called the photoelectric effect.

The photoelectric effect can be divided into external photoelectric effect and internal photoelectric effect. In the external photoelectric effect, the electrons escape from the surface of the irradiated material. The conditions for this effect, the requirements for incident light, and the properties of the escaping electrons have already been discussed in detail in Chapter 1. A typical detector based on external photoelectric effect is introduced in Section 6.2.

Internal photoelectric effect is more obvious in semiconductor materials, and is divided into the photoconductive and photovoltaic types. When light with a sufficient energy to raise electrons across the bandgap strikes a semiconductor, a large number of conductive carriers (electron-hole pairs) are excited, so the conductivity of the material is increased significantly. This phenomenon is called the photoconductive effect, and the detectors based on this working mechanism are called photoconductive detector.

On the other hand, these electron-hole pairs recombine in the diffusion process. However, if there is an electric potential barrier in the semiconductor, the electron-hole pairs within the diffusion length range near the barrier will possibly reach the strong field region before recombining. The electrons and holes are pushed to both sides of the barrier by a strong field region, resulting in charge accumulation. This phenomenon is known as the photovoltaic effect, and the detectors based on this mechanism are called photovoltaic detectors.

In addition to the photoelectric effect, another common phenomenon is photothermal effect. When the material is irradiated, the absorbed energy is converted to the energy of the lattice or molecular thermal motion, which results temperature increases. The detectors based on this mechanism are called photothermal detectors. It is worth noting that the detectors usually do not measure heat or temperature rise directly but instead convert the temperature increase into other physical quantities which are electrical signals in most cases. According to the mechanism of the second conversion, these detectors can be divided into the bolometer, the radiation thermocouple, the thermopile, and the pyroelectric detector, etc. The pyroelectric detector is widely used among these detectors.

https://doi.org/10.1515/9783110500608-006

The photoconductive effect and the photoconductive detector as well as the photovoltaic effect and photovoltaic detector will be discussed in Sections 6.3 and 6.4, respectively. The photothermal effect and photothermal detector will be briefly described in Section 6.5.

6.1 Overview of photoelectric detector performance

This section introduces some performance parameters of photoelectric detectors, which can be used to determine detector performance and provide some suggestions for selecting detectors.

6.1.1 Responsivity

If the detector gives an output y for input x, then we define the responsivity as follows

$$R = \frac{dy}{dx} \tag{6.1}$$

where y is usually current or voltage, the unit of which is A or V; and x is usually the radial flux or luminous flux, the unit of which is W or lm.

For detectors that operate in the linear region, equation (6.1) is simplified as

$$R = \frac{y}{x} \tag{6.2}$$

It is noteworthy that, when there is no external radiation input, the detector will output a finite value y_0 because of internal noise (dark current etc., which will be described in Section 6.1.2). Then unreasonable result $R \to \infty$ will be obtained from equation (6.2). Considering this, we will substitute equation (6.2) by

$$R = \frac{y - y_0}{x} \tag{6.3}$$

Obviously, $y \geq y_0$ is required in equation (6.3), where y_0 is the ability of a detector for detecting weak signals. The smaller y_0 is, the stronger the ability is.

Considering the dependence of the responsivity on wavelength, we define spectral responsivity as follows

$$R(\lambda) = \frac{dy(\lambda)}{dx(\lambda)} \tag{6.4}$$

The wavelength corresponding to the maximum responsivity is called the peak wavelength, which is often expressed as λ_m. $\frac{R(\lambda)}{R(\lambda_m)}$ is referred as the relative spectral responsivity. The wavelength that satisfies $\frac{R(\lambda)}{R(\lambda_m)} = \frac{1}{2}$ is called cutoff wavelength of the detector expressed as λ_∞.

6.1.2 Noise equivalent power

As mentioned in the previous subsection, owing to internal noise, the detector will give a limited output without any external input. To solve the problem, equation (6.3) is used to replace equation (6.2). It is assumed that, when the output induced by the input radiation is exactly equal to the output of the internal noise, the input radiation is called the noise equivalent power, which is usually expressed as NEP.

We suppose the noise signal is x_n, whose RMS value $\sqrt{\langle x_n^2 \rangle}$ is more appropriate to be used as it is a random variable in general. Thus, we can obtain NEP using the above definition

$$NEP = \frac{\sqrt{\langle x_n^2 \rangle}}{R} \tag{6.5}$$

When the noise mostly comes from the dark current i_n, then $NEP = \frac{\sqrt{\langle i_n^2 \rangle}}{R}$.

NEP shows the detection capability of photoelectric detectors. The smaller the value is, the stronger is the ability to detect weak signals.

6.1.3 Detectivity

The reciprocal of NEP is more commonly used to represent the detectability of the detector.

$$D = \frac{1}{NEP} = \frac{R}{\sqrt{\langle x_n^2 \rangle}} \tag{6.6}$$

D is referred to as the detectivity. It is obvious that the larger the detectivity is, the stronger is the detection ability.

As the noise power P_N is proportional to the area of the detector A_D, i.e., $P_N \propto A_D$. Therefore,

$$\sqrt{\langle x_n^2 \rangle} \propto A_D^{\frac{1}{2}}, \; D \propto A_D^{-\frac{1}{2}} \tag{6.7}$$

In addition, the noise of most detectors is dominated by the white noise. The power of the white noise is proportional to the bandwidth Δf, i.e., $P_N \propto \Delta f$, hence $\sqrt{\langle x_n^2 \rangle} \propto \Delta f^{\frac{1}{2}}$, so

$$D \propto \Delta f^{-\frac{1}{2}} \tag{6.8}$$

Combining equation (6.7) and (6.8), we can obtain $D \propto (A_D \Delta f)^{-\frac{1}{2}}$.

Therefore, the detectivity is inversely proportional to the detector area and the square root of bandwidth, which makes it inconvenient to compare detector performance. Therefore, the normalized detectivity is defined as follows:

$$D^* = D(A_D \Delta f)^{\frac{1}{2}} \tag{6.9}$$

D^* is independent of the detector area and the bandwidth, but it does depends on the wavelength and modulation frequency. Substituting equation (6.6) into (6.9), we can get

$$D^* = \frac{R(\lambda)}{\sqrt{\langle x_n^2 \rangle}} (A_D \Delta f)^{\frac{1}{2}} \tag{6.9a}$$

6.1.4 Quantum efficiency

Quantum efficiency η is defined as the ratio of the number of the output signal elements to that of the incident photons. In many cases, the output signal elements are photoelectrons, therefore

$$R = \frac{e}{hv}\eta \quad \text{or} \quad \eta = \frac{hv}{e}R \tag{6.10}$$

6.1.5 Response time

When the detector is irradiated, the output will not immediately reach a steady value, and the output will not be zero immediately when the radiation is stopped. This delay phenomenon is described by the response time τ. The time required to reach $1 - \frac{1}{e} \approx 63\%$ of the steady output value from the beginning of the radiation is defined as τ_{up}; and the time required to drop to $\frac{1}{e} \approx 37\%$ of the steady output value from the radiation stopping is defined as τ_{down}.

6.1.6 Linear region

A linear region is where the output signal of the detector is linearly related to the input signal. The lower limit usually depends on NEP and the upper limit is also related to the external circuit.

6.1.7 Noise

Noise is an important performance of photoelectric detectors. From previous discussion in this subsection, noise clearly influences the responsivity, detectivity, and linear region of the detector. In this subsection, we will describe some common noise types in the detector.

6.1.7.1 Thermal noise

Thermal noise is the most widespread noise in the detector, which is caused by the irregular thermal motion of the charges in the device; the mean square current of noise is expressed as

$$\langle i_T^2 \rangle = \frac{4kT\Delta f}{R} \tag{6.11}$$

where k is the Boltzmann constant whose recommended value is $1.380658 \times 10^{-23}\ \mathrm{JK}^{-1}$ in 1986; T is the thermodynamic temperature; Δf is the bandwidth; and R is the internal resistance.

If thermal noise is expressed using the mean square voltage, it can be expressed as

$$\langle u_T^2 \rangle = \langle i_T^2 \rangle R^2 = 4kTR\Delta f \tag{6.11a}$$

Thus, as long as internal resistance exists in the photoelectric detector, the thermal noise will be generated and the noise expressed by the mean square current or voltage is directly proportional to the operating temperature and the system bandwidth. Therefore, reducing the operating temperature and the system bandwidth is effective to reduce the thermal noise. Moreover, because thermal noise is independent of frequency, it is white noise.

6.1.7.2 Shot noise

Shot noise is caused by irregular fluctuations of the current in the detector, and the mean square current is expressed as

$$\langle i_n^2 \rangle = 2e\langle i \rangle \Delta f \tag{6.12}$$

where e is the electronic charge; $\langle i \rangle$ is the average current passing through the detector; and Δf is the bandwidth.

Therefore, shot noise is also white noise, and the mean square current is proportional to the average current and bandwidth. It is worth noting that equation (6.12) is obtained under the assumption that the electrons flow through the photoelectric detector in a very short time, hence, it is only suitable when frequency is not too low. Equation (6.12) can be applied in the photomultiplier because the time gap for photoelectron coming from the cathode to anode (that is, the time of photoelectron transiting the junction region in the photodiode) is very short.

6.1.7.3 Generation-recombination noise

If the transit time of the carriers is too long in the detector, the carriers recombine in the diffusion process. Because the number and lifetime of carriers are random, the generation and recombination of carriers can cause noise, which is called generation-recombination noise. The mean square current of the noise is expressed as

$$\langle i_{g-r}^2 \rangle = 4g(N_0) \frac{i_d^2}{N_0^2} \frac{\tau^2}{1+\omega^2\tau^2} \Delta f \tag{6.13}$$

where $g(N)$ is the generation rate of the carriers, N is the total number of carriers, $N_0 = \langle N \rangle$ is the statistical average of N; i_d is the bias current caused by an applied electric field, τ is the lifetime of the carriers; ω is the modulation frequency of light; and Δf is the bandwidth of the detector.

If we use voltage to express noise, we get

$$\langle u_{g-r}^2 \rangle = \langle i_{g-r}^2 \rangle R^2 = 4g(N_0) \frac{u_d^2}{N_0^2} \frac{\tau^2}{1+\omega^2\tau^2} \Delta f \tag{6.14}$$

where $\Delta u_d = i_d R$, which is the bias voltage.

6.1.7.4 Current noise or $^1/_f$ noise

$^1/_f$ noise is a type of general noise, whose mechanism is complex, such that there is no exact theoretical expression for it. However, after performing a several experimental studies, we have found the following empirical equation:

$$\left\langle i_{1/f}^2 \right\rangle = \frac{k i^a \Delta f}{f^\beta} \tag{6.15}$$

where k is a constant, and a and β are coefficients. The constant k is related to the manufacturing process of the detector, the contact condition of the electrode, the surface state and the size, and a is a coefficient related to the current through the detector. In most cases, $a \approx 2$, so $\sqrt{\left\langle i_{1/f}^2 \right\rangle} \propto i$, and therefore, it is referred to as current noise. β is a coefficient related to the characteristics of the detector material whose value is in the range of (0.8~1.5). For most materials, $\beta \approx 1$. So $\left\langle i_{1/f}^2 \right\rangle \propto \frac{1}{f}$, and therefore, it is called $^1/_f$ noise. Because it is proportional to $^1/_f$, it has a great impact on the signals in the low frequency area (below the 100 Hz); the impact reduces to a lower level when $f > 200Hz$.

6.2 The working foundation of photodetectors

As mentioned in the above section, photodetectors work by converting an optical signal into an electrical signal. This conversion occurs either by photoelectric effect or by photothermal effect and thermoelectric effect. The photoelectric effect can be divided into two types, namely, the external photoelectric effect and the internal photoelectric effect. The internal photoelectric effect mainly includes photoconductive and photovoltaic effect. This section will introduce these effects and will provide a theoretical basis for the discussion in the following sections.

6.2.1 External photoelectric effect

When a material is irradiated by light, the electrons that escape from the surface of the material are called photoelectrons, and the phenomenon is called the photoelectric effect. Usually, this is referred to as the external optical effect to distinguish it from another photoelectric effect. The photoelectric effect related to the quantum properties of light described in Chapter 1 belong to this kind of photoelectric effect.

In the external photoelectric effect, the energy of electrons must be sufficient for them to escape from the surface. The limit of this energy is called the emergence work, or the work function, which is expressed as W. If the photon energy $h\nu$ is greater than W, the escaping electrons will have the following initial kinetic energy

$$\frac{1}{2}m_e v^2 = h\nu - W$$

Thus, the cutoff frequency of irradiation inducing the photoelectron emission can be obtained by $\nu_{co} = \frac{W}{h}$ or the cutoff wavelength $\lambda_{co} = \frac{hc}{W}$.

The work function of metals and semiconductors is defined as the difference between the vacuum stationary electron energy E_0 (the vacuum energy level) and the Fermi energy level E_F. In metals, because energy levels below the Fermi level are filled by electrons, the work function is approximately equal to the minimum energy required for the electrons to escape from the metal and is a fixed value. For semiconductors, the energy difference between the vacuum energy level E_0 and the bottom of the conduction band is fixed, which is often called affinity energy. However, because of the different dopants and different positions of E_F in the bandgap, the work function W is different.

6.2.2 Photoconductivity effect

The phenomenon that the conductivity of material changes under light irradiation is called the photoconductivity effect. Because conductivity is determined by the concentration of free carriers or excess carriers, the change in conductivity is caused by the change carriers' concentration mentioned above. As the concentration of free electrons in metals remains essentially unaffected by light, photoconductivity effect does not occur in metals. Photoconductivity effect can occur in most semiconductors and insulator, but only obviously occur in some kinds of materials, mainly semiconductors and some organics. The following discussion mainly focuses on semiconductor materials.

Under thermal equilibrium conditions, the concentration of carriers in semiconductors remains constant, and is expressed as n_0 (the electron concentration) and p_0 (the hole concentration), respectively. The equilibrium value of conductivity is expressed as

$$\sigma = e\mu_n n_0 + e\mu_p p_0 = \frac{e}{AL}\left(\mu_n N + \mu_p P\right)$$

hence, the detector conductivity is

$$G = \frac{A}{L}\sigma_0 = \frac{e}{L^2}\left(\mu_n N + \mu_p P\right) \tag{6.16}$$

where μ is the mobility of carriers, e is the electronic charge, and A and L are the sectional area and length of semiconductor, respectively.

When the surface of the semiconductor is radiated by the light whose photon energy hv is larger than the bandgap E_g, the partial electrons of the valence band can be excited to the conduction band and the corresponding holes are left in the valence band. Thus, the electron and hole concentration is increased by Δn and Δp, respectively; hence the conductivity is increased to

$$\sigma = \sigma_0 + \Delta\sigma$$

Where

$$\Delta\sigma = e\mu_n\Delta n + e\mu_p\Delta p = e\left(\mu_n\Delta n + \mu_p\Delta p\right) \tag{6.17}$$

We suppose that, in the total number of excess carriers, electrons and holes are ΔN and ΔP, respectively

$$\Delta n = \frac{\Delta N}{AL}, \Delta p = \frac{\Delta P}{AL} \tag{6.18}$$

Whether ΔN and ΔP are equal depends on the electron-hole recombination mechanism. Substituting equation (6.17) into (6.18), we obtain

$$\Delta\sigma = \frac{e}{AL}\left(\mu_n\Delta N + \mu_p\Delta P\right)$$

The photoconductivity can be expressed as

$$\Delta G = \frac{A}{L}\Delta\sigma = \frac{e}{L^2}\left(\mu_n\Delta N + \mu_p\Delta P\right) \tag{6.19}$$

When $\Delta N = \Delta P$, then

$$\Delta G = \frac{e}{L^2}\Delta N\left(\mu_n + \mu_p\right) \tag{6.19a}$$

The lifetime of residual carriers is defined as

$$\tau = \frac{\Delta N}{\Phi_s(\lambda)\eta(\lambda)A} \tag{6.20}$$

Where $\Phi_s(\lambda)$ is the photon flow at wavelength λ, $\eta(\lambda)$ is the ratio of the number of electron-hole pairs excited by the incident light at λ to the number of incident photons. Substituting Equation (6.20) into (6.19a), we obtain

$$\Delta G = \frac{e}{L^2}\mu_p\tau[\Phi_s(\lambda)\eta(\lambda)A](1+k) \tag{6.21}$$

where $k = \mu_n / \mu_p$.

6.2.3 Photovoltaic effect

As mentioned above, when photon energy is greater than the bandgap, light radiation produces a large number of electron-hole pairs. These electron-hole pairs recombine during the diffusion process. However, if there is an electric potential barrier in the semiconductor, the electron-hole pairs that are within the diffusion length range near the barrier possibly reach the strong field region before recombining. The electrons and holes are pushed to both sides of the barrier by the strong field region, resulting in charge accumulation. This phenomenon is known as the photovoltaic effect. Thus, the photovoltaic effect requires the presence of the electrical barrier in the material. The electric potential barrier can be PN junction, PIN junction, or Schottky barrier. Here, we will first introduce the simplest PN junction.

PN junction is the core of many semiconductor devices. In the semiconductor material, if one part is the N-type region (the majority carrier is electron and the minority carrier is hole), the other part is the P-type region (the majority carrier is hole and the minority carrier is electron); then, the PN junction is formed at the interface of the N-type and P-type region.

We assume that the Fermi energy level on the N-type region is higher than that on the P-type region at one moment, as shown in Fig. 6.1(a).

The electrons in the N region diffuse into the P-type region and form a negative space charge near the interface at the P-type region. On the other hand, the holes in the P-type region diffuse into the N-type region and form a positive space charge near the interface at the N-type region, resulting in the formation of a contact potential difference in the PN junction. The P-type region has a negative potential $-V_d$ relative to the N-type region, which will increase the electrostatic potential energy be increased by eV_d in P-type region. The Fermi level E_{fp} and the electron energy levels in the P-type region are both moved up by eV_d which is exactly equal to the difference between the original Fermi levels of the two regions as

$$eV_d = E_{fn} - E_{fp}$$

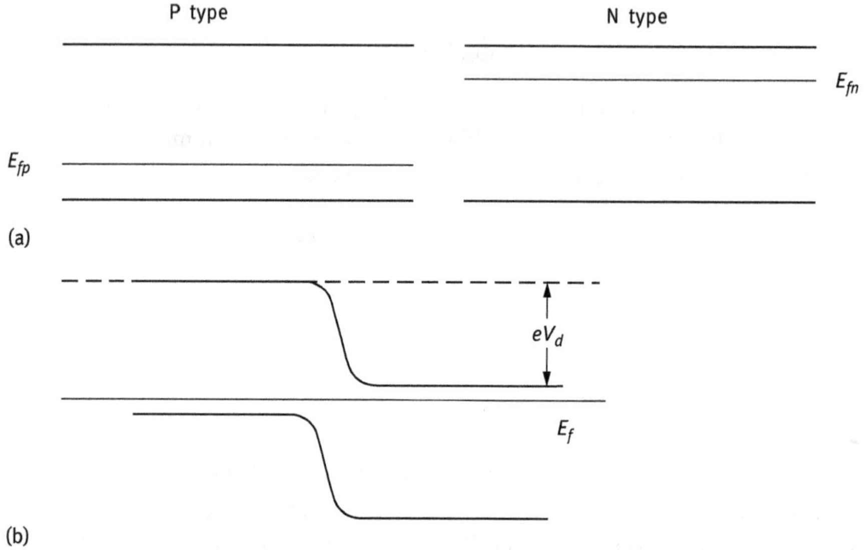

Fig. 6.1: Schematic diagram of Fermi energy in semiconductor materials.

So E_f in the two regions reach the same level, as shown in Fig. 6.1(b). The band-bending zone corresponds to the space charge zone of the PN junction, and there is a strong built-in field, which prevents the electrons in the N-type region and the holes in the P-type region from spreading to each other. Thus, a barrier is formed with a height of eV_d and width of the intrinsic material in the micron range.

From the relationship between the Fermi level and the concentration of carrier, we can obtain

$$eV_d = kTln\frac{n_n^0 p_p^0}{n_i^2} \tag{6.22}$$

where n_n^0 and p_p^0 are the concentration of the majority carrier in N- and P-type region, respectively, when they are in equilibrium, and n_i is the intrinsic carrier concentration. The effects of doping are reflected in n_n^0 and p_p^0 whereas that of E_g is reflected in n_i.

The primary potential maintains the equilibrium Boltzmann relationship between the P- and N-type regions, so there are

$$n_n^0 p_n^0 = n_i^2$$
$$n_p^0 p_p^0 = n_i^2 \tag{6.23}$$

From equation (6.22) and (6.23), we can get

$$n_p^0 = n_n^0 exp\left[- \frac{eV_d}{kT}\right]$$

$$p_n^0 = p_p^0 exp\left[- \frac{eV_d}{kT}\right] \quad (6.24)$$

6.2.4 Light thermal electric effect

The process of converting light radiation into electrical energy can be accomplished not only by the preceding effects but also by the following steps: first, the light signal will be converted into an intermediate quantity, and then the latter will be converted into electrical quantity. For example, when a photothermal detector is irradiated by light, the photosensitive material absorbs the radiation and the temperature increases; this increase in temperature further induces a change in electrical quantity. The rise in temperature leads to the change in electrical quantity, mainly including thermoelectric and pyroelectric effect. The following subsection introduces the pyroelectric effect.

Materials that produce pronounced pyroelectric effect are called pyroelectric materials. Pyroelectric materials have the following two characteristics:

1. They usually have a very poor symmetrical structure, such that the centers of positive and negative charges do not coincide at one direction of the natural state, and a certain amount of polarization charge is formed at the surface perpendicular to this direction. When temperature is constant, the polarization charge also remains constant and is neutralized by the external charge near the surface. This neutralization process takes approximately 1~1000 s.
2. When the temperature of the material temperature, the centers of positive and negative charges shift, leading to change in surface polarization charge. The relaxation time from the change in temperature to the change in polarization charge is very short, and is generally in the ps range.

Thus, if temperature changes rapidly, polarization charge is not neutralized in time but varies rapidly with the temperature. The alternating current can be detected if this kind of material is connected with the electric field.

6.3 Photoelectric emission photodetector (based on external photoelectric effect)

The working mechanism of the photoelectric emission photodetector, which consists of photoelectric tubes, photomultiplier tubes, etc., is the external photoelectric effect. In this section, photomultiplier tube is introduced.

6.3.1 Working process and structure of the photomultiplier tube

The structure of the photomultiplier tube is shown in Fig. 6.2. A photomultiplier tube consists of a photocathode, an electron multiplier, and an anode, and is usually sealed into an evacuated glass tube.

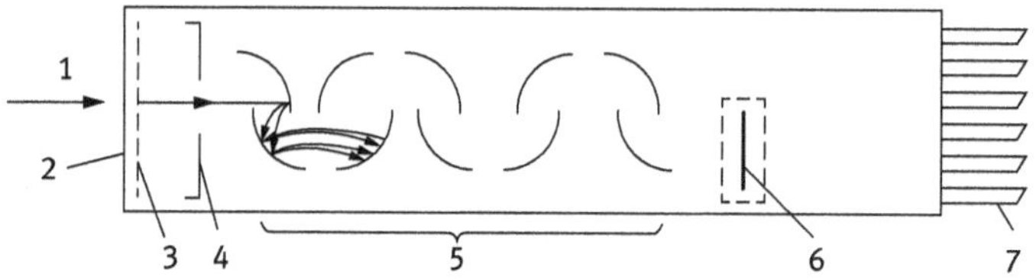

Fig. 6.2: The structure of the photomultiplier tube. 1, incident light; 2, input window; 3, photocathode; 4, focusing electrode; 5, electron multiplier; 6, anode; 7, pin.

When the photomultiplier tube is at work, the incident light 1, which needs to be detected, passes through the input window 2 and excites the electrons in the photocathode 3 such that photoelectrons are emitted into the vacuum. Subsequently, photoelectrons are focused by the focusing electrode 4 onto the first dynode where they are multiplied by secondary electron emission. This process is repeated at each successive dynodes and more and more electrons are excited. Finally, multiplied secondary electrons emitted from the last dynode are collected by the anode to form the anodic photocurrent. When the anodic photocurrent flows through the load resistance, a signal voltage, which is used for detection, is generated.

6.3.1.1. Photocathode

Photocathode is a photoelectron emission unit in the photomultiplier tube. The most important parameter to assess characteristics of the photocathode is sensitivity, including spectral sensitivity and illumination sensitivity.

When a beam of light with wavelength λ, passes through the photocathode, a photocurrent I_k is generated. The ratio between the photocurrent I_k and the radiation power P of the incident light is defined as the radiation sensitivity or spectral sensitivity of the photocathode as

$$S_k(\lambda) = \frac{I_k(\lambda)}{P(\lambda)}$$

The unit of that ratio is $^A/_W$.

This characteristic of the photocathode can also be described by the quantum efficiency η. Under the same condition, the ratio between the number of the emitted photoelectrons $N_e(\lambda)$ and the number of the incident photons $N_p(\lambda)$ is defined as the quantum efficiency as

$$\eta(\lambda) = \frac{N_e(\lambda)}{N_p(\lambda)}$$

Using the relationship between the photocurrent and the number of the photons, as well as the relationship between the radiation power and the number of the photons, we easily obtain

$$\eta(\lambda) = \frac{hc}{\lambda e} S_k(\lambda) = 1240 \frac{S_k(\lambda)}{\lambda}$$

Because the spectral sensitivity is difficult to detect, sensitivity of the photocathode is usually described by the illumination sensitivity or white light sensitivity. The definition of illumination sensitivity is the photocurrent generated by the light whose luminous flux is 1 l m when the photocathode is illuminated by the standard filaments lamp whose color temperature is 2856 K. To a certain extent, there is a correlation between illumination sensitivity and spectral sensitivity in some wavelength ranges.

Except for sensitivity, another important characteristic of the photocathode is the emission of hot electrons. At a certain temperature, although there is no light irradiation, some hot electrons are still emitted from the photocathode because the energy of the thermal motion of these electrons is the same or higher than their work function. In addition to temperature, the number of hot electrons emitted in a unit time is determined by the photocathode material. In addition, the photocathode material is also the key element to determine the response wavelength of the detector. According to the response wavelengths ranging from short to long, some materials usually used in the fabrication of photocathode are described below.

1. Cesium compounds

1. Cs-I

   Cs-I is sensitive to radiation with wavelength shorter than 200 nm and Cs-I is the special-purpose material in the vacuum ultraviolet region. In fact, the lower bound of the response wavelength of a detector whose photocathode is made of Cs-I is determined by its window material. For example, when we use synthetic silica or MgF_2, both of which have excellent transparency over ultraviolet radiation, as the entrance window, the range of the response wavelength is from 115 to 200 nm. Although sensitivity of Cs-I to the radiation, whose wavelength is shorter than 115 nm, is still high, it is difficult to find an appropriate window material. To detect these radiations, a windowless photomultiplier with Cs-I plated directly on the first dynode is chosen.

2. Cs-Te

   Cs-Te is sensitive to radiation whose wavelength is shorter than 300 nm.

2. Antimonide of alkali metal

According to the types of alkali metals in the antimonide, alkali antimonide can be divided into single alkali antimonide, bialkali antimonide, and multialkali antimonide.

1. Single alkali antimonide
   Single alkali antimonide contains antimony and an alkali metal. The most common single alkali antimonide is CsSb with the highest sensitivity ranging from the ultraviolet wavelength range to 700 nm.
2. Bialkali antimonide
   Bialkali antimonide is a compound of antimony and two types of alkali metals such as Sb-K-Cs, Sb-Na-Cs, and Sb-Rb-Cs. Bialkali antimonide has a spectral response range similar to the single alkali antimonide, but with higher sensitivity.
3. Multialkali antimonide
   Multialkali antimonide is a compound of antimony and more than two types of alkali metals. One of the most commonly used multialkali antimonides is Sb-Na-K-Cs. The response wavelength of Sb-Na-K-Cs is from ultraviolet region to near infrared region extending to 930 nm.

3. Ag-O-Cs

Ag-O-Cs has a wider range of spectral response than other materials, whereas the reflection type exhibits a spectral response region from 300 to 1100 nm and the transmission type exhibits a spectral response region from 300 to 1200 nm. Compared to other materials, Ag-O-Cs has lower sensitivity in the visible region, but has higher sensitivity at longer wavelengths in the near infrared region. Thus, Ag-O-Cs is chiefly used for near-infrared detection.

6.3.1.2 Electron multiplier

Electron multiplier can detect extremely weak light depending on the ability of its dynodes to multiply electrons. Thus, dynodes in electron multiplier are key components to determine the sensitivity of the detector. Here, we briefly introduce the working mechanism, the materials, and the dynode structure.

1. Working mechanism

The working mechanism of the dynode is the secondary electron emission, that is, when electrons whose kinetic energy is high enough strike the dynode, new electrons are emitted from the surface of the dynode. The incident electron is referred to as the primary electron while the outgoing electron is called the secondary electron. Usually, the secondary emission coefficient is the main parameter used to evaluate dynode performance. Secondary emission coefficient is expressed as

$$\sigma = \frac{n_2}{n_1}$$

where n_1 is the amount of the primary electrons while n_2 is the amount of the secondary electrons. σ is determined not only by the material of the dynode but also by the kinetic energy of the primary electron and the acceleration voltage between the dynodes. If the interstage voltage of dynodes is U, the relationship between the secondary emission coefficient and the voltage is expressed at

$$\sigma = aU^k$$

where a is the collection efficiency, and in general, is a constant. The parameter k that ranges from 0.7 to 0.8 and is determined by the material and the structure of the dynode. If there are $n - 1$ dynode stages in an electron multiplier, the total current gain can be expressed as:

$$\sigma_t = \left(aU^k\right)^n$$

We suppose that the working voltage V is equally distributed into n dynodes, such that the voltage on each dynode is expressed as

$$U = \frac{V}{n}$$

As a result, we can derive the expression of the total current gain as

$$\sigma_t = \frac{a^n}{n^{kn}} V^{kn} \tag{6.25}$$

2. Materials and structure of the dynode

Dynode can be fabricated with alkali antimonides, for example, CsSb, some oxides, such as MgO and BeO, or some negative electron affinity materials such as GaP. Moreover, there are a variety of dynode structures such as circular-focused structure, linear-focused structure, box-and-grid structure, fine-mesh structure, venetian blind structure, and microchannel plate structure.

6.3.1.3 Anode

The function of the anode is to collect the secondary electrons emitted from the final dynode. Grid structure is the typical type of anode structure. One of the performances of the anode is the illumination sensitivity which is defined as the ratio between the signal current I_p output from the anode and the luminous flux Φ of the radiation with the temperature of 2856 K, when the photomultiplier tube is illuminated. Thus, the expression of illumination sensitivity S_p becomes

$$S_p = \frac{I_p}{\Phi}$$

6.3.2 Main performance of the photomultiplier tube

The performance of the photomultiplier tube is mainly determined by that of the three components mentioned above.

6.3.2.1 Response characteristics

Response characteristics mainly include spectral response, irradiation sensitivity, and time response.

1. Spectral response
   Spectral response is determined by the photocathode and the property of the material of the input window.
2. Irradiation sensitivity
   The sensitivities of the photocathode and the anode, discussed in the preceding subsection, together determine the irradiation sensitivity.
3. Time response
   Transit time, rise time, and fall time constitute the time response. The time interval between the arrival of the incident light at the photocathode and appearance of the output signal at the anode is called the transit time. The rise time is defined as the time for the output pulse to increase from 10 to 90% of the peak pulse height. Conversely, the fall time is defined as the time required by the output pulse to decrease from 90 to 10% of the peak pulse height. The time characteristics are mainly determined by the electrode structure and have a close relationship with working voltage.

6.3.2.2 Current gain

Current gain is relevant to the structure, number, material, and working voltage of the dynode. Moreover, the size of the current gain can even reach 10^8.

6.3.2.3 Noise-to-signal ratio

There are two types of noise in the output signal of a photomultiplier tube: one is the noise itself even without light input; the other is the noise caused by the incident signal. Normally, the former is determined by dark current generated by emission of hot electrons while the latter mainly consists of the shot noise generated by the signal current.

1. Dark current

Dark current is mainly caused by the thermionic emission of the photocathode and the first few dynodes. When the electron absorbs the energy of the photon, which is higher than the work function of the electron, it can escape from the material surface, which is the electron emission discussed earlier. However, some electrons may also escape from the material surface at a certain temperature because their thermal motion energy may be higher than their work function, which is the hot electron emission. Thus it can be seen that the emission of hot electrons is relevant to temperature T and work function Ψ of the material, and the generated dark current can be expressed as follows:

$$i_n \propto T^{5/4} exp\left[-\frac{e\Psi}{kT}\right] \tag{6.26}$$

Equation (6.26) shows that the dark current will be larger if the work function is lower and the temperature is higher. Normally, the longer the radiation wavelength is, the lower the energy of the corresponding emitted quantum will be. Thus, to improve the efficiency of the photoelectric effect, the work function Ψ needs to be lower. Moreover, under the same conditions, the noise of the dark current in the photomultiplier tube used for the detection of infrared radiation is more obvious than that in the ultraviolet photomultiplier tube, and therefore, infrared devices need more cooling.

2. Shot noise

Shot noise is determined by the characteristic of the emitted quanta. The time when the quanta, which consists of radial flow, arrive at the photocathode is random and after the absorption of the photons on the material surface, the time interval of the generation of photoelectron is also random. These two processes appear as the shot noise.

3. Signal-to-noise ratio

Signal-to-noise ratio (SNR) is usually expressed as the ratio between the mean value of the signal (noise component included) and the AC component of the signal (noise component included). However, the mean value of the noise component is normally much less than that of the signal, hence, the SNR can be expressed as:

$$SNR = \frac{\langle i_s \rangle}{i_{s+n}} \tag{6.27}$$

where i_s is the mean value of the signal while i_{s+n} is the AC component of the signal and the noise. Compared with the signal, if the AC component of the noise can be ignored, the expression of the SNR can be simplified as

$$SNR \approx \frac{\langle i_s \rangle}{i_s} \tag{6.28}$$

where i_s is the AC component of the signal.

6.4 Photoconductive detector

6.4.1 Overview

Photoconductive detector is a photodetector based on the photoconductivity effect. Its typical structure is a thin photoconductive material plating on a thick substrate, as shown in Fig. 6.3(a). The photoconductive material, with electrodes connecting to the lead wire at each end, is the core component of the detector and the interface A is the area to receive the illumination. Figure 6.3(b) is the common symbol of the detector in the circuit diagrams.

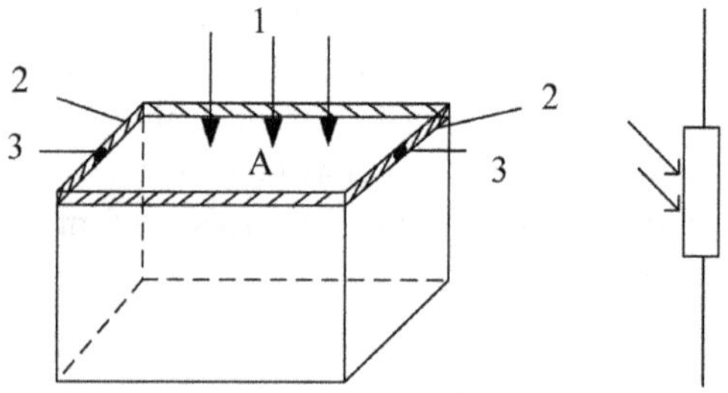

Fig. 6.3: The structure (a) and symbol (b) of the photoconductive detector. 1, Illumination area of the detector; 2, electrodes; 3, lead wire.

There are many types of photoconductive detectors, and some common photoconductive detectors are listed below according to their response wavelengths ranging from short to long.

6.4.1.1 CdS detector

The response wavelength of the CdS detector, which is the most sensitive photoconductive detector in the visible spectrum, ranges from 0.3 to 0.8 μm and the peak wavelength is approximately 0.5 μm. The main characteristics of the CdS detector are high reliability and long service life. However, its response is very low and its response time τ is between 1 and 1000 ms. In addition, the response time of this kind of detector is relevant to the illumination of the incident light, for example, when the illumination is 1001 x, the response time τ is approximately few tens of ms.

6.4.1.2 PbS detector

The PbS detector, whose response wavelength ranges from 1 to 3 μm and peak wavelength is approximately 2 μm, is one of the most sensitive photodetectors in the near-infrared range. Its peak detection rate D^* can reach 10^{11} cm $Hz^{1/2}W^{-1}$. Nevertheless, its response speed is low and the response time is over 10 μs.

6.4.1.3 PbSe detector

The response wavelength of the PbSe detector ranges from 1 to 5 μm and the response time is over 100 μs.

6.4.1.4 InSb detector

InSb detector has a response wavelength ranging from 5 to 7.5 μm and it is another kind of infrared detector with a good performance. At room temperature, its peak detection rate D^* can reach $10^9 \text{cmHz}^{1/2}\text{W}^{-1}$, and when it works at low temperature, with reducing noise, the detection rate D^* has a huge improvement. In addition, the other advantage of this kind of detector is its high speed. The typical response time is merely 10^{-1} to 10^{-2} μs.

6.4.1.5 $Hg_{1-x}Cd_xTe$ detector

The bandgap of $Hg_{1-x}Cd_xTe$ is relevant to the coefficient x, and as a result, the response wavelength of the device, within 1 and 15 μm, changes with the coefficient x. Two different kinds of detectors with the response wavelength ranged from 3 to 5 μm and from 8 to 14, respectively, are widely used. When the value of the coefficient x is 0.19, the peak response wavelength is approximately 10.5 μm, which well matches the output wavelength of the CO_2 laser.

At present, $Hg_{1-x}Cd_xTe$ detector is the best performing photodetector, which is discussed below.

6.4.2 Performance of the $Hg_{1-x}Cd_xTe$ photoconductive detector

6.4.2.1 Response wavelength

Like all semiconductor materials, the response wavelength range of the $Hg_{1-x}Cd_xTe$ is determined by its bandgap E_g. Moreover, there is a simple relationship between the cutoff wavelength λ_{co} and the bandgap E_g, which is expressed as

$$\lambda_{co} = \frac{hc}{E_g}$$

In the equation, bandgap E_g is relevant to the environment temperature T. When the temperature T is higher, The bandgap E_g will be larger, and as a result, the cutoff wavelength λ_{co} will be shorter.

Unlike other materials, the bandgap of $Hg_{1-x}Cd_xTe$ is the function of coefficient x as well as the environment temperature T, so its response wavelength range is also determined by coefficient x in its molecular formula. With the increase in the coefficient x, the bandgap E_g will be larger and the cutoff wavelength λ_{co} will be shorter.

For some typical values of coefficient x and temperature T, the corresponding values of bandgap E_g and cutoff wavelength λ_{co} are shown in Table 6.1.

Table 6.1: Relationship between the bandgap, cutoff wavelength of $Hg_{1-x}Cd_xTe$, and the temperature.

X	0.19		0.20		0.30		0.40		0.55	
T/K	77	170	30	77	77	125	200	200	300	300
E_g / eV	0.079	0.170	0.080	0.094	0.251	0.260	0.275	0.425	0.433	0.656
λ_{co} / μm	15.8	11.6	15.4	13.2	6–9	6–8	6–5	2.9	2.9	1.9

If we know the value of cutoff wavelength λ_{co}, we can estimate the value of the peak response wavelength λ_p using the following equation as

$$\frac{\lambda_{co}}{\lambda_p} \approx 1.1$$

Thus, it is obvious that $\lambda_p(0.19, 170) \approx 10.5\mu m$. Thus, the $Hg_{0.81}Cd_{0.19}Te$ photoconductive detector is suitable for the detection of CO_2 laser and its ideal working temperature is approximately 170 K.

6.4.2.2 Photoconductivity ΔG

Using equation (6.21), we can calculate the photoconductivity. For n-type $Hg_{1-x}Cd_xTe$, $k \gg 1$, so we can get:

$$\Delta G \approx \frac{e}{L^2}\mu_p \tau[\Phi(\lambda)\eta(\lambda)A] \tag{6.29}$$

where τ is the carrier lifetime which can be expressed as equation (6.20).

6.4.2.3 Noise

1. Thermal noise

Using conductance G to replace the expression ${}^1\!/_R$ in equation (6.11a), we can get the expression of thermal noise as

$$\langle u_T^2 \rangle = \frac{4kT\Delta f}{G} \tag{6.30}$$

2. Generation-recombination noise

We can rewrite the equation (6.14) as

$$\langle u_{g-r}^2 \rangle = \frac{4u_d^2}{N_0^2} g\tau \frac{\tau\Delta f}{1+\omega^2\tau^2} \tag{6.31}$$

Here, a crucial assumption for the simplification is that the processes of generation and recombination, caused by heat and light, are independent of each other. The two parts of $g\tau$ can be added together as

$$g\tau = (g\tau)_t + (g\tau)_0$$

and

$$(g\tau)_t = \frac{N_0 P_0}{N_0 + P_0} = \frac{n_0 p_0}{n_0 + p_0} V$$

$$(g\tau)_0 = P_b = p_b V \tag{6.32}$$

where p_b is the hole density caused by the background light while V is the effective volume of the material of the detector.

If equation (6.32) is substituting into equation (6.31) and the result is squared, we will get the RMS of the generation-recombination noise voltage as

$$\sqrt{\langle u_{g-r}^2 \rangle} = \frac{2u_d^2}{n_0 V^{1/2}} \left[\left(p_b + \frac{n_0 p_0}{n_0 + p_0} \right) \frac{\tau\Delta f}{(1+\omega^2\tau^2)} \right]^{1/2} \tag{6.33}$$

The complete equation (6.33) is only used at a small temperature range when the semiconductor changes from the intrinsic type into the extrinsic type. Beyond this range, only one factor, either light or heat, plays a determinative role in this equation. Thus, equation (6.33) can be simplified corresponding to a process. For example, if light is the determinative factor, we obtain

$$\sqrt{\langle u_{g-r}^2 \rangle} = \frac{2u_d^2}{n_0 V^{1/2}} \left(\frac{p_b \tau\Delta f}{(1+\omega^2\tau^2)} \right)^{1/2}$$

On the contrary, if heat is the determinative factor, we obtain

$$\sqrt{\langle u_{g-r}^2 \rangle} = \frac{2u_d^2}{n_0 V^{1/2}} \left[\frac{n_0 p_0 \tau\Delta f}{(n_0 + p_0)(1+\omega^2\tau^2)} \right]^{1/2} = 2u_d^2 \left[\frac{p_0 \tau\Delta f}{n_0 V (n_0 + p_0)(1+\omega^2\tau^2)} \right]^{1/2}$$

3. $1/f$ noise

As mentioned earlier, $1/f$ noise mainly determines the performance of the system at low frequency. However, there are no rigorous theories to describe this. For convenience, $1/f$ noise is usually described by "turning frequency" f_0. At frequency point f_0, the power of the $1/f$ noise is equal to that of the g-r noise at the same frequency point. Consequently, the RMS of the $1/f$ noise voltage at any frequency point f can be expressed as

$$\langle u_{1/f}^2 \rangle = \frac{f_0}{f} \langle u_{g-r}^2(0) \rangle \tag{6.34}$$

where $u_{g-r}(0)$ is the value of u_{g-r} in a smooth region where the frequency is obviously higher than f_0 but lower than the frequency that makes u_{g-r} start to move upward. Both $u_{1/f}$ and u_{g-r} are dependent on the bias, temperature and background signal in possibly different ways. f_0 should be the function of these variables. According to the classical theory of the photoconductivity proposed in 1960s, we can obtain

$$\left\langle u_{1/f}^2 \right\rangle = \frac{C}{T} \frac{L}{W} E^2 \frac{\Delta f}{f} \tag{6.35}$$

where L, W, and T are the length, width, and thickness of the detector, respectively. E is the DC bias field and Δf is the bandwidth of the noise. f denotes the frequency. Furthermore, C with dimension of cm^3, is the coefficient used to express the strength of the $1/f$ noise. Although its dimension is cm^3, it only depends on the carrier concentration and is irrelevant to the size of the detector. For the n-type $Hg_{1-x}Cd_xTe$, we usually assume that

$$n_0 \gg p_0, p_b \text{ and } (\omega\tau)^2 \ll 1$$

Thus, we can obtain

$$f_0 = \frac{cn_0^2}{4(p_0 + p_d)\tau} \tag{6.36}$$

Thereafter, a large amount of the $Hg_{1-x}Cd_xTe$ detector data, obtain under different working conditions, is summarized and analyzed, and the phenomenological theory of the $1/f$ noise of the $Hg_{1-x}Cd_xTe$ photoconductive detector was developed in 1974. Moreover, a simple empirical equation that can express the relationship between $u_{1/f}$ and u_{g-r} is expressed as

$$u_{1/f}^2 \approx \left(\frac{C}{f}\right)\left(\sqrt{\langle u_{g-r}^2 \rangle}\right)^3 \tag{6.37}$$

where C is a constant.

Because u_{g-r} is the current noise and varies inversely with the internal resistance size of the detector, according to equation (6.37), $u_{1/f}$ is the current noise and varies inversely with the internal resistance size as well.

Comparing equation (6.37) with (6.34), we can see:

$$f_0 \approx C\sqrt{\langle u_{g-r}^2 \rangle} \tag{6.38}$$

Equations (6.37) and (6.38) show that $1/f$ noise increases with the increasing $g-r$ noise, which is a new and pivotal conclusion of the phenomenological theory. According to the classical theory mentioned above, these two noise types are essentially independent, and a few connections are built between them only because of their current dependence.

6.5 Photovoltaic detector

6.5.1 Overview

Photovoltaic detector is based on the photovoltaic effect of the junction region in the detector material. Thus, it is named as the junction detector in some papers. According to the characteristics of the junction, photovoltaic detectors can be divided into PN type, PIN type, and Schottky type. In this subsection, we will mainly discuss the PN type device, and the high-performance $Hg_{1-x}Cd_xTe$ is chosen as the study object.

The $Hg_{1-x}Cd_xTe$ junction diode, working as a high-speed detector earlier, is mainly used to detect the radiation of the CO_2 laser whose wavelength is 10.6 μm. At the beginning, it can only work at low temperature of approximately 77 K. Thereafter, to meet the requirements of direct detection at 10.6μm wavelength and heterodyne detection, $Hg_{1-x}Cd_xTe$ junction diode is well developed and can work at approximately 200 K which can be attained by thermoelectric refrigeration conveniently. Later, the interests in $Hg_{1-x}Cd_xTe$ junction diode were mainly concentrated in combination with silica-based CCD chips to fabricate the mixed mosaic focal plane array, which can be used in direct detection at wavelength ranging from 3 to 5 μm and from 8 to 12 μm.

There are two types of mixed mosaic focal plane arrays: plane processing type and back-illuminated type. Irrespective of the type of focal plane array, the $Hg_{1-x}Cd_xTe$ action layer must be very thin. The thickness is approximately 10 to 15 μm, which is approximately the diffusion length of the minority carrier.

6.5.2 Brief introduction of the current characteristic of the PN junction photodiode

The current (I)-voltage (V) characteristic of the PN junction photodiode can determine its dynamic impedance and shot noise. Under the condition of zero bias, the relationship between the dynamic impedance R_0 and the $I-V$ characteristic can be expressed as

$$R_0^{-1} = \left.\frac{dI}{dV}\right|_{V=0} \tag{6.39}$$

If current I is replaced by the current density $J = \frac{I}{A}$, then equation (6.39) can be rewritten as

$$(R_0A)^{-1} = \left.\frac{dJ}{dV}\right|_{V=0} \tag{6.40}$$

where A is the area of the detector.

The product R_0A is independent of the area of the detector, which is why it is used to evaluate the performance of the detector more widely. However, if the detector is so small that its size is comparable with the diffusion length of the minority carrier, we need to notice the lateral diffusion effect of the carrier when R_0A is used to evaluate the performance of the detector.

6.5.2.1 Diffusion current

In PN junction photodiode, basic current is the diffusion current and can be discussed based on equation (6.24). For discussion, we first assume that the material consists of a thicker p-type material and a thinner n-type material. Between the p-type region and the n-type region, it is the space charge region, which is shown as Fig. 6.4.

The appearance of the diffusion current results from the random thermal generation and recombination of the electron-hole pairs in the diffusion region of the minority carrier at both sides of the space charge layer. For convenience, we rewrite equation (6.24) as:

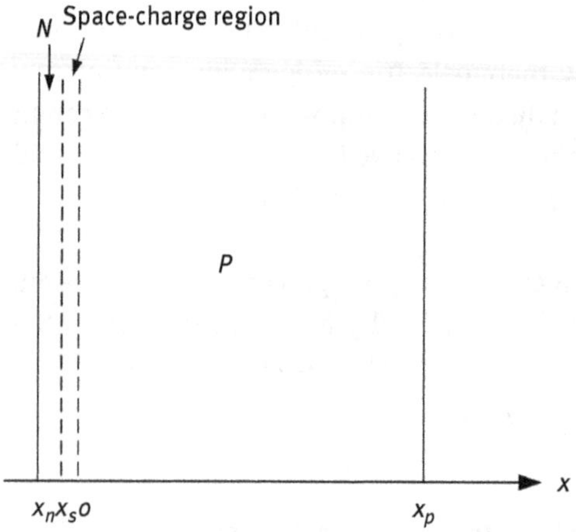

Fig. 6.4: Sectional view of the PN junction photodiode.

$$p(x_p) = p_n^0 exp\left(\frac{eV}{kT}\right)$$

$$n(0) = n_p^0 exp\left(\frac{eV}{kT}\right) \tag{6.41}$$

where p_n^0 and n_p^0 are the thermal balance values of the minority carriers in n-type region and p-type region, respectively. I the space charge region, the relationship of the carrier concentration is as follows

$$n(x)p(x) = n_i^2 exp\left(\frac{eV}{kT}\right)$$

where n_i is given by equation (6.23).

Next, we consider the situation where the thermal balance is destroyed. In the p-type region, carrier concentration is the function of time.

$$n(x,t) = n_p^0 + \Delta n(x,t)$$

$$p(x,t) = p_p^0 + \Delta p(x,t)$$

Because the p-type region is electrically neutral, we obtain

$$\Delta n(x,t) = \Delta p(x,t)$$

If we use $g(n,p)$ and $r(n,p)$ to represent the generation rate and recombination rate of the electrons in unit volume, respectively, the time rate of change of the electron number density can be written as

$$\frac{\partial n}{\partial t} = G_{ex} + g(n,p) - r(n,p) + \frac{1}{e}\frac{\partial J_e}{\partial x} \tag{6.42}$$

where G_{ex} is the external mechanism, for example, the electron-hole pairs generated by the incident radiation. The electron current density can be expressed as

$$J_e = eD_e\frac{\partial n}{\partial x} \tag{6.43}$$

where D_e is the diffusion coefficient of the electron.

Substituting equation (6.43) into equation (6.42), we will get the steady-state equation as

$$D_e\frac{\partial^2 \Delta n}{\partial x^2} - (r-g) = 0 \tag{6.44}$$

If $\Delta n \ll n_p^0,\ p_p^0$, then we can expand $(r-g)$ as Taylor series in the neighborhood of $n_p^0,\ p_p^0$, and retain the preceding two items which are as follows

$$\mathrm{r-g} = \left[r\left(n_p^0,\ p_p^0\right) - g\left(n_p^0,\ p_p^0\right)\right] + \left.\left[\frac{\partial(r-g)}{\partial n} + \frac{\partial(r-g)}{\partial p}\right]\right|_{n_p^0,\ p_p^0}$$

$$= \frac{\Delta n}{\tau_e}$$

where $\tau_e = \dfrac{1}{\left.\left[\frac{\partial(r-g)}{\partial n} + \frac{\partial(r-g)}{\partial p}\right]\right|_{n_p^0,\ p_p^0}}$ is the minority carrier lifetime.

So, equation (6.44) can be simplified as

$$D_e \frac{\partial^2 \Delta n}{\partial x^2} - \frac{\Delta n}{\tau_e} = 0 \tag{6.45}$$

To solve equation (6.45), in addition to equation (6.41), another boundary condition is needed. The boundary condition is that, when $x \to \infty$, there is $\Delta n(x \to \infty) \to 0$. We also assume that, when x_p is big enough, we can deem that $\Delta n(x_p) \to 0$. Under these conditions, the solution solved from equation (6.45) as

$$\Delta n(x) = n_p^0 \left[exp\left(\frac{eV}{kT}\right) - 1 \right] exp\left(\frac{-x}{L_e}\right) \tag{6.46}$$

where

$$L_e = \sqrt{D_e \tau_e} \tag{6.47}$$

L_e is the diffusion length of the minority carrier. The assumption of x_p mentioned above now can be further defined as

$$x_p \gg L_e \tag{6.48}$$

Under this condition, interface characteristics at $x = x_p$ do not influence the diffusion current in p-type region. Substituting equation (6.46) into (6.43), we can get the current density at $x = 0$as

$$J_e = eD_e \frac{\partial \Delta n}{\partial x}\bigg|_{x=0} = eD_e \left(\frac{\Delta n}{L_e}\right)\bigg|_{x=0} = en_p^0 \frac{D_e}{L_e} \left[exp\left(\frac{eV}{kT}\right) - 1 \right] \tag{6.49}$$

The corresponding R_0A can be obtained by equation (6.40) and (6.47) as

$$(R_0A)_p = \frac{kT}{e^2 n_p^0} \frac{\tau_e}{L_e} \tag{6.50}$$

Analogously, if the thickness of the n-type region is thicker than the diffusion length L_p of the holes inside the region, then there is:

$$(R_0A)_n = \frac{kT}{e^2 p_n^0} \frac{\tau_p}{L_p} \tag{6.51}$$

where

$$L_p = \sqrt{D_p \tau_p}$$

D_p and τ_p are the diffusion coefficient and lifetime of the minority carrier in n-type region, respectively.

In fact, many $Hg_{1-x}Cd_xTe$ photodiodes, which are used in the focal plane array, do not satisfy equation (6.48). For example, the diffusion length of the minority carrier in p-type $Hg_{0.8}Cd_{0.2}Te$ is 45 μm while in $Hg_{0.7}Cd_{0.3}Te$, the diffusion length L_e can even reach 100 μm. These lengths mentioned above are obviously longer than the length of the p-type region. Under this condition, equation (6.50) needs to be replaced by

$$(R_0A)_p = \frac{kT}{e^2}\frac{N_a}{n_i^2}\frac{\tau_e}{x_p} \tag{6.52}$$

where N_a is the concentration of the pure acceptor in the p-type region.

6.5.2.2 Photocurrent of the PN junction

When the infrared radiation whose wavelength is shorter than λ_{co} is absorbed by the photodiode, some electron-hole pairs are generated. If the absorption happens in the space charge region, the photogenerated electron–hole pairs will be taken apart by the strong electric field soon and a current will be generated in the external circuit. However, if the absorption happens in the n-type region or p-type region, which are on the side of the space charge region, within the length of L_e, only those photogenerated electron-hole pairs which go into the space charge region by diffusion effect, can be taken apart by the strong electric field and generate photocurrent in the external circuit.

We suppose that Φ is the steady state photon flow shined on the photodiode, so the steady state photogenerated current can be expressed as

$$I_{ph}(\Phi) = \eta e \Phi A \tag{6.53}$$

where A is the photosensitive area of the photodiode, and

$$\eta = \frac{N_e}{N_p}$$

where N_e and N_p are the number of the photogenerated electron and the incident photon, both of which contribute to the generation of the photocurrent, respectively. Furthermore, η is the function of the wavelength of the incident radiation and depends on the diffusion length of the minority carrier in the geometry and quasi-neutral zone of the diode. Unless the photon flow Φ is big enough that the concentration of the photogenerated surplus minority carrier is comparable with that of the majority carrier, η is irrelevant with the incident photon flow and photogenerated current is the linear function of Φ.

6.5.3 Response rate and detection rate

Two important evaluating indicators used to evaluate the performance of the photodiode are detection rate D_λ^* and noise equivalent power NEP_λ. We assume that the diode is illuminated uniformly by the monochromatic radiation whose wavelength is λ and the RMS of the signal photon flow is Φ_s, thus, the RMS of the radiation power of the signal detected by the detector can be expressed as

$$P_\lambda = \frac{hc}{\lambda}\Phi_s A$$

The RMS of the photogenerated current of the signal is as follows

$$I_s = \eta e \Phi_s A$$

Thus, we can get the expression of the response rate of the current listed as follows

$$R = \frac{I_s}{P_\lambda} = \frac{\lambda}{hc}\eta e \tag{6.54}$$

If we use $\sqrt{\langle I_n^2 \rangle}$ to represent RMS of the current noise, then the signal-to-noise ratio can be expressed as

$$\frac{S}{N} = \frac{I_s}{\sqrt{\langle I_n^2 \rangle}} = \frac{RP_\lambda}{\sqrt{\langle I_n^2 \rangle}}$$

If we make $S/N = 1$ and normalize the noise current bandwidth, the expression of noise equivalent power is gotten as follows

$$\mathrm{NEP}_\lambda = \frac{\sqrt{\langle I_n^2 \rangle}}{R}\sqrt{\Delta f}$$

The normalized detection rate of photosensitive area of the detector is

$$D_\lambda^* = R\sqrt{\frac{A\Delta f}{\langle I_n^2 \rangle}} \tag{6.55}$$

To further express D_λ^*, $\langle I_n^2 \rangle$ is discussed in Sub section 6.5.4.

6.5.4 Noise

6.5.4.1 Johnson noise

Under the thermal equilibrium condition without bias and illumination radiation, the mean square noise current of the photodiode is the Johnson-Nyquist noise of the resistance R_0 with zero bias, and is expressed as

$$\langle I_n^2 \rangle = \frac{4kT}{R_0}\Delta f \tag{6.56}$$

The meanings of these parameters in this equation are the same as before.

6.5.4.2 Shot noise

Under nonthermal equilibrium condition, we consider that junction current is generated only by the diffusion effect and background photon flow Φ_b. The junction current generated by the diffusion effect consists of reverse current I_r and forward current $I_r exp\left(\frac{eV}{kT}\right)$. Moreover, there is:

$$I_d = I_r\left[exp\left(\frac{eV}{kT}\right) - 1\right] \tag{6.57}$$

The current generated by the background light can be obtained by equation (6.53), that is:

$$I_{ph}(\Phi_b) = \eta e \Phi_b A \tag{6.58}$$

Thus, we obtain

$$I(V) = I_r\left[exp\left(\frac{eV}{kT}\right) - 1\right] - \eta e \Phi_b A$$

Because the ups and downs of the two currents are independent, the mean square of the total shot noise can be expressed as

$$\langle I_n^2 \rangle = 2e\left\{I_r\left[exp\left(\frac{eV}{kT}\right) + 1\right] + I_{ph}(\Phi_b)\right\}\Delta f \tag{6.58}$$

Under the condition of zero bias, the ups and downs of the diffusion current can be obtained by equation (6.57):

$$\left.\frac{dI_d}{dV}\right|_{V=0} = \frac{eI_r}{kT}$$

Substituting equation (6.57) into (6.39), we obtain

$$I_r = \frac{kT}{eR_0} \tag{6.60}$$

On substituting equation (6.58) and (6.60) into equation (6.59) and supposing that $V = 0$, we can get the expression of mean square of the total noise with zero bias as follows

$$\langle I_n^2(V=0) \rangle = \left(\frac{4kT}{R_0} + 2\eta e^2 \Phi_b A\right)\Delta f \tag{6.61}$$

Combining equation (6.61) with (6.55), we obtain the normalized detection rate as

$$D_\lambda^* = R\frac{1}{\left(\frac{4kT}{R_0 A} + 2\eta e^2 \Phi_b\right)^{1/2}}$$

where R is the response rate, and according to equation (6.54), we can get the further expression as follows

$$D_\lambda^* = \frac{\lambda\eta}{hc}\frac{1}{\left(\frac{4kT}{e^2R_0A} + 2\eta\Phi_b\right)^{1/2}} \tag{6.62}$$

6.5.4.3 $^1\!/_f$ noise

Previous studies have shown that the $^1\!/_f$ noise in $Hg_{1\text{-}x}Cd_xTe$ photodiode is not simply relevant to the total current in the diode but depends on the generation mechanism of the current. Specifically, the current of $^1\!/_f$ noise is basically irrelevant to the photogenerated current and the diffusion current, but is approximately proportional to the surface leakage current I_S. Moreover, it is the function of detection frequency f and noise bandwidth Δf, and can be written as

$$I_{1/f} = CI_S\sqrt{\frac{\Delta f}{f}} \tag{6.63}$$

where C is a constant with the magnitude of 10^{-3}. Relatively, Δf is a small parameter. The surface leakage current I_S is the function of reverse bias and temperature, thus, $I_{1/f}$ is the function of the above-mentioned parameters.

7 Photoelectric imaging and imaging system

7.1 Overview

In our daily life, what we can see in an object depends on what is imaged on the retina. The human eye is an excellent optical imaging system. However, it is only sensitive to radiation within the wavelength range of 400~800 nm, beyond which it becomes difficult to perceive radiation. In addition, direct observation of objects with the naked eye is associated with certain difficulties, for example, the light emitted (or reflected) by the object is too faint, the field angle of the object to the eye is too small (object is too small or too far away from eye), or the object is on the move at a high speed. Photoelectric imaging techniques can help overcome these difficulties and expand the function of the human eye.

Photoelectric imaging system is a key component of a photoelectric detection device. Its role is to convert images that cannot be directly observed by the human eye to a visible image Different detection devices have been described based on different concerns that need to be solved. For example, to an obtain image of an object with radiation wavelengths greater than 800 nm, we need to use an infrared detector, whereas to observe objects that emit (or reflect) extremely faint light detection devices with image enhancing function are needed. Therefore, a wide range of optoelectronic detectors are used in optoelectronic imaging system. Since considerable research has been published in this field, we do not repeat it here; only a simple introduction has been presented in Section 7.2. Another important component of photoelectric imaging system is the optical imaging subsystem. Following sections in this chapter will present some indexes for the evaluation of the performance of several optical systems.

On the system level, the photoelectric imaging system can be divided into scanning imaging systems and staring imaging systems. Scanning imaging system includes thermal imaging system, charge-coupled device (CCD) pickup camera, among others. The camera is an example of a staring imaging system. The imaging principle of these two systems is different, which will be discussed in Section 7.7, and the two kinds of optical imaging systems are introduced in Sections 7.8 and 7.9.

The performance of optoelectronic imaging systems is a very important problem for designers, manufacturers, and users. The relative literature is comparatively sparse, hence, this chapter focuses on the system performance and evaluation of performance.

7.2 Image detector profiles

The main difference between the image detector and the non-imaging detector described in Chapter 6 is the ability to output the visual image. Similar to the latter,

https://doi.org/10.1515/9783110500608-007

image detectors can also be divided into two categories, namely, vacuum devices and solid devices.

7.2.1 Vacuum imaging device

As the name suggests, a vacuum imaging device is a vacuum tube with a built-in photoelectric imaging element. If there is a scanning mechanism in addition to the imaging element within the tube, it is called a camera tube, otherwise, it is called an image tube.

The image tube includes an image converter and image intensifier. The main function of the image converter is spectral transformation, wherein it converts the non-visible radiation image with wavelength the range of 400~800 nm into the visible light range. The main function of the image intensifier is to enhance the light intensity. It enhances an image which cannot be perceived by the human eye.

The basic components of the image tube are the photoelectric cathode and the fluorescent screen. The main function of the former is photoelectric conversion, which transforms the incoming radiation into an image of the photoelectron emission. An electronic optical system is placed between the photoelectric cathode and the fluorescent screen. This system can accelerate photoelectrons emitted from the photoelectric cathode and retain the relative space distribution of the photoelectron.

The spectral response characteristics of the image tube are determined by the response characteristics of the photoelectric cathode. The relevant content has been discussed earlier in Chapter 6. For example, the photocathodes of most common infrared image converters use Ag-o-Cs, which is sensitive for infrared radiation with wavelength less than 1.15 μm; and the photocathodes of common ultraviolet image converters use Sb-Cs which is sensitive for ultraviolet radiation with wavelength greater than 200 nm.

Combining different image convertors forms a cascaded image intensifier, also known as the first-generation image intensifier. The devices with a microchannel plate instead of an ordinary electronic optical system to accelerate cathode emission optoelectronics are called microchannel plate image intensifiers, also referred to as second-generation image intensifiers. Replacing the photocathode of a second-generation image intensifier with a negative electron affinity photocathode results in a third-generation image intensifier, which plays a role in spectral transformation and weak light enhancement simultaneously.

In addition, image tubes with some special features have also been reported in the past, such as image magnification tube and location-sensitive sensor tube. Here, we do not discuss these in detail.

When the camera tube works, first, the incident optical image is converted into a charge image, and then the electrical signal is converted back to the optical image signal output. Between these steps, it is necessary to accumulate and store the potential image, which is read out by electron beam scanning.

Since the 1970s, solid automatic-scanning-imaging technology and devices have been rapidly developed and widely used, therefore, the vacuum tube is not discussed further in this chapter.

7.2.2 CCD imaging device

In 1969, W. S. Boyle and G. E. Smith (Baer Laboratory, United States) first proposed the concept of CCD, and soon developed CCD devices with a variety of features. Because this type of solid-state device can realize all functions of optical image conversion, storage, and sequentially output screen signals, it has the advantages of small volume, light weight, low power, high reliability, and long service life. With the development of semiconductor integrated-circuit technology, especially, metal-oxide semiconductor (MOS) integrated circuit, solid-state imaging device and technology have also developed rapidly. In addition to CCD, the more commonly used devices are self-scanned, photodiode-device (SSPD) and charge injection device (CID). This section describes the most widely used CCD, and the next section will briefly introduce CID.

The main component of the CCD is a shift register which is composed of an MOS capacitor arranged in certain pattern. The typical structure of a basic MOS unit is shown in Fig. 7.1. It consists of a growth layer of SiO_2 on a substrate of a semiconductor single crystal Si, to which the metal layer is added to form the MOS structure.

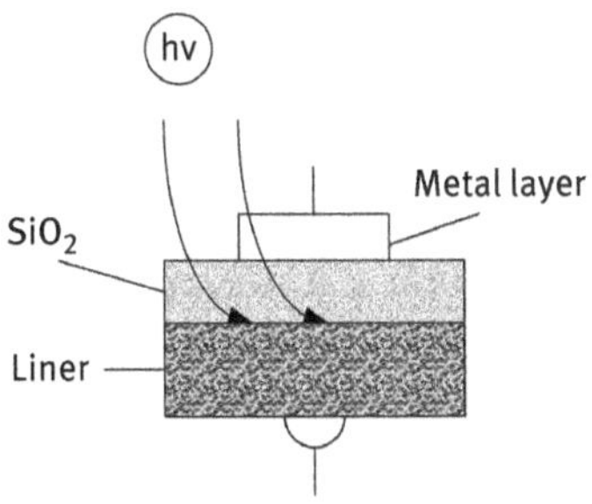

Fig. 7.1: Typical structural of an MOS unit.

The working principle of a CCD is charge storage and transfer. When the gate is biased, a potential well is formed between Si and SiO_2. The function of this potential well is to store signal charge, and the change in the applied voltage on the gate results in changes in the depth of the potential well of the same law, which leading to the transfer of trap signal charges along the surface of the semiconductor; finally, screen signal is transmitted by the output diode. Figure 7.2 represents the above

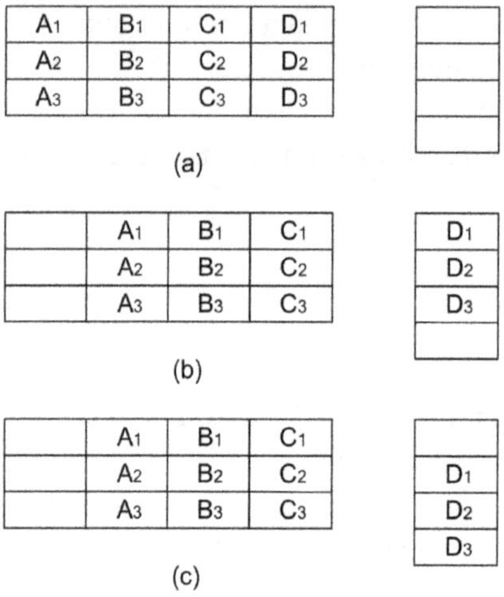

Fig. 7.2: Signal charge transfer in the CCD schematic.

process in a 4×3 unit CCD. The capital letters on the left side of the figure represent different charges, and the empty column on the right side of the distal end is the shift register.

Figure 7.2(a) Represents the stage that the exposure is just completed but has not yet read. When the CCD reads out, charge shifts first from a file of the "trap" to the next file. Each transfer makes the signal charge of the right file get into the shift register (Fig. 7.2b), then this column shifts downward and enters the output state (Fig. 7.2c).

7.2.3 CID imaging device

The structure of the CID photosensitive unit is shown in Fig. 7.3. In the figure, poly1 and poly2 are two metal gates that are isolated from the doped N-type Si region by the SiO_2 insulating layer (not shown).

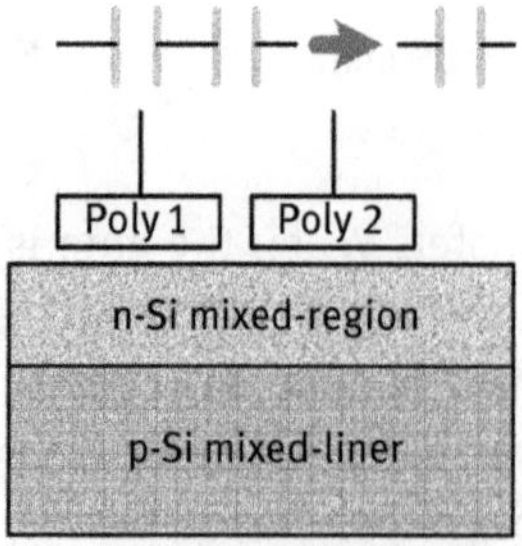

Fig. 7.3: CID photosensitive element.

Compared with CCD, the structure and working principle of the CID are different. First, the photosensitive unit of the latter corresponds to two MOS capacitors of the former, where poly1 and poly2 and doped N-type silicon region are the electrodes and SiO_2 is the medium. For certain CIDs, the charge of equivalent capacitance C of two series capacitors remains constant. Hence, the storage charge in the photosensitive element can be obtained from the voltage V between the two electrodes as

$$Q = CV$$

In addition, with regards to the detection principle, while CCD detects the photo-generated charge after it is transferred, CID directly detects each detection unit and photo-generated charge is stored in the MOS capacitor during measurement.

7.3 Point-spread function and performance index based on the point-spread function

7.3.1 Point-spread function

The image of the point source formed by optical system is called the point-spread function (PSF) of the system. PSF is also known as impulse-response function, Green function, Fraunhofer diffraction function, among others.

The foundation of modern image quality analysis is based on both the object and image can be understood as the light intensity distribution on both the object plane and the image plane. Suppose geometric aberrations are completely eliminated, only diffraction dictates the characteristics of the PSF. An optical system with an optical clean round exit pupil image for point source, the intensity distribution is expressed as

$$I(r) = I_0 \left[\frac{2J_1(kr)}{kr} \right]^2 \tag{7.1}$$

where

$$k = \frac{\pi}{\lambda F}$$

and F is the F-number of the system; λ is the wavelength of the monochromatic radiation; r is the radial distance on the image plane from the center of the PSF; and J_1 is the first-class, first-order Bessel function.

If the pupil central has a circular block, the ratio of the blocked part with the total diameter is ε, thus

$$\frac{I(R)}{I_0} = \frac{1}{(1-\varepsilon^2)^2}\left[\frac{2J_1(kr)}{kr} - \varepsilon^2\frac{2J_1(k\varepsilon r)}{k\varepsilon r}\right]^2 \tag{7.2}$$

For rectangular holes,

$$\frac{I(R)}{I_0} = \left[\frac{\sin(k_x r_x)}{k_x r_x}\right]^2\left[\frac{\sin(k_y r_y)}{k_y r_y}\right]^2 = \text{sinc}_x^2\text{sinc}_y^2 \tag{7.3}$$

where

$$k_x = \frac{\pi D_x}{\lambda f_x}, k_y = \frac{\pi D_y}{\lambda f_y}$$

and f is the effective focal length; and D is the length of the pupil.

In extreme cases of $D_y \ll D_x$, rectangular hole changes into the slot, leading to

$$\text{sinc}_x^2 \gg \text{sinc}_y^2$$

Thus, the linear PSF is obtained, with its longitudinal direction being perpendicular to the length of the slit. The linear PSF generated by point objectives through the slit should not confused with line spread function (LSF) resulting from a linear object imaged by the optical system. The latter is the strength superposition result of numerous independent PSFs along the longitudinal direction of a linear object.

$(1-\varepsilon^2)^{-2}$ in equation (7.2) is required for normalization to ensure that the PSF peak intensity at the center of the loop system is one.

The above equation represents the energy distribution in an image plane based on PSF. To quantify this distribution, we need to calculate the ratio of the energy contained within a given region to the total energy. Consider a circular hole as an example, to calculate the energy inside the circle with a radius R (circle around energy EE) and to normalize it by ensuring $R = \infty$, in the case of unobstructed holes, we have

$$EE(R) = 1 - J_0(kR) - J_1(kR) \tag{7.4}$$

By calculating of Bessel function, the energy contained within the first ring region is 84% of the total energy, and the energy contained within the first three rings is more than 95% of the total energy.

For a circular hole with a cover and a rectangular hole, there is no analytical expression similar to equation (7.4). To derive an energy distribution, numerical integration is required. For the sheltered circular hole, calculation shows that energy dispersion becomes severe when ε increases. For example, when $\varepsilon = 0.3$, the relative energy within the first and first three-ring is 68% and 93%, respectively. When $\varepsilon = 0.9$, the corresponding values are 8% and 38%, respectively.

7.3.2 Strehl ratio

PSF is one of the two most complete image-quality-appraisement functions used for scanning (the other is the optical transfer function, OTF, described below). However, the calculation of OTF is complicated, and therefore, some relatively simple functions have been derived which are more commonly used. One of them is the Strehl ratio (SR).

SR was proposed by K. Strehl more than a hundred years ago. It is defined as the ratio of the peak intensity of the actual system to that of the diffraction limit system and can be expressed as

$$\mathrm{SR} = \frac{1}{\pi^2}\left|\int_0^1\int_0^{2\pi} e^{jk\phi(\rho,\theta)}\rho d\rho\, d\theta\right|^2 \tag{7.5}$$

where $\phi(\rho, \theta)$ is an aberration function of the optical system, and (ρ, θ) is the polar coordinate on the image plane. To calculate SR use using equation (7.5), we need to know the exact form of the aberration function on the pupil coordinates. This condition is not satisfied for an optical system with random errors during operation. Therefore, a number of approximations have been described, with the mean square value σ^2 of the wave front error representing SR.

First, on expanding the exponential of equation (7.5) into Maclaurin series the SR series can be expressed as

$$\mathrm{SR} = \frac{1}{\pi^2}\left|\int_0^1\int_0^{2\pi}\left[1 + ik\phi(\rho,\theta) + \frac{1}{2}(ik\phi)^2 + \cdots\right]\rho d\rho\, d\theta\right|^2 \tag{7.6}$$

Then, define the mean of the wave front error in the pupil as

$$\langle\phi^n\rangle = \frac{\int_0^1\int_0^{2\pi}\phi^n(\rho,\theta)\rho d\rho\, d\theta}{\int_0^1\int_0^{2\pi}\rho d\rho\, d\theta} = \frac{1}{\pi}\int_0^1\int_0^{2\pi}\phi^n(\rho,\theta)\rho d\rho\, d\theta$$

or

$$\pi\langle\phi^n\rangle = \int_0^1\int_0^{2\pi}\phi^n(\rho,\theta)\rho d\rho\, d\theta \tag{7.7}$$

Substituting equation (7.7) into (7.6) gives

$$\mathrm{SR} = 1 - k^2\left[\langle\phi^2\rangle - \langle\phi\rangle^2\right] + \frac{1}{4}k^4\langle\phi^2\rangle^2 + \cdots$$

where

$$(\Delta\phi)^2 = \langle\phi\rangle^2 - \langle\phi\rangle^2$$

and

$$\sigma = k\Delta\phi$$

represents the phase standard deviation. SR can be expressed approximately as

$$SR \approx e^{-\sigma^2}$$

or

$$SR \approx e^{-(2\pi\sigma)^2} \tag{7.8}$$

In the above two equations, the units of σ are rad and λ, respectively .

To calculate SR, Equation (7.8) has a higher degree of approximation, allowing a greater value of σ, for instance, $\sigma \sim \frac{\lambda}{6}$ may be sufficiently small. According to the literature, σ is permitted even up to 2 rad, corresponding to or slightly less than $\frac{\lambda}{3}$.

There are some simple approximate formulas for SR. On developing the right side of equation (7.8) as the series and retaining only the first two terms gives

$$SR \approx 1 - \sigma^2 \tag{7.9}$$

The condition that expression (7.9) can be approximately established is $\sigma^4 \ll \sigma^2$, normally we expect to be $\sigma < \frac{\lambda}{10}$. A similar expression was derived by Marechal in 1947 as

$$SR \geq \left(1 - \frac{1}{2}\sigma^2\right)^2 \tag{7.10}$$

Equation (7.10) is applicable for $R \geq 0.8$, corresponding to the wave front error $\sigma \sim \frac{\lambda}{14}$.

Assuming that there are other factors that result in a reduction of SR, while these factors are independent of each other, the final SR becomes

$$SR = \prod_{i=1}^{N} SR_i$$

As an example, for a round pupil with central occlusion and an occlusion ratio of ε

$$SR_\varepsilon = e^{-\varepsilon^2}$$

If random jitter exists in images, SR is decreased as

$$SR_{jit} \approx \left[1 + 4.93\left(D\sigma_{jit}\right)^2\right]^{-1}$$

the total SR can then be expressed as

$$SR = \frac{\exp\left\{-\left[(2\pi\sigma)^2 + \varepsilon^2\right]\right\}}{1 + 4.93\left(D\sigma_{jit}\right)^2}$$

7.3.3 Relationship between circle surrounding energy and spatial frequency

An important drawback of SR is that it does not reflect the relationship between energy distribution and spatial frequency of the system. A circle around energy can compensate for this limitation.

Consider a grating equation

$$\sin\theta + \sin i = \pm n\frac{\lambda}{d}$$

where i and θ are the incidence and exit angles, respectively, and d is the grating period. Under normal incidence condition, $i = 0$, such that

$$\sin\theta = \pm n\frac{\lambda}{d}$$

At peak principal diffraction, n = 1, when there is a small angle diffraction

$$\theta \approx \pm n\frac{\lambda}{d}$$

For a cylinder of diameter D, the spatial frequency becomes

$$F_s = \frac{D}{d} = \frac{D}{\lambda}\theta \tag{7.11}$$

or

$$\theta = \frac{\lambda}{D}F_s \tag{7.12}$$

Equation (7.12) shows that the higher the spatial frequency, the greater the corresponding diffraction angle. Typically, spatial frequency F_{sc} is defined corresponding to the detectable minimum angle θ_{min} as the critical value between low and medium spatial frequency. According to this definition, for example,

$$D = 2\text{m},\ \lambda = 0.5\mu\text{m},\ \theta_{min} = 0.75\mu\text{rad}$$

results in

$$F_{sc} = 3\left(\frac{c}{a}\right)$$

Therefore, for spatial frequencies of less than three cycles, each aperture is considered to be a low frequency. At the same time, for spatial frequencies of greater than three cycles, each aperture is considered to be a medium spatial frequency.

Let us calculate EE corner radius ϕ_e as

$$\phi_e = n\frac{\lambda}{D}F_s \tag{7.13}$$

When λ and D are known, for ϕ_e, n and F_s, while the first two are known and the third can be obtained by equation (7.13). Commonly, when ϕ_e is known, for the main peak (n = 1) to fall within ϕ_e, F_s can be obtained by

$$F_s = \frac{D}{\lambda}\phi_e$$

7.4 OTF

PSF is one of the two most complete functions to evaluate the quality of imaging optical systems. However, it is mainly applied to point source imaging. To describe the imaging of a continuous distribution and surface radiation object, OTF is more reliable.

OTF of an optical imaging system is defined as ratio of the spectral distribution intensity $I_i(\nu)$ on the image plane to that on the object plane $I_O(\nu)$, using $T(\nu)$ it can be expressed as

$$T(\nu) = \frac{I_i(\nu)}{I_O(\nu)} \tag{7.14}$$

"Transfer function" is derived from the linear network theory. Figure 7.4 is a linear four-terminal network. On applying a harmonic signal $U_{in}(\nu)$ with frequency ν, the output will be at the same frequency signal $U_{out}(\nu)$. The ratio of the two signals can be expressed as

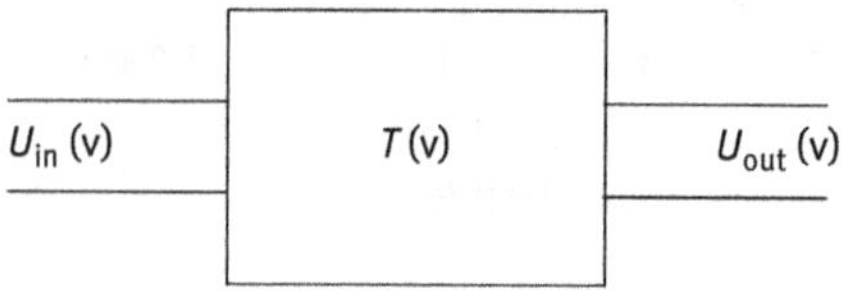

Fig. 7.4: Voltage transfer of the network.

$$T(\nu) = U_{in}(\nu)/U_{out}(\nu) \tag{7.15}$$

$T(\nu)$ is the voltage transfer function of the network. On comparing equation (7.14) with (7.15), the similarities become obvious. The difference is that the voltage transfer function is a time-frequency response function of the circuit system, whereas the OTF is the spatial-frequency response function of the optical imaging system.

OTF plays a key role in the theoretical evaluation and optimization of an optical system. The modulation transfer function (MTF) is the magnitude and the phase transfer function (PTF) is the phase of the OTF. When an ideal system is experiencing incoherent illumination, the OTF is real-valued and is equal to MTF. The MTF is a primary parameter used for system design, analysis, and specifications.

7.5 Modulation transfer function

7.5.1 Modulation

Modulation (M) is the variation of a sinusoidal signal about its average value (Fig. 7.5), and is defined as

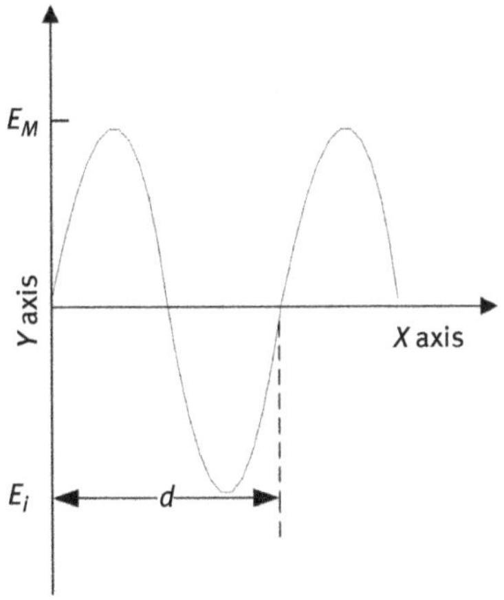

Fig. 7.5: Modulation of optical system.

$$M = \frac{E_M - E_m}{E_M + E_m} \tag{7.16}$$

where E_M and E_m are the maximum and minimum signal levels, respectively.

An example of modulation is shown in Fig. 7.5. For system analysis performed in object space, d is measured in mrad and spatial frequency $\nu = \frac{1}{d}$ in cycles/mrad. Figure d is measured at an angle of space and is called the spatial frequency. For optical system, d is measured in image space (mm) and the spatial frequency $\nu = \frac{1}{d}$ in cycles/mm.

7.5.2 Modulation transfer function

The MTF is the output modulation (M_0) of a system divided by the input modulation (M_i) at the same frequency:

$$MTF = M_0 / M_i$$

MTF is a function of spatial frequency ν and is usually normalized at $= 0$, i.e., MTF (0) = 1. When ν increases, MTF decreases. Figure 7.6 schematically illustrates this

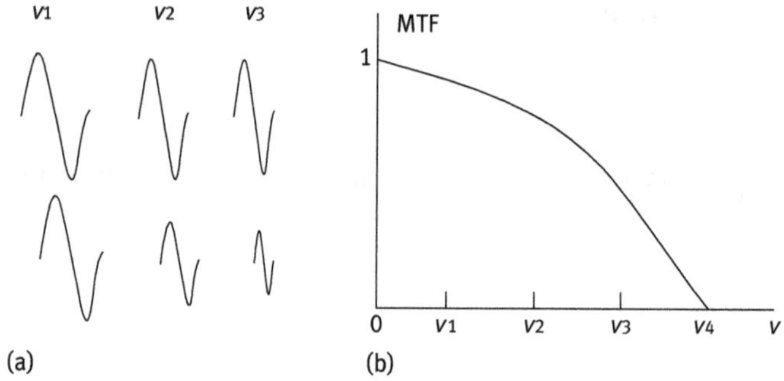

Fig. 7.6: MTF changes with spatial frequency v corresponding to output–input signals with different spatial frequency $MTF - v$ curves.

decline. The top line of Fig. 7.6(a) is the input signal of three different spatial frequencies, and the following line is the corresponding output signal. Figure 7.6(b) shows the schematic MTF changes with spatial frequency. The frequency which makes MTF = 0 is called the cutoff frequency and is denoted by v_c. The signal with frequency $v > v_c$ cannot pass through the optical system. In this sense, the optical system is similar to the low-pass linear network.

Because MTF is a function of spatial frequency, during the design, evaluation, or selection of an optical system, we must consider the spatial frequency characteristics of practical application. If the two curves in Fig. 7.7 represent the MTF of two optical systems, we cannot simply say which system is better only based on the curves but need to be consider the intended use of the system. If the spatial frequency of input information is below v_0, the first system is better; otherwise, the second system is a better fit.

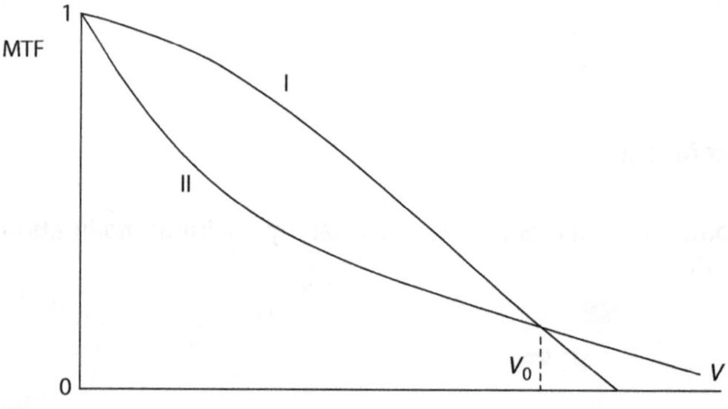

Fig. 7.7: MTF and system performance.

7.6 MTF of the optical system

MTF_O of an optical system mainly includes three parts, namely the diffraction-limited MTF_d, aberration-caused MTF_a, and defocus-caused MTF_{def}. As they are independent of each other, they can be expressed as

$$MTF_O = MTF_d \cdot MTF_a \cdot MTF_{def} \tag{7.17}$$

7.6.1 Diffraction-limited MTF

The diffraction-limited MTF_d for a circular aperture is

$$MTF_d(\nu_n) = \begin{cases} \frac{2}{\pi}[\cos^{-1}\nu_n - \nu_n(\cos^{-1}\nu_n)], \nu_n \le 1 \\ 0, \nu_n > 1 \end{cases} \tag{7.18}$$

where

$$\nu_n = \frac{\nu}{\nu_c}$$

is the normalized spatial frequency, and

$$\nu_c = D/\lambda$$

is the cutoff frequency, D is the diameter of the pupil, and λ is the wavelength.

Therefore, equation (7.18) can be only applied to monochromatic radiation. For application to polychromatic radiation, the mean wavelength λ_{ave} is required to replace λ. Under certain conditions,

$$\lambda_{ave} \approx \frac{\lambda_M + \lambda_m}{2}$$

where λ_M and λ_m are the maximum and minimum wavelength, respectively, in the polychromatic radiation.

For a rectangular aperture,

$$MTF(\nu_{nx}, \nu_{ny}) = (1 - \nu_{nx})(1 - \nu_{ny})$$

where ν_{nx} and ν_{ny} are normalized spatial frequencies in x and y directions, respectively.

7.6.2 Aberrations effect

When there is aberration, MTF of the system declines. Shannon developed an empirical relationship that encompasses most aberrations of real lens systems.

$$\mathrm{MTF_{abe}} \approx 1 - 30.86\sigma^2 \left[1 - 4\left(\nu_n - \frac{1}{2}\right)^2\right], \nu_n \leq 1 \tag{7.19}$$

where σ is the rms wave front error whose unit is λ.

However, when $\nu_n > 1/2$, $\mathrm{MTF_{abe}}$ increases with increase in ν_n and approaches 1 at $\nu_n = 1$ which is clearly unreasonable. Moreover, as noted in the literature, for large σ, equation (7.19) cannot be applied. It is generally considered that $\sigma \leq 0.14$ is the condition of the available equation and when $\sigma \leq 0.07$ system is approximately satisfied for the diffraction-limited condition. It is clear that $\mathrm{MTF_{abe}}$ is approximately 0.4 and 0.87, respectively, for the two cases.

7.6.3 Defocus

The MTF for a circular, non-aberration, defocused lens is calculated by different approximations according to the size of the defocus. When the peak–peak value of the wave-front error caused by defocus satisfies $W_{pp} \leq 2.2\lambda$, we have

$$\mathrm{MTF_{def}} \approx \frac{2J_1(Z)}{Z}$$

Here, J_1 is the first-order Bessel function, and its independent variable is

$$Z = 8\pi W_{pp}\nu_n(1 - \nu_n)$$

When W_{pp} is small, the Bessel function can be developed into a power series. For example, in the case of $W_{pp} \leq 0.5\lambda$

$$\mathrm{MTF_{def}} = 1 - \frac{Z^2}{8} + \frac{Z^4}{192} - \frac{Z^6}{‘9216}$$

7.7 Introduction to optical imaging system

The subsections 7.7.1 and 7.7.2 introduce the staring array imaging systems and scanning imaging systems, respectively. Sections 7.7.3 and 7.7.4 introduce performance evaluation methods commonly used in the two systems.

7.7.1 Staring array optical imaging system

A staring array optical imaging system consists of four basic subsystems, namely optical subsystem, detectors and auxiliary circuit subsystem, analog–digital conversion subsystem, and image reconstruction subsystem. The optical subsystem images the radiation of the scene on the detector, and the image is then converted into a measurable electrical signal. An electronic image is produced as a result of secondary amplification and signal processing circuitry, with the voltage difference of the electronic image depicting the strength difference of different objects in the field-of-view or different parts of an object. The signal is digitized because it is easy to handle.

In each of the above subsystem, more attention is paid to those parts that may add noise or change the image fidelity. Because detectors have already been discussed, the following section briefly describe optical system and detector and auxiliary circuits.

Most optical systems are composed of individual lenses or mirrors, with each element having a different refractive index and shape to minimize the aberration. However, for ease of analysis, the optical system is generally treated as a single element with an equivalent focal length f.

As described above, the optical subsystem images radiation of an object on the detector. The received radiation is the strongest when the detector is on the optical axis of the optical subsystem. Otherwise, the intensity of the radiation reaching the detector attenuates in $\cos^m\theta$ according to the law of radial symmetry, where θ is the included angle between the line linking the pupil of the optical subsystem with the detector and the optical axis. m is a constant with a possible value of 2.4 in the actual system. A single-detector system is always put in optical axis; the detector of a staring system is typically a pluralism array and it is optical design decisions that which sensing element on the optical axis or off-axisx. Therefore, $\cos^m\theta$ is determined depending on the circumstances.

A complete staring system includes no scanner system, and its field-of-view is determined by the optical design. If the total array size is Sa, the field-of-view is

$$\mathrm{FOV} = 2\tan^{-1}(S_a/2f) \tag{7.20}$$

Appropriate electronic amplifier circuit can compensate for the effects of $\cos^m\theta$ to obtain a high-contrast image. However, any amplifier circuit will also increase the noise, such that the signal–noise ratio remains constant.

7.7.2 Scanning optical imaging system

In comparison to the staring system, scanning imaging system includes a scanning device which may be a rotary polygon mirror, a reflection prism, or a galvanometer.

Scan mechanism can either be within or external to the optical subsystem. Let the whole scanning angle be θ_S, then the scan range provided by the former is $\pm\theta_S/2$, whereas the latter provides a scan range of $\pm M\theta_S/2$, where M is the magnification of the defocus telescope.

The field-of-view of a scanning imaging system is determined by both the instantaneous field-of-view and the scanning angle θ_s at the moment. For a scanning mechanism external to the optical subsystem, we have

$$\mathrm{FOV} = \theta_s + \mathrm{IFOV}$$

While the opposite case,

$$\mathrm{FOV} = \tan^{-1}(\theta_s + \mathrm{IFOV})$$

When $\theta_s + \mathrm{IFOV}$ is small, the above two equations are approximately equal.

7.7.3 Optical imaging system performance

The performance of an optical imaging system is often described by the sensitivity, resolution, and signal-to-noise ratio. The resolution includes the temporal resolution, spatial resolution, spectral resolution, and grayscale resolution. Spatial resolution is of main concern as it reflects the system's ability to identify the smallest details of the target.

The ratio of the output signal to noise is called signal-to-noise ratio. When the signal-to-noise ratio of the input signal is 1, the resolution is sensitive and can reflect the system's ability to detect small signals.

There are many different forms of resolution which are followed with interest by researchers in different fields. For example, for optical designers, these include Rayleigh criterion, Sparrow criterion, and Airy spot diameter. A common starting point is that the radiation spot diameter on the image plane of the optical system is equal to the size of the detector. The main concern for detectors is the number and size of the detection element. For system analysts, if geometric analysis is used, resolution is often expressed by the detector aperture angle detector angular subtense (DAS); and if MTF analysis is used, instantaneous field-of-view is used as resolution indicators. It is worth mentioning that the resolution represented by DAS is only suitable for limited detector systems, in particular, cutoff spatial frequency f_{oco} of optical system and cutoff spatial frequency f_{dco} of detector should satisfy the relationship $f_{oco} \geq 2.44 f_{dco}$. Otherwise, the resolution of the system may be limited by the cutoff spatial frequency f_{oco} of the optical system or Nyquist frequency. At this time, DAS only reflects the performance of the detector subsystem but not as the resolution of the whole imaging system.

The detector as the main noise source of optical imaging systems has been discussed Chapter 6. Here, we discuss introduced noise in combination of detector elements to array, as well as background noise and the noise produced by amplifying circuits.

When many detector elements are composed of numerous detector arrays, non-uniform noise is generated that that does not correspond to each detector element or different gains of amplifiers. This noise mainly occurs in the staring imaging system, however, it may be present in scanning imaging system having multiple detecting elements in a direction vertical to the scanning direction.

In principle, all noise apart from random noise can be eliminated or at least reduced to below a measurable value. When random noise is caused only by random events associated with the photon detector, the system has a background-limited performance (BLIP). Thermal imaging system can be operated in the BLIP mode.

Noise current value of the amplifier is provided by the manufacturer, usually depicted as noise current i_a per unit bandwidth, and the total noise current rms value is expressed as

$$\sigma_a = \left(\langle i_a^2 \rangle \Delta f_e\right)^{1/2}$$

where Δf_e is the noise-equivalent bandwidth (NEBW) of amplifiers in Hz.

7.8 Performance of staring array imaging system

7.8.1 Field of view

Field-of-view of staring array imaging system is represented by equation (7.20), where Sa is the total size of the detector array. Suppose in an array direction, for example, the x direction, there is Nx sensing element, the size in the x direction of each detector element is dx, and the distance between the centers of two adjacent detector units is dccx, then

$$S_a = (N_x - 1)d_{ccx} + d_x$$

If the detector elements are enough, then

$$N_x \gg 1$$

thus

$$S_a \approx N_x d_{ccx}$$

On substituting it into equation (7.20) gives

$$FOV_x \approx 2\tan^{-1}\frac{N_x d_{ccx}}{2f}$$

Similarly, in the y direction,

$$FOV_y \approx 2\tan^{-1}\frac{N_y d_{ccy}}{2f}$$

where the meanings of Ny and d_{ccy} are self-evident.

7.8.2 Noise and signal-to-noise ratio

Noise of a staring array imaging system includes shot noise, fixed pattern noise, and multiplier noise. The corresponding noise-currents are i_{shot}, i_{nu}, and i_{mux}, respectively, and the signal current is the signal to noise ratio expressed as

$$SNR = \frac{i_s}{\sqrt{\langle i_{shot}^2 \rangle + \langle i_{nu}^2 \rangle + \langle i_{mux}^2 \rangle}}$$

Here, sources of non-uniformity noise include non-linear response rate of detection elements, different spectral response of the sensing element, and 1/f noise. The first two do not substantially change over time, and the 1/f noise with a lower frequency, and thus non-uniform noise, also does not change with time, and is known as fixed pattern noise. In fact, it causes a slow change in the output signal due to changes in the 1/f noise. The degree of influence is dependent on the 1/f noise characteristic and data acquisition time after the last calibration. If the system is corrected only once after it starts working, noise accumulates slowly over time.

7.9 Further description of the scanning imaging system performance

7.9.1 Scanning imaging system

Scanning optical imaging systems generally require wide-angle imagery for airborne reconnaissance. The imagery can be obtained either with push-broom systems or line scanner.

The operation principle of line scanners is shown in Fig. 7.8. A line scanner is an imaging system that comprises detectors, optical subsystems, and components for mechanical scanning. The line scanner is mounted on a flying platform such as aircraft or satellites. For each mirror rotation, the detector array senses a narrow swath on the ground. The forward motion of the aircraft provides vertical extent. The data obtained is continuously displayed in a waterfall manner, that is, the imagery

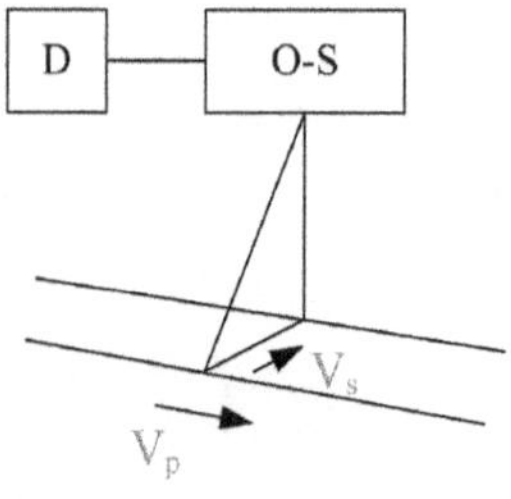

Fig. 7.8: A line scanner functional electro-optical block diagram. D: Detectors; O-S: optical and scanning subsystems; Vs: line scanning direction; Vp: platform moves direction

constantly moves down the monitor screen as the platform moves forward. The scan time T of each row is equal to the ratio of the instantaneous field-of-view and the floor projection speed of platform motion. If each line scanner samples (reading) n times, the integration time of each resolution element is T/n.

The basic diff`erence of push-broom optical imaging system (Fig. 7.9) and the line scanning imaging system is the absence of mechanical scanning components. Line scanning function is performed by self-scanning inside the detector (as CCD). Thus, the optical system images once for one row of radiation of ground resolvability element. Platform motion provides extended vertical direction. If the detector is in the mxn array, the detection element number m is equal to the pixel number imaging one row of ground resolvability element radiation. The integration time on each pixel is theoretically m times of line scanning.

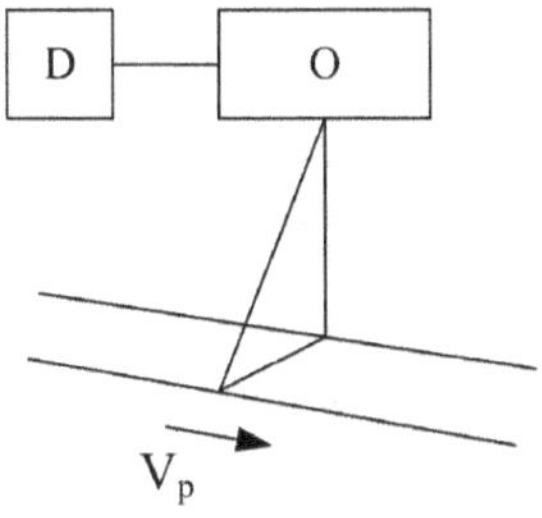

Fig. 7.9: A push-broom scanner functional electro-optical block diagram. D: Detector; O: optical subsystem; Vp: platform moves direction

7.9.2 System noise of scanning imaging

Noise of the scanning imaging system includes photon noise (Ip), Johnson noise (IJ), and amplifier noise (Ia). The mean square value of noise current in unit bandwidth is expressed as

$$\langle i_n^2 \rangle = \langle i_p^2 \rangle + \langle i_J^2 \rangle + \langle i_a^2 \rangle$$

Suppose that noise equivalent bandwidth is Δf_e, then the total noise current rms value is

$$\sigma_n = \sqrt{\langle i_n^2 \rangle \Delta f_e}$$

8 Fundamental of Nonlinear Optics

Nonlinear optics is an important branch of modern optics that emerged after the invention of laser. It covers the second, third, and higher-order nonlinear optic effects, and also studies some phenomena such as photorefractive effect and photon echo. In Section 8.2, we introduce the basic knowledge of nonlinear optics phase conjugate. Section 8.3 consider three-wave mixing as an example to discuss the second-order nonlinear optics effect. Sections 8.4 to 8.6 successively discuss the four-wave mixing, near-degenerate four-wave mixing, as well as harmonic oscillation four-wave mixing. In Section 8.8, we describe two kinds of elastic scattering that belong to the third order of nonlinear optics. Photon echo is a transient process along with a special nonlinear optics effect, Photorefractive phenomenon will be descripted in Sections 8.7 and 8.9.

Phase conjugate waves can be produced in several nonlinear optics processes. Because phase conjugate wave can be used to correct distorted wavefront, it has important applications in laser nuclear fusion, lens-less imaging, and high-power, and high-light beam quality laser. These topics are also an important part of this chapter.

8.1 Introduction

In this section, we briefly review some basic knowledge about nonlinear optics, including nonlinear wave function and its slow varying approximation form, as well as nonlinearity of materials and its coupling with light wave.

8.1.1 Nonlinear wave function

We know that, when light travels in a medium, it reacts with atoms, molecules, and other particles inside the medium. First, the electric field of light polarizes the medium, and then the electric field generated as a result of this polarization in turn changes the original optical field. These mutual interactions can be described by Maxwell equations and other related functions. Assume that the medium is homogeneous, nonmagnetized, and nonconductor and does not contain free electrons, in SI units, these equations can be expressed as

$$\nabla \cdot \boldsymbol{D} = \rho$$

$$\nabla \cdot \boldsymbol{B} = 0$$

$$\nabla \times \boldsymbol{H} = J + \frac{\partial \boldsymbol{B}}{\partial t}$$

https://doi.org/10.1515/9783110500608-008

$$\nabla \times \boldsymbol{E} = -\frac{\partial \boldsymbol{B}}{\partial t} \tag{8.1}$$

and

$$\boldsymbol{D} = \varepsilon_0 \boldsymbol{E} + \boldsymbol{P}$$
$$\boldsymbol{J} = \sigma \boldsymbol{E} \tag{8.2}$$

Here, each quantity has the same meaning as in any ordinary electrodynamic textbook. The only difference is the polarization of matter, which should be

$$\boldsymbol{P} = \boldsymbol{P}_L + \boldsymbol{P}_{NL} \tag{8.3}$$

Here, P_L and P_{NL} are linear and nonlinear parts, respectively. On combining the above equations, we get the plane wave equation that propagates in a nonlinear medium.

$$\nabla^2 \boldsymbol{E} = -\mu_0 \varepsilon \frac{\partial^2 \boldsymbol{E}}{\partial t^2} = \mu_0 \frac{\partial^2 \boldsymbol{P}_{NL}}{\partial t^2} \tag{8.4}$$

Equation (8.4) shows that, when nonlinear polarization exists, new electric field is generated. On neglecting the nonlinear polarization, right-hand side of Equation (8.4) becomes zero and the equation is reduced to a standard linear wave equation. Many waves can travel independently in this kind of nonboundary media without generating any new waves. Coupling between waves and new wave generation can only occur through interaction of nonlinear polarization.

In this chapter, we treat electromagnetic wave as classical wave and use quantum mechanics to deal with nonlinear media. Therefore, we use semi-classical study interaction between radiation and media.

To simplify our discussion without losing generality, we assume all electric fields and polarization vectors are polarized in the same direction. Hence, the vector symbols can be omitted. Moreover, we also treated nonlinear tensors as scalars.

8.1.2 Slowly varying envelope approximation (SVEA) of equation

Solving second-order nonlinear wave equation is tedious and usually not required. Often times, we simplify it to a first-order equation. For this, express the plane wave in z direction as

$$E(z,t) = \frac{1}{2}\varepsilon(z,t)exp[j(\omega t - kz)] + c.c. \tag{8.5}$$

Here, $k = n_0\omega/c$ is a wave vector, n_0 is a linear refraction coefficient, and c.c. is a complex conjugate of the former term, and

$$\varepsilon(z,t) = a(z,t)exp[-j\delta\phi(z,t)]$$

represents the envelope of the complex amplitude. $\delta\varphi$ is the phase modulation function, and $a(z, t)$ is a real amplitude.

Nonlinear susceptibility that can couple with the above wave field should have the same frequency ω and wave vector k as the field. Therefore, the nonlinear susceptibility can be expressed as

$$P_{NL}(z,t) = \frac{1}{2}F(z,t)exp[j(\omega t - kz)] + c.c. \tag{8.6}$$

Slow variation means characters (envelope and instant phase) of light field change very little in one period or in one wavelength range. In mathematics

$$\left|\omega^2\varepsilon\right| \gg \left|\omega\frac{\partial\varepsilon}{\partial z}\right| \gg \left|\frac{\partial^2\varepsilon}{\partial t^2}\right| \tag{8.7a}$$

or equivalently

$$\left|k^2\varepsilon\right| \gg \left|k\frac{\partial\varepsilon}{\partial z}\right| \gg \left|\frac{\partial^2\varepsilon}{\partial z^2}\right| \tag{8.7b}$$

On substituting equation (8.5) and equation (8.6) into equation (8.4) using equation (8.7), we get

$$\left(\frac{\partial}{\partial z} + \sqrt{\mu_0\varepsilon}\frac{\partial}{\partial t}\right)\varepsilon = j\frac{\omega}{2}\sqrt{\frac{\mu_0}{\varepsilon}}F \tag{8.8}$$

Equation (8.8) is the required first-order equation. It shows that, for a given nonlinear polarization, an electric field with the same frequency and wave vector similar to coming wave will be induced.

The above result can be extended to a situation wherein multiwave train existed in a medium. In this situation, the electric field can be expressed as

$$E(r,t) = \frac{1}{2}\sum_i \varepsilon_i(r,t)exp[j(\omega_i t - k_i \cdot r)] + c.c.$$

This is true for all electric field components of each wave. Similarly, nonlinear polarization is

$$P_{NL}(r,t) = \frac{1}{2}\sum_i F_i(r,t)exp[j(\omega_i t - k_i \cdot r)] + c.c.$$

Note that propagation direction is arbitrary (not necessary follow z direction). Equation (8.9) shows the first-order equation in i direction.

$$\left(\frac{\partial}{\partial z} + \sqrt{\mu_0\varepsilon}\frac{\partial}{\partial t}\right)\varepsilon_i = j\frac{\omega}{2}\sqrt{\frac{\mu_0}{\varepsilon}}F_i \tag{8.9}$$

Equation (8.9) indicates that a wave with a certain frequency and wave vector is defined only by nonlinear polarization which has the same frequency and wave vector and has nothing to do with other components in polarization with a different frequency.

8.1.3 Nonlinearity of material and its coupling with light wave

The permeability of a nonlinear material can be expressed as a series

$$\chi(E) = \chi^{(1)} + \chi^{(2)}E + \chi^{(3)}E^2 + \dots \tag{8.10}$$

The nonlinear polarization is

$$P(E) = E\chi(E) = \chi^{(1)}E + \chi^{(2)}E^2 + \chi^{(3)}E^3 + \dots \tag{8.11}$$

E in equation (8.10) and equation (8.11) represents total electric field. It can be composed of many wave fields of different frequencies, wave vector, and polarization status. Term $\chi^{(1)}$ corresponds to linear optical effects, namely, refraction, absorption, gain, and birefringent. These topics are studied by classical optics. The other components of χ correspond to nonlinear optical effects.

In equation (8.10), high order of χ components are much smaller than the lower ones, the relationship between adjacent components is approximately

$$\chi^{(2)}/\chi^{(1)} \approx \chi^{(3)}/\chi^{(2)} \approx 10^{-10}$$

From equation (8.11), we can see that, only when the field strength of incident radiation is in the order of 10^{10} V/m, the nonlinear optical effect reach the same level as the linear optical effects.

Phenomena which belong to second-order nonlinear optical effects mainly include

- Second-harmonic wave generation $\chi^{(2)}(2\omega, \omega, \omega)$
- Optical rectification $\chi^{(2)}(0; \omega, -\omega)$
- Parametric mixer $\chi^{(2)}(\omega_1 \pm \omega_2; \omega_1, \omega_2)$
- Pockels effect $\chi^{(2)}(\omega, \omega, 0)$

Usually, these processes are called three-wave mixing. In parentheses, ω_i before semicolon are the frequencies of output waves and ω_i after semicolon are the frequencies of incident waves. For instance, $\chi^{(2)}(\omega_1 \pm \omega_2; \omega_1, \omega_2)$ represents the nonlinear interaction between the two waves with frequencies ω_1 and ω_2 producing waves with frequencies $\omega_1 + \omega_2$ or $\omega_1 - \omega_2$.

The second-order effect can occur only in materials that are not inversion symmetry. Second-harmonic wave generation and parametric mixing conversion effect are dependent on the conditions of phase matching. Just as shown in equation (8.9), only when the frequency and wave vector are tuning in certain values, does effective conversion can be realized (we will discuss this in later sections). These conditions can be reached by adjusting the wave frequency, incident angle, or changing the material's temperature.

$\chi^{(3)}$ corresponds to the third-order, nonlinear effect. Important examples are:
- Third harmonic generation $\chi^{(3)}(3\omega;\omega,\omega,\omega)$
- DC Kerr effect $\mathrm{Re}\left[\chi^{(3)}(\omega,\omega,0,0)\right]$
- Degenerate four-wave mixing (instantaneous AC Kerr effect) $\mathrm{Re}\left[\chi^{(3)}(\omega,\omega,\omega,-\omega)\right]$
- Nondegenerate four-wave mixing $\chi^{(3)}(\omega_1+\omega_2\pm\omega_3;\omega_1,\omega_2,\pm\omega_3)$
- Brillouin scattering $\chi^{(3)}(\omega\pm\Omega;\omega,-\omega,\omega\pm\Omega)$
- Raman scattering $\chi^{(3)}(\omega\pm\Omega;\omega,-\omega,\omega\pm\Omega)$
- Dual photon absorptiometry $\mathrm{Im}\left[\chi^{(3)}(\omega,\omega,-\omega,\omega)\right]$
- DC-induced harmonic generation $\chi^{(3)}(2\omega;\omega,\omega,0)$

The above processes involved four-wave interaction. ω in parentheses has the similar meaning as in the three-wave mixing. Third-order effect is independent of whether the material is inversion symmetry.

8.2 Optical phase conjugate

In this section, we mainly discuss two problems: one is the definition of phase conjugate wave; and the other one is the comparison of phase conjugate mirror (PCM) and common plane mirror (CPM).

8.2.1 Definition of phase conjugate wave

Consider a plane wave with frequency ω travels in the positive z direction.

$$\mathrm{E}=\mathrm{Re}\{\varepsilon(r)exp[j(\omega t-kz)]\}\equiv Re\left[\psi(r)e^{j\omega t}\right] \tag{8.12}$$

Its phase conjugate wave is defined as

$$\mathrm{E_c}=\mathrm{Re}\{\varepsilon^*(r)exp[j(\omega t+kz)]\}=Re\left[\psi^*(r)e^{j\omega t}\right] \tag{8.13}$$

That means phase conjugate wave only contains the space conjugate, and the part related to time remains constant. If we consider a complex variable has the same real part as its conjugate, then we can write

$$E_c=Re\left\{\left[\psi^*(r)e^{j\omega t}\right]^*\right\}=Re\left[\psi(r)e^{-j\omega t}\right] \tag{8.14}$$

Therefore, phase conjugate wave is the time inverse wave while keeping the space part uniform, we obtain

$$E_c(r,t) = E(r, -t) \quad (8.15)$$

Based on this, sometimes "time reversal reproduction" has been used to describe phase conjugate wave in literature.

8.2.2 Comparison of PCM and CPM

All the devices that can change incidence wave to phase conjugate wave are called PCM. Compared to CPM, PCM has the following special characteristics.

1. Reflected light returns in the same route as the incident route

Figure 8.1 shows the obvious difference between PCM and CPM. The latter only changes the sign of the normal component of wave vector, while keeping the tangent component unchanged. PCM inverses wave vector **k**, implying that PCM changes sign of both normal and tangential components of the wave vector. The result is the phase conjugate wave returns exactly along the route of the incident wave, which is independent to the orientation of the incident wave. Obviously, this characteristic of PCM can change a converging wave to a diverging wave and vice versa. When the incident wave comes from a point source, all reflected phase conjugate waves from PCM focus on the origin, as showed in Fig. 8.2(b). Figure 8.2(a) shows that the diverging wave reflected from CPM is still diverging.

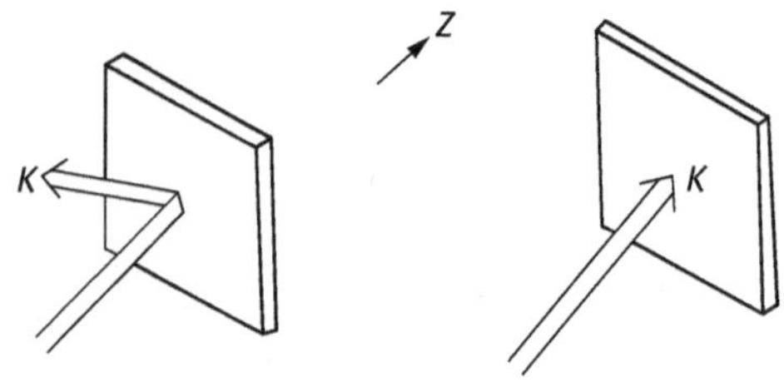

Fig. 8.1: Directions of reflected waves from CPM and PCM. (a) Reflection from CPM, $k_{in} = k_x x + k_y y + k_z z$, $k_{out} = k_x x + k_y y - k_z$. (b) Reflection from PCM, $k_{out} = -k_{in}$.

The above discussion is one of the most important properties of PCM, and is also one of the main reasons for its wide applications. This property can be described more accurately from a mathematical viewpoint.

Assume reflection mirror is in the xy plan, its normal is in the z direction. An incident wave has wave vector $k_{in} = (k_x, k_y, k_z)$ strikes on the mirror, and after being reflected from a CPM, the wave vector becomes $\mathbf{k}_{out} = (k_x, k_y, -k_z)$; whereas after being reflected from a PCM, the wave vector changes to $\mathbf{k}_{out} = (-k_x, -k_y, -k_z)$.

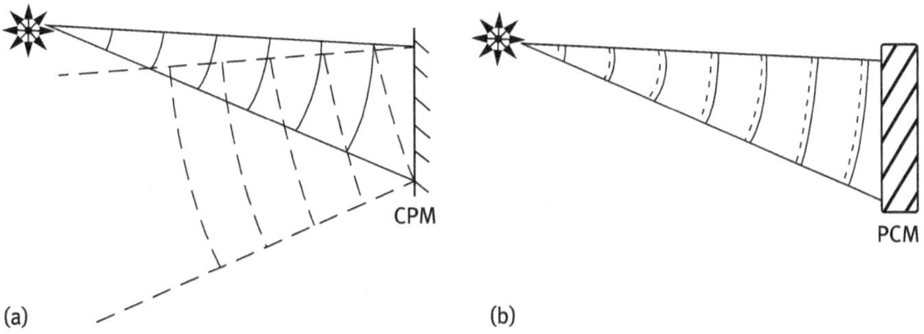

Fig. 8.2: Divergent characteristics of CPM and PCM. (a) Reflected from CPM. (b) Reflected from PCM.

2. Inverse polarization

When an ideal PCM reverses the direction of wave propagation, it also reverses the polarization vector of the incident wave. Therefore, a right-hand circular polarization light (RHCP) reflected from PCM will travel back along the income route and still remains RHCP. If the same light is reflected from a CPM, the returning light will be a LHCP. Figure 8.3 shows these two situations.

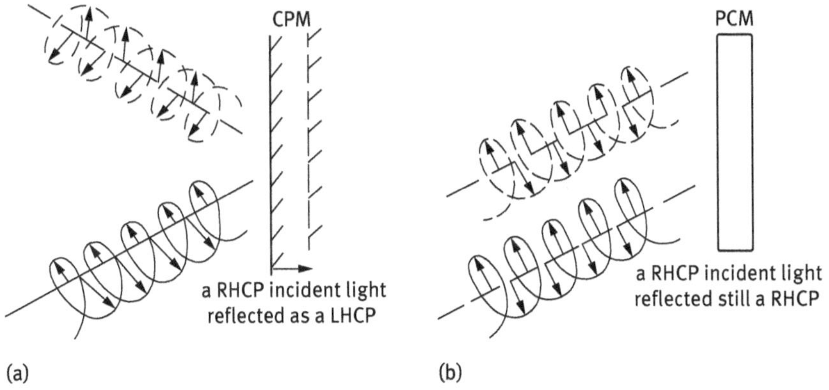

Fig. 8.3: Polarization properties of CPM and PCM. (a) an RHCP incident light reflected as a LHCP; (b) an RHCP incident light reflected still a RHCP.

Fundamentally, all quantum numbers of incident photons reflected from an idea lossless PCM are the same such as linear momentum and angular momentum. Consequently, PCM is not affected by photon radiation pressure and torque, and therefore, there is no recoil. On the other hand, when a CPM illuminated by light, it bears zero radiation pressure and has a recoil.

In brief, the conjugated reproduction E_{pc} is a field which has the same equiphasic surface, but the opposite propagation direction still satisfies Maxwell equations. Assume a PCM is located at $z = z_0$, then the above statement holds for the entire region of $z \le z_0$ (reflective conjugate mirror).

3. Eliminate wavefront distortion

Figure 8.4 shows the influence of phase distortion on the propagation of a plane wave reflected from CPM and PCM. Assuming that wavefront distortion happens after a monochromatic plane wave travels through a phase medium, and the distorted wave is reflected from CPM and goes through the same medium once again, the phase difference is doubled (Fig. 8.4a). In contrast, PCM effectively inverses the time of the incident field. As a result, when the wave passes the same medium once again, the wave front recovers well (Fig. 8.4b).

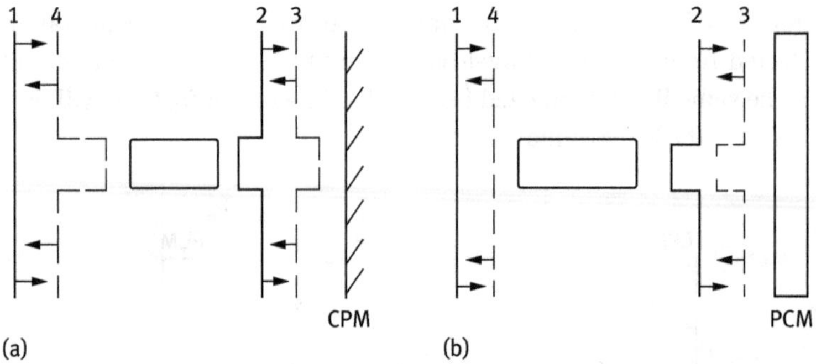

Fig. 8.4: Remove distortion of reflection from CPM and PCM. (a) Reflection from CPM; (b) reflection from PCM.

The main function of adaptive optics imaging system is to perform wavefront correction. Therefore, the property of the phase conjugate wave can remove wavefront distortion is the foundation that phase conjugate wave could be used widely in adaptive optics.

Property of phase conjugate wave eliminates wavefront distortion and can be proven using mathematics. For this, consider a monochromatic plane wave that passes through a medium whose permittivity can be expressed as a real variable $\varepsilon(r)$ (for more complex situation, such as nonmonochromatic plane wave or permittivity is a complex number, please refer to related literature). If $\varepsilon(r)$ is a real number, indicates that there exist passive linear elements (which include lenses, optical wedge) or a distortion medium (such as atmospheric turbulence) in the optical path. A light wave traveling along +z direction can be expressed as

$$E(r,t)=\frac{1}{2}\varepsilon(r)\exp[j(\omega t-kz)]+c.c. \tag{8.16}$$

Under slow variation approximation, it satisfies the following scalar wave equation

$$\nabla^2\varepsilon+\left[\omega^2\mu_0\varepsilon(r)-k^2z\right]\varepsilon+2jk\frac{\partial\varepsilon}{\partial z}=0 \tag{8.17}$$

Taking conjugates on both sides of equation 8.17 gives

$$\nabla^2\varepsilon^*+\left[\omega^2\mu_0\varepsilon(r)-k^2z\right]\varepsilon^*+2jk\frac{\partial\varepsilon^*}{\partial z}=0 \tag{8.18}$$

Equations (8.18) and 8.17 are the same wave equations, and describe that when wave travels in −z direction, the solution is

$$\frac{1}{2}a\varepsilon^*(r)exp[j(\omega t+kz)]+c.c.$$

Except an unimportant constant a, this is the exact phase conjugate wave of equation (8.5), which implies that phase conjugate wave travels in a direction opposite to the initial wave and both wavefronts coincide everywhere in the route.

The above conclusion is true for real number $\varepsilon(r)$, but for lossy and amplified media, $\varepsilon(r)$ is a complex number, so the above proof is not necessary unless loose and gain are independent of r.

8.3 Three-wave mixing

In this section, we first introduce elastic photon scattering. Elastic photon scattering is a process in which the medium stays in the same quantum state before and after light reaction on it. The nonlinear optics processes that belong to this category include three-wave mixing (TWM), four-wave mixing (FWM), photon echo, situation effect, plasma, nonlocal effect, thermal effect, and surface effect. Due to space limitations, we mainly discuss TWM and FWM and briefly introduce photon echo. Here, we concentrate on FWM.

TWM can only produce forward conjugate wave, but FWM can generate both forward and backward conjugate waves. For TWM and FWM, forward processes are limited by severe phase match conditions, and their applications are also restricted. Backward reaction are not restricted by the same conditions, therefore, it gets has applications in imaging and space propagation.

In this discussion, we ignore polarization effect, therefore, reaction fields can be expressed as scalars.

8.3.1 Phase matching three-wave mixing

Assume there are two monochromatic waves, namely pump wave

$$E_1 = \frac{1}{2}\varepsilon_1(r)exp[j(\omega_1 t - k_1 z)] + c.c. \tag{8.19}$$

and probe wave

$$E_p = \frac{1}{2}\varepsilon_p(r)\exp\left[j\left(\omega_p t - k_p z\right)\right] + c.c. \tag{8.20}$$

strike on a nonlinear, noninversion symmetry material, nonlinear polarization will be induced. If we want to obtain a conjugate wave E_c and its amplitude is a conjugate of E_p , then the interested polarization vector is

$$P_{NL} = \frac{1}{2}\chi^{(2)}\varepsilon_1(r)\varepsilon_p^*(r)exp\{j[(\omega_1 - \omega_p)t - (k_1 - k_p)z]\} + c.c. \tag{8.21}$$

Assume $\omega_1 = 2\omega_p$, then P_{NL} will radiate a wave

$$E_c \propto \chi^{(2)}\varepsilon_1(r)\varepsilon_p^*(r)exp\left[j(\omega_p t - k_c z)\right] + c.c. \tag{8.22}$$

with frequency $\omega_c = \omega_1 - \omega_p = \omega_p$

This implies that E_c is wavefront inversion of E_p, and travels along the z direction. This gives us a forward conjugate wave with wave vector

$$\begin{aligned} \boldsymbol{k}(\omega_1 - \omega_p) &= \boldsymbol{k}_1(\omega_1) - \boldsymbol{k}_p(\omega_p) \\ &= \boldsymbol{k}(2\omega_p) - \boldsymbol{k}(\omega_p) \end{aligned} \tag{8.23}$$

Equation (8.23) is called a phase match condition. An example of this matching is shown in Fig. 8.5(a). When we choose $\omega_1 \approx 2\omega_p$ in the graph, plane wave $\varepsilon_p(r)$

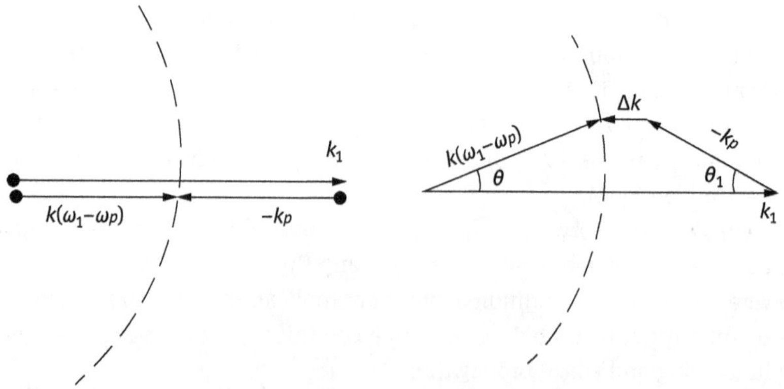

Fig. 8.5: Phase conjugate generated by TWM. (a) Phase match; (b) Phase mismatch.

propagation direction is parallel to $\varepsilon_1(\mathrm{r})$, thus an ideal phase matching is obtained. However, till now, in all possible applications, all pump waves are not plane waves and can be treated as successive superpositions of plane waves. All these plane waves have the same frequency ω_p, their wave vectors occupy certain sloid angles called the angular field view of incident wave. To find the complex conjugate of total incident wave, a necessary and sufficient condition is to first find complex conjugate of each plane wave, such that each conjugate has the same ratio as the new total conjugate as the original ratio for each component to the original total light.

8.3.2 Phase mismatching three-wave mixing

If pump wave is a plane wave with frequency of ω_p and its propagation direction is not parallel to k_1, then the phase condition is not satisfied, the mismatching quantity

$$\Delta\boldsymbol{k} = \boldsymbol{k}(\omega_1 - \omega_p) - [\boldsymbol{k}_1(\omega_1) - \boldsymbol{k}_p(\omega_p)]$$

takes the minimum value when $\theta = \theta_1$. In phase mismatch, conjugate wave gets greatly affected. To reduce this, generally

$$\left|[\boldsymbol{k}_c(\omega_1 - \omega_p) - \boldsymbol{k}_1(\omega_1) - \boldsymbol{k}_p(\omega_p)] \cdot L\right| \ll 1$$

is required. Here, L is the reaction distance. This requirement brings many restrictions on incident receiving angles.

8.4 Degenerate four-wave mixing

In the phase conjugation process of four-wave mixing, three waves are incident on a nonlinear medium and the fourth wave can be generated due to elastic photon scattering. Its amplitude is proportional to the conjugate of one of the incident waves. If two of three incident waves are stronger, they are called pump wave can be written as

$$E_1(\omega_1, \boldsymbol{r}) = \frac{1}{2}\varepsilon_1(\omega_1, \boldsymbol{r}) exp[j(\omega_1 t - \boldsymbol{k}_1 \cdot \boldsymbol{r})] + c.c.$$

$$E_2(\omega_2, \boldsymbol{r}) = \frac{1}{2}\varepsilon_2(\omega_2, \boldsymbol{r}) exp[j(\omega_2 t - \boldsymbol{k}_2 \cdot \boldsymbol{r})] + c.c.$$

The third weaker wave is called a probe wave, that is

$$E_p(\omega_p, \boldsymbol{r}) = \frac{1}{2}\varepsilon_p(\omega_p, \boldsymbol{r}) exp[j(\omega_p t - \boldsymbol{k}_p \cdot \boldsymbol{r})] + c.c.$$

The polarization generated in a medium is

$$P_{NL} = \frac{1}{2}\chi^{(3)}\varepsilon_1(\boldsymbol{r})\varepsilon_2(\boldsymbol{r})\varepsilon_p^*(\boldsymbol{r}) exp\{j[(\omega_1 + \omega_2 - \omega_p)t - (\boldsymbol{k}_1 + \boldsymbol{k}_2 - \boldsymbol{k}_p) \cdot \boldsymbol{r}]\} + c.c. \quad (8.24)$$

P_{NL} radiates a conjugate wave with frequency

$$\omega_c = \omega_1 + \omega_2 - \omega_p \tag{8.25}$$

If three incident waves have the same frequency ω, from equation 8.25, we know the conjugate waves also have frequency ω . This process is called degenerate four-wave mixing (DFWM). In this situation, equation 8.24 becomes

$$P_{NL} = \frac{1}{2}\chi^{(3)}\varepsilon_1(\boldsymbol{r})\varepsilon_2(\boldsymbol{r})\varepsilon_p^*(\boldsymbol{r})exp\{j[\omega t - (\boldsymbol{k}_1 + \boldsymbol{k}_2 - \boldsymbol{k}_p)\cdot \boldsymbol{r}]\} + c.c.$$

Based on the conjugate wave propagation direction relative to the direction of the probe wave, there are forward conjugate waves and backward conjugate waves. We will discuss these two situations later and emphasize on the backward conjugate wave.

8.4.1 Forward conjugate wave generated by FWM

In the process of generating forward conjugate wave from DFWM, two pump waves travel in the same direction, such they have the same wave vector, that is

$$\boldsymbol{k}_1(\omega_1 = \omega) = \boldsymbol{k}_2(\omega_2 = \omega)$$

From this, we obtain

$$P_{NL} = \frac{1}{2}\chi^{(3)}\varepsilon_1(r)\varepsilon_2(r)\varepsilon_p^*(r)exp\{j[\omega t - (2k_1 - k_p)\cdot r]\} + c.c.$$

So, the wave vector of conjugate wave is

$$\boldsymbol{k}_c = 2\boldsymbol{k}_1 - \boldsymbol{k}_p$$

The choice of beam direction, as shown in Fig. 8.6, conjugate wave k_c travels in near forward. Similar to TWM, severe phase matching condition restricts the receiving angles of probe wave, therefore, its application is limited.

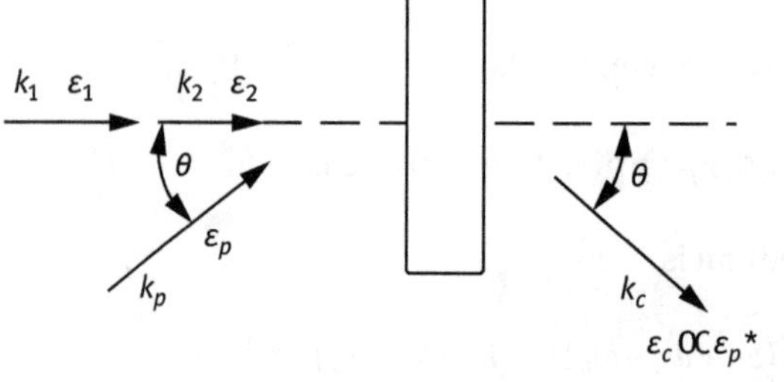

Fig. 8.6: Forward conjugate wave generated by FWM.

8.4.2 Backward conjugate wave generated by FWM

Figure 8.7 shows backward conjugate wave generated by FWM. Here, two plane pump waves travel in opposite directions, therefore (more strictly) we have

$$k_1 + k_2 = 0$$

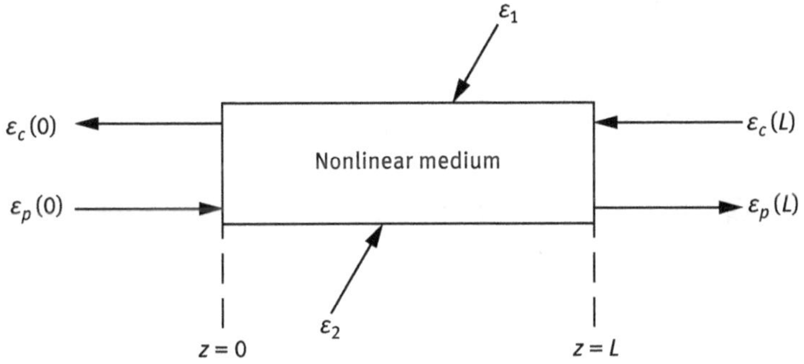

Fig. 8.7: Schematic of FWM that generates backward cognate wave.

Thus, degenerate, nonlinear polarization is expressed as

$$P_{NL} = \frac{1}{2}\chi^{(3)}\varepsilon_1(\boldsymbol{r})\varepsilon_2(\boldsymbol{r})\varepsilon_p^*(\boldsymbol{r})exp\{j(\omega t + \boldsymbol{k}_p \cdot \boldsymbol{r})\} + c.c. \tag{8.26}$$

The resulting conjugate wave $\varepsilon_c(L)$ inverses both in the wavefront and propagation direction, implying the conjugate wave $\varepsilon_c(L)$ travels in the opposite direction of incident wave $\varepsilon_p(0)$, and is called the backward conjugate wave.

It is not hard to see that the process of generating a backward conjugate wave is not affected by the incident angle of probe wave. Its function is very close to an ideal PCM. Therefore, it has wide applications, which we will discuss in this section.

Substituting equation (8.26) into equation (8.8) under steady conditions gives

$$\frac{d\varepsilon_c}{dz} = j\kappa^*\varepsilon_p^*$$

and

$$\frac{d\varepsilon_p}{dz} = -j\kappa^*\varepsilon_c^* \tag{8.27}$$

Here

$$K^* = \frac{\omega}{2}\sqrt{\frac{\mu_0}{\varepsilon}}\chi^{(3)}\varepsilon_1\varepsilon_2 \tag{8.28}$$

For boundary conditions

$$\varepsilon_p(z=0)=\varepsilon_p(0)$$

$$\varepsilon_c(z=L)=\varepsilon_c(L)$$

solutions of equation (8.27) are

$$\varepsilon_p(z)=-j\frac{|\kappa|\sin(|\kappa|z)}{\kappa\cos(|\kappa|L)}\varepsilon_c^*(L)+\frac{\cos[|\kappa|(z-L)]}{\cos(|\kappa|L)}\varepsilon_p(0)$$

and

$$\varepsilon_c(z)=\frac{\cos(|\kappa|z)}{\cos(|\kappa|L)}\varepsilon_c(L)+j\frac{\kappa^*\sin(|\kappa|(z-L))}{\cos(|\kappa|L)}\varepsilon_p(0)$$

In reality, conjugate wave is zero when, $z=L$, that is

$$\varepsilon_c(L)=0$$

So we have

$$\varepsilon_p(z)=\frac{\cos[|\kappa|(z-L)]}{\cos(|\kappa|L)}\varepsilon_p(0)$$

and

$$\varepsilon_c(z)\frac{\kappa^*\sin(|\kappa|(z-L))}{\cos(|\kappa|L)}\varepsilon_p(0) \tag{8.29}$$

Usually, people are interested in the conjugate wave at z = 0 and probe wave at z = L. That is

$$\varepsilon_c(0)=-j\left[\frac{\kappa^*}{|\kappa|}\tan(|\kappa|L)\right]\varepsilon_p^*(0)$$

$$\varepsilon_p(L)=\sec(|\kappa|L)\varepsilon_p(0) \tag{8.30}$$

Equation (8.30) is the main result of this section, which shows that the reflection phase conjugate field is proportional to the conjugate of the incident field. Power reflection and transmission coefficients R and T, respectively, are defined as

$$R=\left|\frac{\varepsilon_c(0)}{\varepsilon_p(0)}\right|^2 \text{ and } T=\left|\frac{\varepsilon_p(L)}{\varepsilon_p(0)}\right|^2$$

Applying equation (8.30) gives

$$R=tan^2(|\kappa|L) \tag{8.31a}$$

and

$$T=sec^2(|\kappa|L) \tag{8.31b}$$

For a given medium, when frequency of light is fixed, from equation (8.28) we know that κ is proportional to light intensity. Equation (8.31) shows reflection and transmission coefficients defined by the product of the intensity of pump wave and interaction length. When

$$\left(m+\frac{1}{4}\right)\pi < |\kappa|L < \left(m+\frac{3}{4}\right)\pi,$$

$$R > 1.$$

and $T > 1$ is true for all $|\kappa|$

$$L \neq m\pi, m = 0, 1, 2,$$

Therefore, in these ranges, power of reflection and transmission waves are higher than the power of incident wave. So, DFWM can give amplified phase conjugate wave (Fig. 8.8). Obviously, the extra energy is coming from pump light. The property that DFWM can produce amplified conjugate wave has great potential for conditions wherein a phase conjugate of a very weak signal is desired.

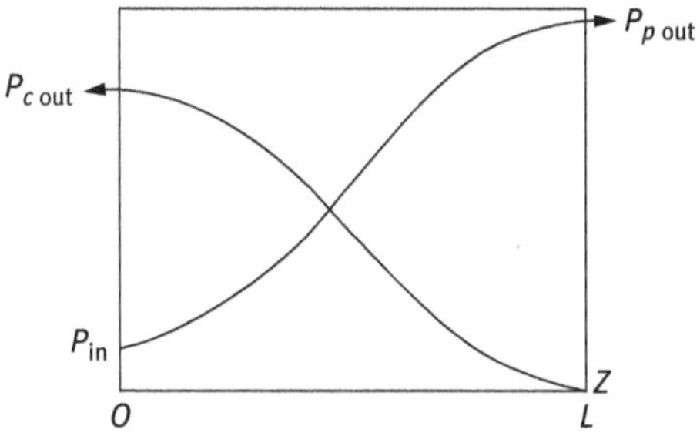

Fig. 8.8: Amplification character of FWME.

Especially, when

$$|\kappa|L = \left(m+\frac{1}{2}\right)\pi$$
$$R \to \infty,\ T \to \infty.$$

That means FWM can realize mirrorless oscillations. Even the strength of incident wave is zero, at z\mequal0 and z\mequalL limited output can still be found.

DFWM can generate an amplified conjugate wave and oscillation can be proven employing quantum optics. As this book is limited to the semi-classical theory, we do not discuss this in detail, but only use a simple graph to illustrate it. In Fig. 8.9, two counter-propagating pump photons and a probe photon (moving in z direction) strike on a nonlinear medium. After interaction, two pump photons annihilate and generate two new photons moving in z direction. As a result, the probe photon beam gets

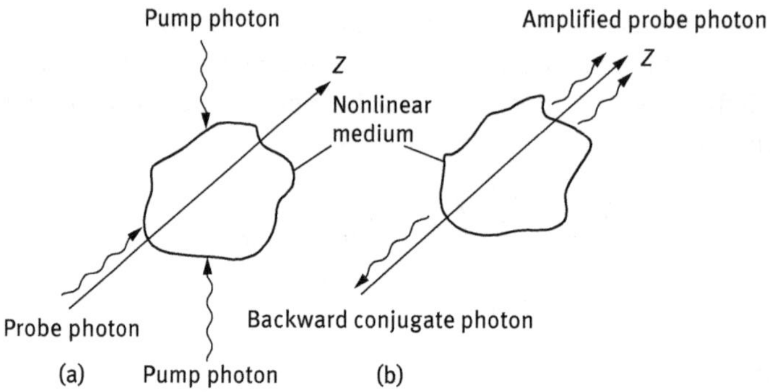

Fig. 8.9: Photon view of amplification reaction of backward degenerate four-wave mixing. (a) Before reaction; (b) after reaction.

amplified and generates a backward conjugate photon. Interested readers can refer to previously published research.

8.4.3 Experimental study of DFWM phase conjugate

Here, we will briefly introduce the experimental method of studying phase conjugate. Typical experiments are used to show the dependence of phase conjugate wave properties on some parameters. These parameters can be strength of pump wave and interaction length. Comparison between experimental results and the above-mentioned theory is also discussed.

Typical experimental set ups for verifying phase conjugate distortion property are showed in Figs. 8.10 and 8.11. A laser beam (continuous or pulsed) is split into three beams ε_1, ε_2, and ε_p by beam splitter BS_1, BS_2, and BS_3, respectively. These beams incident into a nonlinear medium NLM. For simplicity, assume that

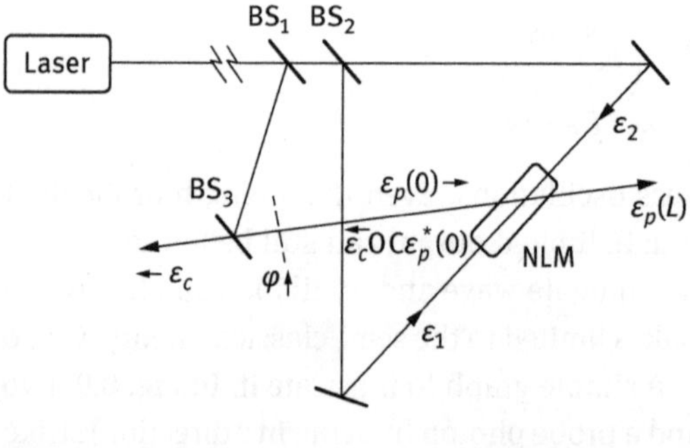

Fig. 8.10: Principle of phase conjugate DFWM: pump light comes from splitters. BS, beam splitter; NLM, nonlinear medium.

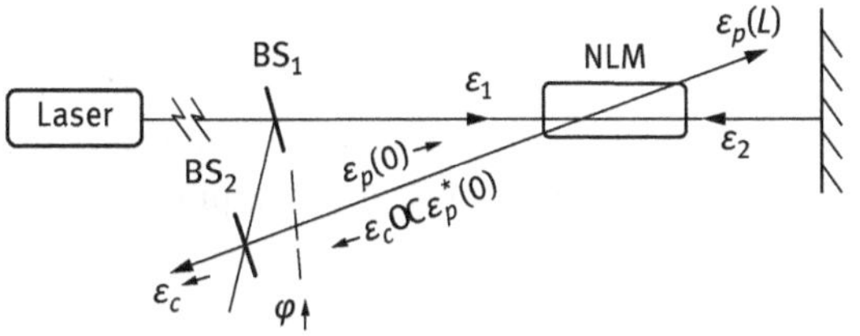

Fig. 8.11: Principle of phase conjugate DFWM: pump waves comes from reflection. BS, beam splitter; NLM, nonlinear medium.

polarization directions of these three beams are parallel. Counter-propagating ε_1, ε_2 form a strong pump field. The third weaker wave ε_p works as a probe wave, where conjugate reproduction is expected.

The difference between apparatus in Figs. 8.10 and 8.11 is how to obtain pump wave ε_2. In Fig. 8.10, ε_2 comes from the splitter while in Fig. 8.11 ε_1 passes through the medium reflected from a plane mirror to form ε_2. To decide which set-up to use in the experiment dependent on the coherence length of laser and the linear loss α in the medium. As an example, if $exp[\alpha L] > 1/4$, Fig. 8.11 is a better choice.

During the experiment, phase conjugate wave generated travels in a direction opposite to ε_p, which is detected after splitter. Probe wave ε_p incidents into nonlinear medium with a limit angle with respect to pump wave. Doing so will make phase conjugate wave separated from pump wave with the same angle and will make each field distinguishable. The disadvantage is it reduces the overlap region of interaction field and limited the reaction length. This eventually reduces nonlinear reflection coefficient.

Another popular way used to separate each field is based on their different polarization. In this situation, each field follow the same axis, this will increase interaction overlap region and overcome the shortcomings of the above method.

To prove the wavefront compensation ability of a phase conjugate, put an object which can generate phase conjugate in the route of ε_p (such as an etched glass). A famous typical experiment using above set up was done by Jain and Lind in 1983. They used a semiconductor coated glass sheet as a nonlinear medium and a Q-switched rube laser as light source for providing interacting fields. Three pictures taken were probe wave without passing the glass, probe wave passes the glass one time and phase conjugate wave. Result showed clearly that the second picture was greatly expanded and fuzzy and the third picture were very close to diffraction limit image of the first picture. The experiment proved that phase conjugate has ability to eliminate wavefront distortion.

Another typical experiment was done by Bloom et al. In the experiment, they used sodium vapor as nonlinear medium, a dye laser tuned to the vicinity of Na-D line as pump light source. They measured relationship between R and T versus intensity of pump light. Comparison of the experimental data with theoretical one is plotted in Fig. 8.12. The solid line is a theoretical gain curve of forward conjugate wave, while the dot lines are experimental data; dashed line is theoretical gain curve of backward conjugate wave. This graph also showed using DFWM can get an amplified phase conjugate wave, this agreed with the former discussed theory.

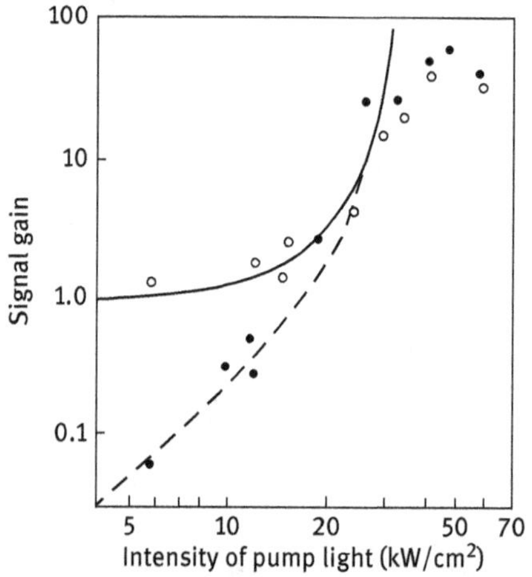

Fig. 8.12: Signal gain vs. intensity of pump light.

It is worth to notice that the characteristics of time and space of a conjugate wave. Assume nonlinear respond is instant, since nonlinear polarization of generating time-reversed wave is product of three incident waves, so in the case of pulse input, pulse width of conjugate wave should be narrower than the one of input wave.

The space character of conjugate wave can also disclose information about nonlinear reaction. Assume three conjugate waves are space Gaussian type waves, then the near-field spatial extension (i.e. light spot) of conjugate wave will be reduced. It can be explained as phase conjugate mirror works like a space Gauss wave sharpener formed by a pair of Gauss pump wave.

Before we finish this section, we would like to talk briefly about what requirements of experimental conditions should have before to get an effective time reverse wave.

First all, lasers should work in single-longitudinal and transverse mode. This will maximize coherence length therefore longer interaction distance can be achieved, it also extends the restrictions on the length of the path. Secondly, pump wave should be close to plane wave so that effective and valuable conjugate wave can be obtained. Last, prevent from any light beam comes back to laser, because it will affect the unimodality and coherence properties of laser.

Other factors related to the experimental arrangement must also be considered. First, propagation directions of two pump waves must be parallel, so that the conjugate wave reflection coefficient will be maximized. That is because if $k_1 + k_2 \neq 0$ (that is $\omega_1 = \omega_2$), phase mismatch will reduce the efficiency of the PCM. More important, this may make the conjugate wave cannot propagate precisely along the opposite direction of incident wave and resulting in inadequate compensation. Second, in most pulse light experiments, time overlap of three waves should be the maximum. Last, the ratio of pump amplitudes $\varepsilon_1/\varepsilon_2$ and the maximum intensity of pump wave must be considered. The former affects R value while the later can lead to unwanted nonlinear effect. Such as nonlinear phase move, self-focusing and self-excitation effect, etc. Above requirements can be fulfilled in most cases, this was proved by many successful experiments.

8.5 Near-Degenerate four wave mixing

Our discussion so far is limited in the situation in which only three input waves with same frequency is considered $\omega_1 = \omega_2 = \omega_p = \omega$. Thus the output frequency $\omega = \omega_1 + \omega_2 - \omega_p$ is also ω. This is called "degenerate" frequency interaction. But in actual applications often encounter are, pump source is provided by the same laser, so $\omega_1 = \omega_2 = \omega$, while probe wave may come from "noncooperated source". In this case, even if you work hard in tuning, may still cannot get $\omega = \omega_p$ precisely. In this case, assume $\omega_p = \omega + \delta$, then the output frequency is $\omega_c = \omega - \delta$. This situation is called "nondegenerate" frequency interaction. We are not going to discuss this kind of problem in general. Instead, we assume $|\delta/\omega| \ll 1$ always satisfied, so the problem becomes near degenerate four-wave mixing.

The geometry relationship between near degenerate four-wave mixing beams is the same as degenerate four-wave mixing showed in Fig. 8.11. Their frequency relationship as mentioned above. From this, the nonlinear polarization is

$$P_{NL} = \frac{1}{2}\chi^{(3)}\varepsilon_1\varepsilon_2\varepsilon_p^* \times exp\{j(\omega - \delta)t - (k_1 + k_2 - k_p) \cdot r\} \tag{8.32}$$

This leads to a set of coupled mode equation

$$\frac{d\varepsilon_p}{dz} = -j\kappa_p^*\varepsilon_c^* exp(j\Delta kz)$$

$$\frac{d\varepsilon_c}{dz} = j\kappa_c^*\varepsilon_p^* exp(j\Delta kz) \tag{8.33}$$

Here

$$\kappa_l^* = \frac{\omega_l}{2}\sqrt{\frac{\mu_0}{\in}}\chi^{(3)}\varepsilon_1\varepsilon_2, l = p, c$$

defined the complex conjugate coefficient. As there exists a frequency difference $2|\delta|$ between ε_p and ε_c, so the phase mismatch Δk is not zero any more, its value is

$$|\Delta k| = 2n\frac{|\delta|}{c} = 2n\pi\left(\frac{\Delta\lambda}{\lambda^2}\right),\ |\Delta\lambda| = |\lambda_p - \lambda_c|$$

Applying boundary conditions

$$\varepsilon_p(z=0) = \varepsilon_p(0) and\ \varepsilon_c(z=L) = \varepsilon_c(L)$$

The solutions of equation 8.33 can be written as

$$\varepsilon_p(z) = [exp(j\Delta kz/2)/D]\left\{\begin{array}{l} -j\kappa_p^* exp(j\Delta kL/2) \times \sin(\beta z)\varepsilon_c^*(L) + (\beta\cos[\beta(z-L)] \\ -(j\Delta k/2)\sin[\beta(z-L)])\varepsilon_p(0) \end{array}\right\} x$$

$$\varepsilon_c(z) = [exp(j\Delta kz/2)/D]\left\{\begin{array}{l} exp(-j\Delta kzL/2) \times [\ \beta\cos(\beta z) - (j\Delta k/2)\sin(\beta z)] \\ \varepsilon_c(L) + j\kappa_c^*\sin[\ \beta(z-L)]\varepsilon_p^*(0) \end{array}\right\} \quad (8.34)$$

here

$$D \equiv \beta\cos(\beta L) - (j\Delta k/2)\sin(\beta L)$$

and

$$\beta \equiv \left[\kappa_p\kappa_c^* + (\Delta k/2)^2\right]^{\frac{1}{2}}$$

Assume $\kappa_p = \kappa_c = \kappa$,then

$$\beta = \left[|\kappa|^2 + (\Delta k/2)^2\right]^{\frac{1}{2}} \quad (8.35)$$

Once again assume $\varepsilon_c(\mathrm{L}) = 0$, only some weak signals exist in $z = 0$ plane which is input of probe wave. Therefore equation (8.29) defines a nonlinear reflection wave in the input plane.

$$\varepsilon_c(0) = \frac{1}{D}\left[-j\kappa_c^*\sin(\beta L)\right]\varepsilon_p^*(0)$$

Substitute D into the above equation, get

$$\varepsilon_c(0) = \frac{-j\kappa_c^*\tan(\beta L)}{\beta - (j\Delta k/2)\tan(\beta L)}\varepsilon_p^*(0) \quad (8.36)$$

Power reflection coefficient

$$R = \frac{|\varepsilon_c(0)|^2}{|\varepsilon_p(0)|^2} = \frac{|\kappa|^2 tan^2(\beta L)}{\beta^2 + (\Delta k/2)^2 tan^2(\beta L)} \quad (8.37)$$

Using equation (8.35) for β, we finally get

$$R = \frac{|\kappa L|^2 tan^2(\beta L)}{|\kappa L|^2 + (\Delta kL/2)^2 sec^2(\beta L)} \quad (8.38)$$

Wavelength detuning parameter defined as

$$\Psi = \frac{\Delta\lambda}{2}\frac{2nL}{\lambda^2} = \frac{\Delta kL}{2\pi}$$

A curve of power reflection coefficient R varies with Ψ is plotted in Fig. 8.13. Different curve represents different nonlinear gain $|\kappa|L$.

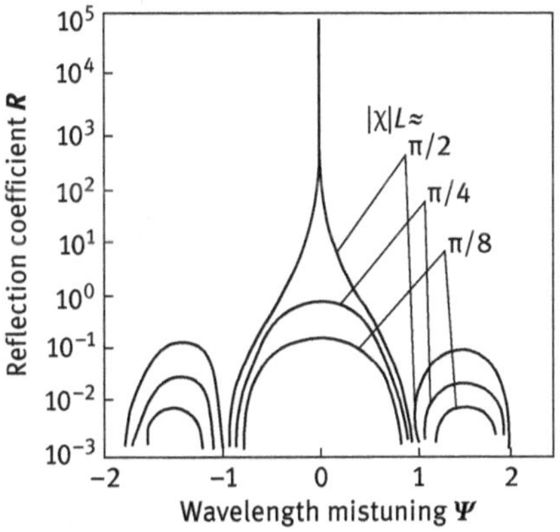

Fig. 8.13: Reflection coefficient *R* changes with respect of Ψ.

From equation (8.38) or Fig. 8.13, it can be seen that the existence of mismatch Δk cause power reflection coefficient R of conjugate wave reduces rapidly. Thus the behavior of a conjugate medium just like a narrow bandpass mirror or an optical filter. It has the following characters. First, $\varepsilon_c(0) \propto \varepsilon_p^*(0)$, that means the filter output has near time inversion character. So after space filter, the signal to noise ratio will be improved. For instance, input wave signal to be measured after passing different optical elements (space filter and lenses etc.), becomes time inverse filter wave field. By contrary, unwanted noise will be reduce greatly after passing the above optical elements. Furthermore, for a certain range of κ and $\Delta\kappa$, amplitude of conjugate wave can be bigger than the amplitude of input wave. That means the filter has amplification effect. Finally, the amount of frequency of input wave shifts up relative to pump wave is equal to the same amount of frequency of conjugate wave shifts down relative to pump wave and vice versa. So, the wavelength difference between probe wave ε_p and pump wave $\varepsilon_{1,2}$ is $\Delta\lambda/2$.

From Fig. 8.13, we can see, as gain $|\kappa|L$ increases, band pass curves become steeper. When $|\kappa|L > \pi/2$, the reflect coefficient of filter in the band pass region is bigger than 1.

If we normalize the power reflect coefficient showed in Fig. 8.13, we will have Fig. 8.14. The later makes the wavelength response characteristic become more apparent. As the nonlinear gain $|\kappa|L$ increases, the band width plummeted first and then the sidelobe structure reduced too. Which results to a sharp band pass curve. From physics point of view, these properties are based on the similarity between filter

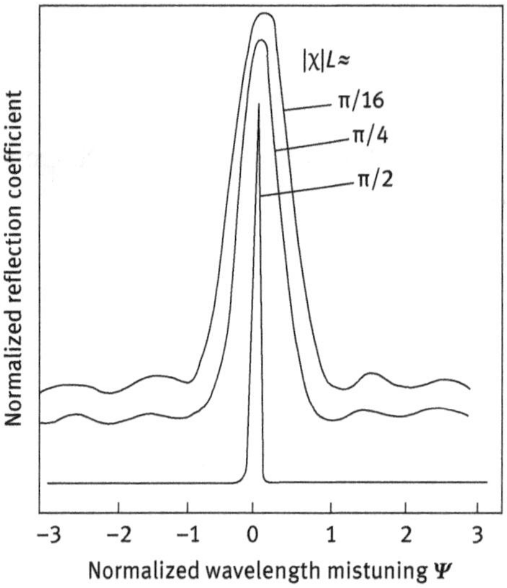

Fig. 8.14: Normalized reflection coefficient vs. wavelength mistuning.

and the Bragg reflector resonator which has real-time distribution medium intrinsic gain. For Bragg reflector resonator, increasing the gain of medium in chamber, the sharpness of wavelength responding curves of resonator or Q value sharpness will be increased. Especially, when $|\kappa|L$ is big enough so that the oscillation condition is reached, band width closes to zero, and eventually restricted by the line width or the coherence length of pump source. Use an elongated nonlinear medium with proper nonlinear coupling coefficient κ such as a long fiber optics, can get large $|\kappa|L$ value then an extremely narrow passband filter can be obtained.

It is also noticed that if enhanced resonant nonlinear medium is used, sharper filter characteristic can be obtained. The reason is involved transition linewidth. Due to the limited space, we won't talk more detail here.

In conclusion, from near degenerate four-wave mixing we can obtain an active narrow band pass optical filters. Interaction has a wide viewing angle range and frequency response, for a giving medium, it depends on interaction length and pump strength. However, for a certain device, real optical view field is restricted by optical explore method and degree of space filtering.

Equation (8.37) gives the reflection coefficient of conjugate wave, it can be used to analysis many other applications of optical phase conjugate. More specific, it contributes in understanding deeply in phase conjugate instantaneous effect and estimating some applications of optical phase conjugate.

The above formula can be used to analyze conjugate procedure of multi-wavelength. Assume pump wave $\varepsilon_{1,2}$ or probe wave ε_p contains a group of components whose frequencies are close (such as laser waves from multi-line molecule laser or chemical laser). Under this circumstance, as nonlinear interaction coupling these three input fields (multi-frequency), it will exist sum and difference frequency fields. Thus combination of a series of degenerate and near degenerate conjugate waves will be obtained. Generally speaking, as nondegenerate mode has noncomplete phase reversal and components with nonrequired frequency, therefore, it is undesirable. Applying the above theory ratio of desirable conjugate output to undesirable components can be calculated.

8.6 DFWM Resonance

Up to now, the media we have discussed whose nonlinear behavior only appears when there exists a linear dependence of reflection on wave intensity. These kind of media are called Kerr-like media. In this section, we will study FWM in resonance or near resonance media. This will involve more complex nonlinearity.

8.6.1 Qualitative description

In the most general sense, phase conjugate wave obtained from material's nonlinear respond to light stimulating. Polarization P was used as source term in Maxwell equations in most study in nonlinear optics. In DFWM, the physical meaning of P reflects that refraction and absorption can vary with light intensity. Saturated absorption and dispersion are the most obvious examples.

First consider qualitatively the process in Fig. 8.15. Assume $\varepsilon_{1,2}$ are plane waves and $\omega_{1,2} = \omega_p = \omega$. Their interference will form spatial intensity modulation structure. Saturation of atomic resonance will lead to material complex refraction spatial modulation. When ε_p is also a plane wave, there are two important modulations existed as showed in Figs. 8.16 and 8.17. In Fig. 8.16(a), ε_p and ε_1 form a grating with larger period, as satisfy the phase match, this grating will scatter ε_2 on to ε_c (see Fig. 8.16b). On the other hand, in Fig. 8.17(a), ε_p and ε_2 form a denser grating and scatter ε_1 onto ε_c (see Fig. 8.17b). The third grating formed by interference of ε_1 and ε_2 makes no contribution to this process, because the scattering of ε_p and ε_c from this grating does not satisfy phase matching condition.

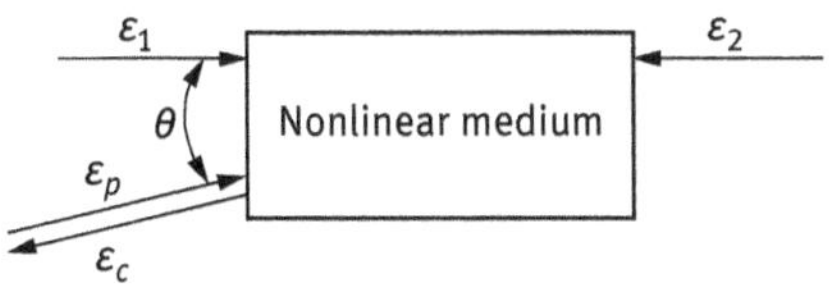

Fig. 8.15: Schematic diagram of DFWM.

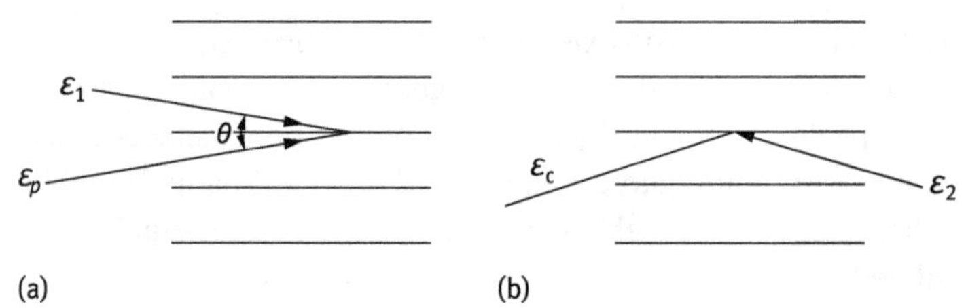

Fig. 8.16: Dual grating I. (a) Formation of sparse grating; (b) Scattering of sparse grating.

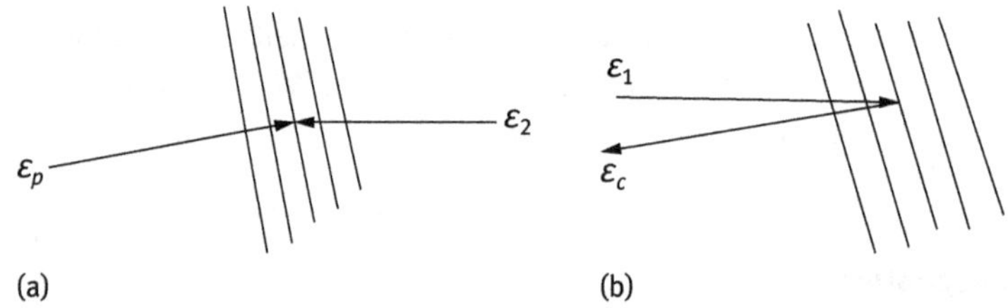

Fig. 8.17: Dual grating II. (a) Formation of dense grating; (b) Scattering of dense grating.

8.6.2 Quantitative discussion

In the resonance system, there are many other more subtle effects that can lead to or change nonlinear polarization. Due to the limited space, we only discuss the simplest situation that is saturated absorption and dispersion. The simplest situation has the following meaning. First the working substance to be two energy levels of saturated absorption system. $|1\rangle$ and $|2\rangle$ to represent ground state and excited states respectively. Second, we ignore the motion of atom and molecule. Last, assume the medium is homogeneous broadening.

The approach is to introduce a density matrix $\rho(r, v, t)$ from quantum mechanics. It can be used to descript response of macro medium and consider the statistics character of the system quantum states. If we use μ to express induced dipole moment, then the trace of matrix $\{\mu\rho\}$ is the polarization P.

Consider

$$\rho = \sum_n c_n |\Psi_n\rangle\langle\Psi_n|$$

Here c_n is statistical weight state function $|\Psi_n\rangle$. From Schrödinger equation one can find a quantum mechanical transport equation (QMTE)

$$j\left(\frac{\partial}{\partial t}+v\cdot\nabla\right)\rho=[H_0,\rho]+[V,\rho]-\frac{j}{2}\{\Gamma,\rho\}+j\Lambda;$$

If we ignore operator $v\cdot\nabla$, the above equation simplifies as

$$j\frac{\partial\rho}{\partial t}=[H_0,\rho]+[V,\rho]-\frac{j}{2}\{\Gamma,\rho\}+j\Lambda; \tag{8.39}$$

Here H_0 is nonperturbation electron Hamiltonian, its eigenequation is

$$H_0|n\rangle=\omega_n|n$$

Operator Γ is used to descript relaxation process such as spontaneous radiation decay, therefore

$$\Gamma|n\rangle=\gamma_n|n$$

Λ represents pumping stimulation of each energy level. So we have

$$\Lambda|n\rangle=\lambda_n|n$$

and

$$V=\ -\mu\cdot E(r,t)$$

expresses foreign field coupling electro dipole moment generated from classical radiation field. Finally, in equation (8.39), square brackets represents commutative and curly bracket represents anti-commutative.

The above operators have the following matrix forms

Nonperturbation Hamiltonian

$$H_0=\begin{bmatrix}\omega_1 & 0\\ 0 & \omega_2\end{bmatrix}$$

Relaxation rate

$$\Gamma=\begin{bmatrix}\gamma_1 & 0\\ 0 & \gamma_2\end{bmatrix}$$

Noncoherence pumping rate

$$\Lambda=\begin{bmatrix}\lambda_1 & 0\\ 0 & \lambda_2\end{bmatrix}$$

Dipolar coupling interaction energy

$$V=\begin{bmatrix}0 & V_{12}\\ V_{21} & 0\end{bmatrix}$$

here

$$V_{ij} = -\frac{1}{2}\mu_{ij}\cdot\left(Ee^{i\omega t} + c.c.\right)$$

Equation (8.39) can be solved by perturbation theory. For this, assume ω_p and ε_c are weak, and have no obvious effect on particle number in each energy level. Let

$$V = V^{(0)} + qV^{(1)}$$

$$\rho = \rho^{(0)} + qV^{(1)}$$

here

$$V^{(0)} = -\frac{1}{2}\mu\cdot(E_1 + E_2)e^{i\omega t} + c.c.$$

$$V^{(1)} = -\frac{1}{2}\mu\cdot\left(E_p + E_c\right)e^{i\omega t} + c.c.$$

Substitute into equation (8.38), and make the coefficients of the same power of q equal to each other on both sides, get zero and the first order equations

$$j\frac{\partial\rho^{(0)}}{\partial t} = \left[H_0, \rho^{(0)}\right] + \left[V^{(0)}, \rho^{(0)}\right] - \frac{j}{2}\left\{\Gamma, \rho^{(0)}\right\} + j\Lambda$$

and

$$j\frac{\partial\rho^{(1)}}{\partial t} = \left[H_0, \rho^{(1)}\right] + \left[V^{(0)}, \rho^{(0)}\right] + \left[V^{(1)}, \rho^{(0)}\right] - \frac{j}{2}\left\{\Gamma, \rho^{(0)}\right\}$$

The zero order equation can be solved by rotation wave approximation (RWA). For stable state, nondiagonal elements are

$$\rho_{ij} = \rho_{ij}e^{j\omega t}$$

The number of particle difference and the atomic coherence can be express as

$$\rho_{22}^{(0)} - \rho_{11}^{(0)} = \frac{-\Delta N_0}{1 + \left(\frac{\gamma_{12}}{\Gamma_0}\right)\left(\mu_{12}\cdot\frac{E_0}{\hbar}\right)^2 / \left(\gamma_{12}^2 + \Delta^2\right)} \tag{8.40}$$

and

$$\rho_{12}^{(0)} = -\frac{1}{2}\frac{\mu_{12}\cdot E_0}{j}\cdot\frac{\rho_{22}^{(0)} - \rho_{11}^{(0)}}{\gamma_{12} + j\Delta}$$

here

$$\Delta N_0 = \frac{\lambda_1}{\gamma_1} - \frac{\lambda_2}{\gamma_2}$$

is the particle difference when no external field.

$$\gamma_{12} = \frac{1}{2}(\gamma_1 + \gamma_2)$$

is atomic coherence decay rate, while particle decay rate is given by

$$\Gamma_0^{-1} = \frac{1}{2}(\gamma_1^{-1} + \gamma_2^{-1})$$

and

$$E_0 = E_1 + E_2$$

Represents the sum of two pump fields. and

$$\Delta = \omega - \omega_{21}.$$

$\Delta N_0 > 0$ corresponding to medium absorption, while $\Delta N_0 > 0$ represents amplifying.
Use RWA again to get the solution of first order equation, which is

$$\rho_{22}^{(1)} - \rho_{11}^{(1)} = -\frac{1}{j\Gamma_0}\left[\left(\mu_{12}\cdot E_0^*\rho_{12}^{(1)} - \mu_{12}\cdot E_0^*\rho_{21}^{(1)}\right) + \left(\mu_{12}\cdot E_{cp}^*\rho_{12}^{(0)} - \mu_{12}\cdot E_{cp}\rho_{21}^{(0)}\right)\right] \quad (8.41)$$

$$\rho_{12}^{(1)} = -\frac{1}{2j}\frac{1}{\gamma_{12} + j\Delta}\left[\mu_{12}\cdot E_0\left(\rho_{22}^{(1)} - \rho_{11}^{(1)}\right) + \mu_{12}\cdot E_{cp}\left(\rho_{22}^{(0)} - \rho_{11}^{(0)}\right)\right]$$

Here $E_{cp} = E_c + E_p$, $\rho_{12} = \rho_{21}^*$. Substitute the second equation of equation (8.40) into the first equation of equation 8.41, we find,

$$\rho_{12}^{(1)} - \rho_{11}^{(1)} = \Delta N_0 \frac{\frac{\gamma_{12}}{\Gamma_0}}{\gamma_{12}^2 + \Delta^2} \times \frac{\left(\mu_{12}\cdot\frac{E_0^*}{\hbar}\right)\left(\mu_{12}\cdot\frac{E_{cp}}{\hbar}\right) + \left(\mu_{12}\cdot\frac{E_0}{\hbar}\right)\left(\mu_{12}\cdot\frac{E_{cp}}{\hbar}\right)}{\left[\frac{1+\left(\frac{\gamma_{12}}{\Gamma_0}\right)\left(\mu_{12}\cdot\frac{E_0}{\hbar}\right)}{\left(\gamma_{12}^2+\Delta^2\right)}\right]} \quad (8.42)$$

equation 8.42 represents the first-order particle number difference generated under a situation in which a strong pump field and a weak probe field plus conjugate wave acting at the same time. Substitute it back to the second equation of equation 8.41, will get nondiagonal element:

$$\rho_{12}^{(1)} = \frac{N_0}{2j}\frac{1}{\gamma_{12} + j\Delta}\frac{1}{\left[1+\left(\frac{\gamma_{12}}{\Gamma_0}\right)\left(\mu_{12}\cdot\frac{E_0}{\hbar}\right)^2/\left(\gamma_{12}^2+\Delta^2\right)\right]^2}$$
$$\times\left[\left(\mu_{12}\cdot\frac{E_{cp}}{\hbar}\right) - \frac{\gamma_{12}}{\Gamma_0}\frac{\left(\mu_{12}\cdot\frac{E_0}{\hbar}\right)^2\left(\mu_{12}\cdot\frac{E_{cp}^*}{\hbar}\right)}{\left(\gamma_{12}^2+\Delta^2\right)}\right] \quad (8.43)$$

From $P = T_y\{\mu, \rho\}$, we obtain the total polarization which is the source in Maxwell equations

$$P = \frac{\Delta N_0}{2} \frac{j\gamma_{12} + \Delta}{\gamma_{12}^2 + \Delta^2} |\mu_{12}|^2 \times \frac{E_{cp} e^{j\omega t}}{\left[1 + \left(\frac{\gamma_{12}}{\Gamma_0}\right)\left(\mu_{12}\frac{E_0}{\hbar}\right)\left(\gamma_{12}^2 + \Delta^2\right)^{-1}\right]^2} + \frac{\Delta N_0}{2^3} \frac{\gamma_{12}}{\Gamma_0} \frac{j\gamma_{12} + \Delta}{\left(\gamma_{12}^2 + \Delta^2\right)^2} |\mu_{12}|^4$$

$$\times \frac{E_0^2 E_{cp}^* e^{j\omega t}}{\left[1 + \left(\frac{\gamma_{12}}{\Gamma_0}\right)\left(\mu_{12}\frac{E_0}{\hbar}\right)^2\left(\gamma_{12}^2 + \Delta^2\right)^{-1}\right]^2} + c.c. \tag{8.44}$$

This polarization contains complete response to DFWM in the case where existing of arbitrary strong pump and weak signal waves as well as probe wave. In equation (8.44), the first term corresponds to saturated absorption and nonlinear dispersion; the second term is the responds of phase conjugate signal, its real and virtual parts corresponding to dispersion and absorption of conjugate signal, respectively.

Substitute equation (8.44) to wave function, we can get SVEA equations

$$\frac{d\varepsilon_p}{dz} = -\alpha\varepsilon_p - j\beta\varepsilon_c^*$$

$$\frac{d\varepsilon_c}{dz} = \alpha\varepsilon_c + j\beta\varepsilon_p^* \tag{8.45}$$

Once again these equations show that ε_c and ε_p related to each other through their complex conjugates. It means nonlinear interaction creates phase conjugate of incident wave. The attenuation (or gain) coefficient is

$$\alpha = \alpha_0 \frac{1}{1+\delta^2} \frac{1 + \frac{(I_1 + I_2)}{I_{sat}}}{\left\{\left[1 + \frac{(I_1 + I_2)}{I_{sat}}\right]^2 - \frac{4I_1I_2}{I_{sat}^2}\right\}^{3/2}}$$

Nonlinear coupling coefficient is

$$\beta = \alpha_0 \frac{j+\delta}{1+\delta^2} \frac{2\left(I_1I_2/I_{sat}^2\right)^{1/2}}{\left\{[1 + (I_1 + I_2)/I_{sat}]^2 - 4I_1I_2/I_{sat}^2\right\}^{3/2}}$$

Here

$$\alpha_0 = \left(\frac{\omega}{2nc}\right)\Delta N_0 |\mu_{12}|^2 E_0 \gamma_{12}$$

The frequency dependent saturated light intensity is

$$I_{sat} = \frac{1}{2}\varepsilon_0 c\left(\frac{\hbar^2\gamma_{12}\Gamma_0}{|\mu_{12}|^2}\right)(1+\delta^2)$$

The light intensity dependent refraction index is

$$n^2 = 1 + \frac{\Delta N_0 |\mu_{12}|^2}{h\gamma_{12} E_0} \frac{\delta}{1+\delta^2} \times \frac{1 + \frac{(I_1+I_2)}{I_{sat}}}{\left\{\left[1 + \frac{(I_1+I_2)}{I_{sat}}\right]^2 - \frac{4I_1I_2}{I_{sat}^2}\right\}^{3/2}}$$

Here,

$$\delta = \frac{\Delta}{\gamma_{12}}$$

Once again, we assume the boundary conditions are $\varepsilon_p(z=0) = \varepsilon_p(0)$, $\varepsilon_c(z=L) = 0$, the conjugate wave power reflection coefficient can be obtained as

$$R = \left|\frac{\varepsilon_c(0)}{\varepsilon_p(0)}\right|^2 = \left(\frac{\beta \tan \gamma L}{\gamma + \alpha \tan \gamma L}\right)^2 \tag{8.46}$$

Here

$$\gamma^2 = |\beta|^2 - \alpha^2$$

If $|\beta|^2 > \alpha^2$, then γ is positive real number. The oscillation condition becomes

$$\tan \gamma L = -\frac{\gamma}{\alpha} \tag{8.47}$$

For absorbing medium, $(\alpha_0 > 0)$, equation (8.47) means $\gamma L > \pi/2$, for gain medium $(\alpha_0 < 0)$, equation (8.47) suggests $\gamma L < \pi/2$; if $\alpha_0 = 0$, the oscillation condition requires $\gamma L = \pi/2$, this corresponding to the discussion in previous section.

Several different situation of reflection defined by equation 8.46 are showed in Figs. 8.18, 8.19 and 8.20. Figure 8.18 corresponding to a situation in which pump wave

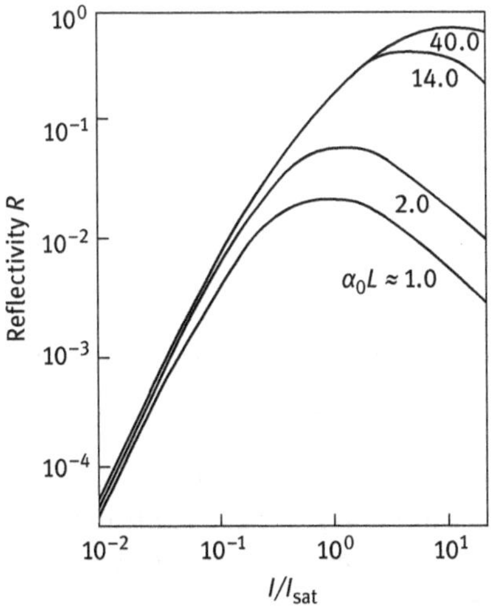

Fig. 8.18: Reflection of absorbing medium ($\delta = 0$).

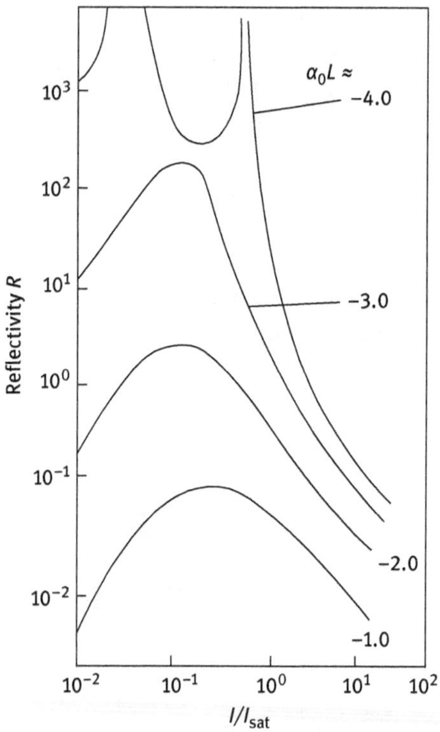

Figure 8.19: Reflection of gain medium ($\delta = 0$).

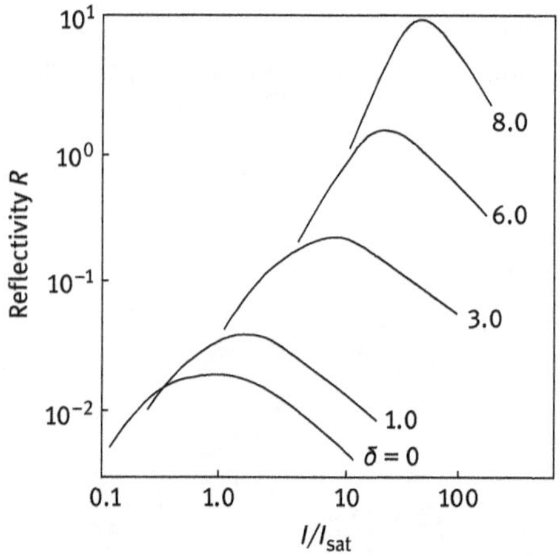

Figure 8.20: Reflection of absorbing medium.

frequency ω equals to the atomic center frequency of an absorbing medium ($\delta = 0$). In this situation, the reflection is produced by pure absorption grating. Usually the peak value of R appears near $I = I_{sat}$. When $\alpha_0 L$ small,R increases linearly with $\alpha_0 L$ increases. After R reaches a certain level, it approaches to saturate. When $\alpha_0 L$ and

I/I_{sat} are bigger, $R \to 1$. In physics the saturation can be explained as such, on one hand, $\varepsilon_C(0)$ increases as $\alpha_0 L$ increases; on the other hand, when probe wave and conjugate wave travel in a medium, due to the absorption of medium, $\varepsilon_C(0)$ will getting smaller. There are two ways to make R>1, one is to choose a gain medium, ($\alpha_0 < 0$, as is in Fig. 8.19) It is very easy to satisfy R >1, and when $\alpha_0 L = -4$, oscillation will appear.

Figure 8.20 shows another way to do the experiment in which the differences between pump frequency and spectrum center ($\delta \neq 0$) are used as a parameter for different cures. From the figure, we can see that when δ is bigger enough, R>1 can be reached. But in this case, $\alpha_0 L$ must be increased accordingly and will keep $\beta L = \alpha_0 L/(1+\delta^2) = 1$. In addition, as I_{sat} increases as $1+\delta^2$ increases, so the intensity of pump should be increased proportionally. For instance, when the difference between pump frequency and center frequency is big enough to make $\delta = 8$, to increase R as much as possible, the intensity of pump wave has to be 60 times as the saturate intensity. This example aims to obtain the large power reflection coefficient, choose proper parameter is necessary.

8.7 Photon echo

Photon echo studies transient optical phenomena which happens when a resonance medium interact with a strong coherence optical field.

From pervious discussion, we know a medium interacts with a continuous or pulsed pump light field will form a phase grating. The existing time of this grating is long enough that allowing a time delay between "read out" pulse (ε_2) and "writing in" (or reference) pump pulse (ε_1) as well as probe pulse ε_p is shorter than this existing time. However, the two pulses (ε_1 and ε_p) which form the grating have to incident onto the medium at the same time.

This section will discuss a different situation. Light field pulse is shorter than the atom quantum states memory cycle. In this period of time particle grating formed by incident pulses who do not incident onto the medium at the same time. Conjugate light pulse radiated spontaneously late on from medium in the form of photon echo. The echo wave front can travel forward as well as backward depends on the wave vector direction of excitation pulse.

8.7.1 Qualitative description of photon echo of two-energy level system

Photon echo is optical modification of nuclear spin echo. When a medium is irradiated by two successive pulses which separated by time τ, the medium radiates a third radiation at time τ after the second pulse reached – photon echo. Echo phenomenon can happen in a nonhomogeneous broadening system in which there exist finite number of resonance coupling interaction quantum states. To simplify, two-energy level system is used here as an example. Isolated atom in gas or impurity ions in solids such as chromium ions in ruby crystal can be treated as two- energy level system. These systems can be stimulated by interaction between electric dipole moments and external light field. A special feature of a nonhomogeneous broadening media is their atoms have different resonance frequencies so that can be distinguished. The nonuniform distribution of resonance frequencies come from Doppler shift (in atomic gas) caused by thermal movement or local crystal field Stark shift (solid with impurity). In a medium, a small unit volume λ^3 at vicinity of r, if there are enough atoms so that a continuous distribution function $g(\Delta\omega)$ can be used to express the frequency variation inside this volume. Here, $\Delta\omega$ is the offset of resonance away from the center frequency ω_0. Macro value of any quantity at r is an average of microvalues distribution $g(\Delta\omega)$ of the same quantity over the volume λ^3. For example, polarization generated by light radiation at r over a unit volume λ^3 is equal to a weighted summation of dipole moments of all atoms in that volume.

A two-level system can be descripted completely by its wave function $\Psi(t)$. There are three variables included: the probability amplitudes and relative phase difference between these two states. Suppose a two-level atom was initially in a pure quantum state, after irradiated by a resonance or near resonance electromagnetic wave, the interaction makes atomic wave function become the superposition of eigenstates of two atoms. But the phase relation between two probability amplitudes keep the same:

$$\Psi(t) = a_1(t)|1\rangle + a_2(t)|2\rangle$$

Time development of wave functions obey Schrödinger equation

$$i\frac{d}{dt}\Psi(t) = (H_0 + V)\Psi(t)$$

Here H_0 is Hamiltonian without external field, V is potential energy caused by external field. Phase factor of incident light field is kept in the wave function until it is eliminated by some irreversible elimination phase mechanism such as atomic collision. Therefore in the transient period of superposition of two basic states, the function of atom likes a storage medium, it can remembers phase distribution of external field. Thus, if the pulse interval is short than the irreversible elimination time T_2, interference of two discrete pulses will form granting in medium.

Photon echo generation mechanism can be understood through vector model of two-level energy Schrödinger equation. To do so, introduce vector P, his three components along coordinator are real function of probability amplitude:

$$P_1 = 2Re\{a_1 a_2^*\}$$

$$P_2 = 2Im\{a_1 a_2^*\}$$

$$P_3 = |\alpha_1|^2 - |\alpha_2|^2$$

From normalization condition

$$|P|^2 = \left(|\alpha_1|^2 + |\alpha_2|^2\right)^2 = 1$$

Equation of motion can be obtained from Schrödinger equation

$$\frac{dP}{dt} = \mathbf{\Omega} \times \mathbf{P} \tag{8.48}$$

Here Ω is a vector in the same space. Its three components are

$$\Omega_1 = \frac{(V_{12} + V_{21})}{\hbar}$$

$$\Omega_2 = \frac{i(V_{12} - V_{21})}{\hbar}$$

$$\Omega_3 = \omega_0$$

and

$$V_{kj} = \langle k|V|j\rangle, k, j = 1, 2$$

expresses transition energy and ω_0 is resonance frequency.

Assume an external circular polarized pulse

$$E(t) = \varepsilon_0(t)(x \cos \omega_0 t + y \sin \omega_0 t) \tag{8.49}$$

stimulates electro dipole moment transition. In a system which rotates around z axis with angular velocity ω_0, the equation of motion of P (in new coordinator, P') becomes

$$\frac{dP'}{dt} = \left(-2\mu\frac{\varepsilon_0}{\hbar}\right)\left(x'^0 \times P'\right) \tag{8.50}$$

Here x'^0 is the unit vector in x' direction. equation 8.50 shows, when external field is absent, P' is a constant. If a circular polarized external field existed, P' makes a precession about a stable vector which has a magnitude $-2\mu\varepsilon_0/\hbar$ and parallel to new x axis. At t = 0, atom in a lower energy state has $P' = (0, 0, -1)$ as showed in Fig. 8.21(a). Under the action of a pulse defined by equation (8.49), P' starts to rotate about x', Fig. 8.21(b) shows when external field disappear, P' rotates to its final position (in "−Y' ") axis. The external pulse in this case is called $\pi/2$ pulse, since it makes P' perform a 90° turn. Vector P' of all atoms which have same initial states in a medium will have same direction, so a huge macro polarization is generated. If all atoms have same resonance frequency, then their P' vectors keep parallel to each

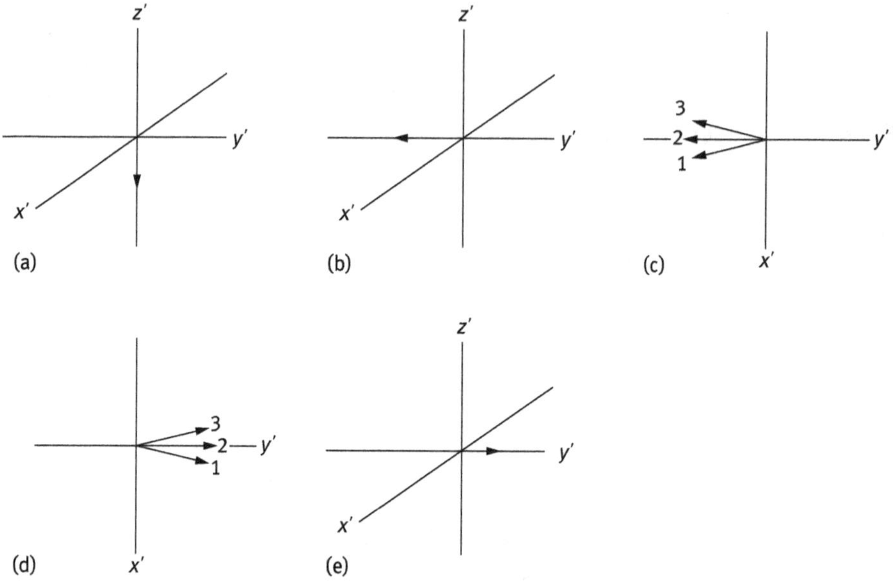

Fig. 8.21: P' in rotational coordinator. (a) $t = 0$, before $\pi/2$ pulse incident; (b) the moment $\pi/2$ pulse incident; (c) $t = \tau$, before π pulse incident; (d) the moment π pulse incident; (e) $t = 2\tau$.

other. And because external field is not existed, P' makes procession about z axis with the atomic resonance frequency. For nonhomogeneous broadening medium, the difference in atomic resonance frequency leads to slightly different procession rate of P' for each atom. In a coordinator that rotates with a fixed angular frequency ω_0, the motion of P' of atoms whose resonance frequency larger than ω_0, will be ahead; the motion of P' of those atoms whose resonance frequency smaller than ω_0, will be behind. The situation is shown in Fig. 8.21(c), where $\omega_1 > \omega_2 = \omega_0 > \omega_3$. The fan-shaped distribution means induced micro dipole moment of each atom will not be in the same direction any more. As a result, a macro polarization won't be co-produced again. Light radiation from polarization source will be behind with nonuniformly line width about $1/T_2^*$. This phenomenon is called free induction delay. Even though, each atom is still in coherent superposition of its own eigenvectors. Therefore the phase information did not lost at all, because it is stored in each atom. However phase difference between each micro dipole moment appears and reciprocal of nonuniform line width T_2^* is shorter than irreversible phase shift time T.

τ After the first pulse passed, the second pulse arrived. Vector P' of every atom will turn about x' again. Especially, assume the second pulse turn the vector an angle π (and thus called π pulse), these vectors' final positions are showed in Fig. 8.21(d). From the figure, one can see that before π pulse comes, behind vector becomes ahead; in turn the ahead vector becomes behind. After π pulse passes, each P' continues procession with their own resonance frequency. The vector behind will catch up due to its bigger procession rate; while the vector ahead will slow down due to its smaller procession rate. The time required for them overlap is exact τ. Therefore, starts from the first pulse arrives, after time interval $t = 2\tau$ a big macro polarization will be formed again. Each dipole moment will be synchronized and in coherence superposition, as showed in Fig. 8.21(e). Radiation from this delayed macro polarization formed photon echo. When the procession continues, Vectors' phases will be mismatch again, photon echo then disappeared. So echo output time approximately equals to the reciprocal of nonlinear linewidth T_2^*, the total duration time is 2τ, between T_2^* and T_2. In this period, information of external field phases are kept in atomic wave functions. Any practical interested external light field has little variation in a space within a wavelength range. Electromagnetic field in small unit volume λ^3 at r is a constant, radiation echo carries amplitude and phase information of incident filed at that point. When incident field varies in a medium, its whole apace distribution is stored first and will be reconstruction by the echo emitted by medium.

8.7.2 Qualitative results of photon echo phase conjugate

In the last section, photon echo phenomenon in atomic medium was descripted quantitatively. Examples of $\pi/2$ and π pulses incident successively was discussed. Here we are going to derive a wave front qualitative expression of radiated photon echo induced by any incident pulse. To accomplish this, first find out macro dipole moment produced by a stimulate pulse, then treat this dipole moment as radiation source of echo.

For simplify the process, consider a thin optical sample formed by stable atoms which have finite energy levels. Among these energy levels, there are two levels which can be coupled through dipole transition with external monochromatic field. Assume at time t, there is a pulse train

$$E_\alpha(r,t) = \frac{1}{2}\varepsilon_\alpha(r,t)exp\{j[\varphi_\alpha(r) + k_\alpha \cdot r - \omega t]\} + c.c. \tag{8.51}$$

incidents onto the above sample, pulse amplitude $\varepsilon_\alpha(r,t)$ is a spatial and temporal slow changing real function. The phase information is carried by $\Phi_\alpha(r)$, it closes to resonance with atomic medium, that is $\omega \approx \omega_0$.

Further assume the pulse width Δt is very narrow, so that $|\omega - \omega_0|\Delta t \ll 1$, that means Rabi frequency $\mu\varepsilon_\alpha/$ is much bigger than $|\omega - \omega_0|$. For gas medium with atoms in motion, this condition also means pulse period is short enough so that during the pulse duration time, atoms can be treated as stationary. The Hamiltonian of above reaction between light field and medium atoms is given by $V_{ij} = -(\mu \cdot E_\alpha)_{ij}$. Wave function solved from Schrödinger equation is

$$\Psi(r,t) = exp[(t - t_\alpha)]U_\alpha(r)\Psi(r,t_\alpha) \tag{8.52}$$

Here $U_\alpha(r)$ can be expressed as

$$U_\alpha(r) = \begin{bmatrix} \cos\left[\frac{\theta_\alpha(r)}{2}\right] & j\sin\left[\frac{\theta_\alpha(r)}{2}\right] exp\{j[\varphi_\alpha(r) + k_\alpha \cdot r - \omega t_\alpha]\} \\ j\sin\left[\frac{\theta_\alpha(r)}{2}\right] exp\{-j[\varphi_\alpha(r) + k_\alpha \cdot r - \omega t_\alpha]\} & \cos\left[\frac{\theta_\alpha(r)}{2}\right] \end{bmatrix} \tag{8.53}$$

and

$$\theta_\alpha(r) = (\mu/\hbar)\varepsilon_\alpha(r,t')dt'$$

phase information is stored in the nondiagonal elements of $U_\alpha(r)$, that is stored in the probability amplitude of final state that is reached by an atom after completes a transition under the action of incident field. Photon echo can be generated by the interaction of two pulses and a medium. Or can be generated by the action of three pulses. We will discuss this two situations in the following.

1. Echo of two pulses

Consider an atom near r, at $t = t_p$ the atom is in its low energy state. Assume at this moment, a short and strong light pulse $E_p(r,t)$ illuminates on it, then at sometime late, say $t = t_1$, its wave function is given by equation 8.52

$$\Psi(r,t_1) = exp\left[-jH_0\left(t_1 - t_p\right)\right]U_p(r)\Psi\left(r,t_p\right)$$

If at this moment, this atom is irradiated by another pulse€, then at $t > t_1$ its wave function is

$$\Psi(r,t) = exp[-jH_0(t - t_1)]U_1(r)\Psi(r,t_1)$$

combine both equations, we have

$$\Psi(r,t) = exp[-jH_0(t - t_1)]exp\left[-jH_0\left(t_1 - t_p\right)\right]U_1(r)U_p(r)\Psi\left(r,t_p\right) \tag{8.54}$$

at time t, the induced dipole moment is given by its expectation value

$$\langle\mu\rangle = \psi^*(r,t)|\mu|\psi(r,t)\rangle$$

Since U is a 2×2 matrix, so above expectation value contains four terms, among them the term who plays the major role for generating echo is

$$\langle\mu\rangle = \left(-\frac{j\mu_{12}}{2}\right)\sin\theta_p(\mathbf{r})sin^2\left(\frac{\theta_1(r)}{2}\right)$$

$$\times exp\{j[2\phi_1(\mathbf{r}) - \phi_p(\mathbf{r}) + (2k_1 - k_p)\Delta \cdot r - \omega t - \Delta\omega(t - 2t_1 + t_p)]\} + c.c \tag{8.55}$$

Equation 8.55 is true for any θ_p and θ_1. It is not hard to see, when $\theta_p(r) = \pi/2$ and $\theta_1(r) = \pi$ that is $E_p(r,t)$ and $E_1(r,t)$ are $\pi/2$ pulse and π pulse respectively, dipole moment reaches its maximum.

To obtain polarization density at r, an average of equation 8.55 over $g(\Delta\omega)$ f for all atoms at r needed to be calculated. Assume at $t = t_p$ total atom number in the vicinity is N, among them, N_2 atoms are in up energy state $|a_2$ and N_1 atoms are in lower energy state. Then from Boltzmann distribution we can get the difference of N_2 and N_1

$$N_2 - N_1 = -N \tan h\left(\frac{h\omega_0}{2K_B T_{th}}\right) \tag{8.56}$$

Here K_B is Boltzmann constant, and T_{th} is the thermal equilibrium temperature. Thus, the polarization density which radiates photon echo is

$$\begin{aligned} P = {} & [jN \tan h(h\omega_0/2k_B T_{th})\mu_{12}/2] \\ & \times \sin\theta_p(\mathbf{r}) sin^2[\theta_1(\mathbf{r})/2] \\ & \times exp\{j[2\phi_1(\mathbf{r}) - \varphi_p(\mathbf{r}) + (2\mathbf{k}_1 - \mathbf{k}_p)\cdot \mathbf{r} - \omega t]\} \\ & \times G(t - 2t_1 + t_p) + c.c \end{aligned} \tag{8.57}$$

Here

$$G(t) = \int_{-\infty}^{\infty} d(\Delta\omega) g(\Delta\omega) exp(-\Delta\omega t) \tag{8.58}$$

If we define the time when the first pulse arrives as $t\backslash mequal0$, and assume the time interval between these two pulses pass the coordinator origin as t_{1p}, we then have

$$t_p = \mathbf{k}_p^o \cdot \frac{r}{c}$$

$$t_1 = t_{1p} + \mathbf{k}_1^o \cdot \frac{r}{c}$$

here $\mathbf{k}_p^o$ and $\mathbf{k}_1^o$ are unit wave vectors for $\mathbf{E}_p(r,t)$ and $\mathbf{E}_1(r,t)$ respectively. From the discussion above, we know if the time when $\mathbf{E}_1$ reaches point $\mathbf{r}$ is t_1, then t_1 after $\mathbf{E}_1$ taking an action, micro dipole moment will line up again. That means the macro dipole polarization becomes maximum. It is also the moment when echo has its maximum intensity. Time between this moments to the time when the first pulse comes to the atom is

$$t = 2t_1 - t_p = 2t_{1p} + \left(2\mathbf{k}_1^0 - \mathbf{k}_p^0\right) \cdot \frac{r}{c} \tag{8.59}$$

From equation 8.57, the total electric field of echo at time t at point R outside of medium is

$$\mathbf{E}_c(R,t) = \frac{1}{4\pi\varepsilon_0} \nabla \times \nabla \times \int_V \frac{P_{echo}\left(\frac{|R-r|}{c}\right)}{|R-r|} d^3 r \tag{8.60}$$

Integration done over the whole area.

Assume $R \gg r$ substitute equation 8.57 into equation 8.60, we get

$$\mathbf{E}_c(R,t) \propto \int_V d^3\mathbf{r}G\left[\left(t-\frac{R}{c}-2t_{1p}\right)+\frac{\mathbf{r}}{c}\cdot\left(\mathbf{k}_c^0-2\mathbf{k}_1^0+\mathbf{k}_p^0\right)\right]$$

$$\times sin\theta_p(\mathbf{r})sin^2[\theta_1(\mathbf{r})/2]$$

$$\times exp\left\{j\left[2\phi_1(\mathbf{r})-\phi_p(\mathbf{r})+(2\mathbf{k}_1-\mathbf{k}_p-\mathbf{k}_c)\cdot\mathbf{r}-\omega(t-R/c)\right]\right\}+c\cdot c \tag{8.61}$$

This means the conditions for the maximum value is

$$\left.\begin{aligned} t &= 2t_{1p}+\tfrac{R}{c} \\ \mathbf{k}_c^0 &= 2\mathbf{k}_1^0\mathbf{k}_p^0 \end{aligned}\right\} \tag{8.62}$$

The first equation of equation 8.62 gives the time for echo pulse to reach point R; while the second equation is the condition for phase matching. Transfer energy from induced polarization vector to echo field can only be possible when these two vectors are in the same direction. So echo pulse propagates forward. Especially, if the second pulse $\boldsymbol{E}_1(r,t)$ is a plane wave with uniform amplitude envelope ε and uniform phase φ and the first pulse $\boldsymbol{E}_p(r,t)$ has uniform amplitude with phase changing in space (a plane wave can satisfy this condition when passes a phase distortion medium), then $\boldsymbol{E}_c$ is a phase conjugate inversion of $\boldsymbol{E}_p$. The reason for having above restrictions on $\boldsymbol{E}_p$ is because $\boldsymbol{E}_c$ dependents on $\sin\theta_p(\boldsymbol{r})$. When $\theta_p(\boldsymbol{r})$ is very small,so that $\sin\theta_p(\boldsymbol{r})\approx\theta_p(\boldsymbol{r})$, even if the amplitude of $\boldsymbol{E}_p$ varies in space, $\boldsymbol{E}_c$ is still its phase conjugate wave.

Phase matching condition requires $\boldsymbol{k}_c^0=\boldsymbol{2k}_1^0-\boldsymbol{k}_p^0$. However, to detect echo, usually making an angle β between k_1 and k_p, and the echo emission angle relative to k_p is 2β. But this will make the echo amplitude reduced as sample thickness and angle β increase. Furthermore, the finite thickness of sample may also lead to wavefront distortion of forward echo. So usually $\beta^2 l/\lambda \ll 1$ is required. To overcome these weak points of the forward echo, some people suggested using standing wave as the second pulse to get backward echo. Interested readers can read reference [8.8].

2. Three-pulse echo

More practical is three-pulse echo. Still assume when $t<t_p$, atom is in its low energy state. That is $|a_2(t<t_p)=0$ and $|a_2(t<t_p)=1$.. When $t<t_p,t_1,t_2$, pulses $\mathbf{E}_p(r,t)$, $\mathbf{E}_1(r,t)$ and $\mathbf{E}_2(r,t)$ incident on to the atom. Then when $t>t_2$, wave function of atom is

$$\psi(\mathbf{r},t)=exp[-jH_0(t-t_2)U_2(\mathbf{r})]$$

$$exp[-jH_0(t_2-t_1)U_1(\mathbf{r})]$$

$$\times\left[-jH_0(t_1-t_p)U_p(\mathbf{r})\right]\psi(\mathbf{r},t_p) \tag{8.63}$$

And the portion which has contribution to the radiation photon echo of induced dipole momentum expectation value (for stationary atom) is

$$\langle \mu \rangle = \left(-\frac{j\mu_{12}}{8} \right) sin\theta_p(\mathbf{r}) sin\theta_2(\mathbf{r})$$

$$\times exp\left\{ j\left[\phi_2(\mathbf{r}) + \phi_1(\mathbf{r}) - \phi_p(\mathbf{r}) + (\mathbf{k}_2 + \mathbf{k}_1 - \mathbf{k}_p) \cdot \mathbf{r} - \omega t - \Delta\omega(t - t_2 - t_1 + t_p) \right] \right\} + c \cdot c \tag{8.64}$$

and the polarization density

$$P_{echo} = [-jNtanh(h\omega_0/2K_B T_{th})\mu_{12}/8] \times sin\theta_p(\mathbf{r}) sin\theta_1(\mathbf{r}) sin\theta_2(\mathbf{r})$$

$$\times exp\left\{ j\phi_2(\mathbf{r}) + \phi_1(\mathbf{r}) - \phi_p(\mathbf{r}) + (\mathbf{k}_2 + \mathbf{k}_1 - \mathbf{k}_p) \cdot \mathbf{r} - \omega t \right\}$$

$$\times G(t - t_2 - t_1 + t_p) + c \cdot c \tag{8.65}$$

It is easy to see from the expression of G, when $t = t_2 + t_1 - t_p$, that is after a time interval equals to the time spacing between the previous two pulses, P_{echo} will reaches its maximum. Here t_p and t_1 are giving by equation 8.59. In equation

$$t_2 = t_{2p} + \mathbf{k}_2^0 \cdot \frac{\mathbf{r}}{c}$$

t_{2p} is the time spacing of pulses $\mathbf{E}_2(\mathbf{r}, t)$ and $\mathbf{E}_p(\mathbf{r}, t)$. When $\mathbf{E}_1(\mathbf{r}, t)$ and $\mathbf{E}_2(\mathbf{r}, t)$ are plane waves with uniform amplitude envelop and phase, echo radiation field

$$\mathbf{E}_c(\mathbf{R}, t) \propto \int_V d^3\mathbf{r} sin\theta_p(r) exp\left\{ j\left[-\phi_p(\mathbf{r}) + \mathbf{r} \cdot (\mathbf{k}_2 + \mathbf{k}_1 - \mathbf{k}_p - \mathbf{k}_c) - \omega\left(t - \frac{R}{c} \right) \right] \right\}$$

$$G\left[t - t_{2p} - t_{1p} - \frac{R}{c} + \left(\mathbf{k}_c^0 - \mathbf{k}_2^0 - \mathbf{k}_1^0 + \mathbf{k}_p^0 \right) \cdot \frac{\mathbf{r}}{c} \right] + c \cdot c \tag{8.66}$$

The condition that an echo can be observed at R is

$$t = t_{2p} + t_{1p} + \frac{R}{c}$$

$$\boldsymbol{k}_c^0 = \boldsymbol{k}_2^0 + \boldsymbol{k}_1^0 - \boldsymbol{k}_p^0$$

As in the two-pulse echo situation, the first equation above expresses the time required for echo to reach point $\boldsymbol{R}$; while the second equation states the phase matching condition. It can be satisfied by two ways: (1) $k_p = k_1 = k_2$, in this case, we get the forward echo. (2) $-k_1 = k_2, k_c = -k_p$, k, in this case, we get the backward echo. The later one has more significant meaning. Especially when $\mathbf{E}_p$ has uniform amplitude, or $\theta_p(\mathbf{r})$ is very small, $\mathbf{E}_c$ will be the back propagation complex conjugate reproduction of $\mathbf{E}_p$.

Now we will end up this section by comparing photon echo and DFWM. First, photon echo and DFWM both can descript real time holographic process. $\mathbf{E}_1$ and $\mathbf{E}_2$ act like pump waves,while $\mathbf{E}_p$ equivalent to an object wave. Besides, as we mentioned before, photon echo can only generate phase conjugate under the condition of weak stimulating ($\sin\theta_p \approx \theta_p$). Similar restriction also existed for DFWM, it reflects that the third term in a perturbation series starting to be ignored. The major difference between photon echo and DFWM is the former don't require one of $\mathbf{E}_p$ and $\mathbf{E}_{1,2}$ incident in the same time. As to the pulse signal DFWM, because conjugate pulse is a result of product of pump wave and probe wave, so its pulse width is smaller than any incident wave. The wave width of photon echo is defined by nonuniform linear Fourier transformation. It has nothing to do with the width of stimulation pulse.

8.8 Stimulated scattering

In this section, we are going to discuss another kind of physical mechanism of producing NOPC, inelastic photon scattering process. Stimulated Raman scattering (SRS) and stimulated Brillouin scattering (SBS) are belong to this kind of interaction. All these interactions have a common property that is incident photons transfer part of their energy in certain forms to nonlinear medium. Molecular vibration is an example of these form. Therefore, unlike the nonlinear process discussed before, the quantum states of nonlinear medium won't be the same before and after the interaction. In these interactions, nonlinear medium can produce "replica" of incident wave with lower frequency, opposite traveling direction and phase conjugate. As stimulation properties of inelastic process, it requires the intensity of incident light stronger than a critical value to maintain the process. This threshold is not a requirement for elastic scattering will add some restrictions on laser source and its structure. The attraction of inelastic process may due to its passive performance which means it does not need additional pump source. Therefore high power pulse laser system, such as laser fusion can be used in these processes. Finally, SBS conjugate mirror can produce reflection polarization wave as traditional reflection mirror.

8.8.1 Stimulated Raman scattering

Stimulated scattering is called to distinguish from spontaneous scattering. In 1962, when Woodbury and other people carried on some experiments using Q-switched lasers, they found besides the normal spectrum components, there accomplished some frequency shift in the spectrum. When laser cavity has some kind of nonlinear medium, the frequency shift can be observed. The total amount of frequency shift is equal to the natural frequency or its integer multiples. This phenomenon has related to light inelastic scattering from molecules discovered by Raman in 1928 when he was doing liquid scattering experiment. The late was named spontaneous Raman effect, it happens in action of weaker light.

After Woodbury, people did a series of studies regarding this kind of scattering. Many materials including ordered or disordered systems in microscopic structure have been found showing this phenomenon when irradiated by strong stimulating (laser) source. The scattering radiation exhibits different characteristics than radiation of spontaneous Raman effect, people called this scattering as stimulated Raman effect.

The basic differences between these two Raman effects are first the radiation waves in stimulated Raman effect can have interference but there is no interference in spontaneous Raman effect. Furthermore, in spontaneous Raman effect the intensity of scattering light is proportional to the excitation intensity, while in stimulated Raman effect it is not the same. Therefore, the later obviously belongs to nonlinear optical phenomenon.

In addition to molecules' vibration, light can also be scattered by phonon and polariton. This related to the interaction between real elementary particle (photon) to quiz-particles (phonon and pole). The phenomenon that electromagnetic wave experiences acoustic scattering in condense matter is called stimulated Brillouin scattering, it is the discussion focus in this section.

Stimulation scattering starts from low level spontaneous scattering and then growths rapidly. In a typical condition, stimulation scattering can be observed in saturation region. Saturation appears when gain total $G|\varepsilon_C|^2L \approx 30$. If we use MW/cm^2 to express corresponding threshold intensity $|\varepsilon_C|^2$, cm for the length of gain region L, then unit for G is cm/MW. SBS has large G constant (G ~10^{-1}~10^{-2} cm/MW), small damping constant τ_S ~10^{-8}~10^{-9} s, and small frequency shift ($\Omega \leq$ 1~10^{-2} cm^{-1}), this make it is widely used in wavefront inversion.

8.8.2 Stimulating Brillouin scattering

When a light which intensity is bigger than a threshold travels in a nonlinear medium, it will produce corresponding sound wave. This happens due to electrostrictive effect in which medium's density increases proportionally to the increase of electric field. The sound wave propagation direction is the same as the incident wave. Here, the medium acts similar to a moving reflect mirror, or laminated dielectric plate assembly. The incident wave reflected from the nonlinear medium and generates Doppler scattering with frequency drift, the amount of frequency drop is equal to the frequency of sound wave.

The process in which scattering and incident waves propagate opposite collinear probably has the highest gain and only is the SBS process that can be observed. Under certain conditions, scatter wave is complex conjugate of incident. Thus through SBS a distortion incident wave will generate a equal distortion sound wave with phase surface exactly match the incident wave. Let us imagine that the nonlinear process forms a deformation mirror, its function is to inverse the phase of reflection wave related to the phase of incident wave. Therefore, when reflected wave travels back along the same route as the incident wave, it cancels the phase errors generated by the incident wave passes through the inhomogeneous medium. If a movie of the incident wave propagation can be taken, then reverse the film screening will get the behavior of conjugate wave.

Solving wave function can qualitatively descript SBS process.

Consider a stimulated wave $E_c(\rho, z)$ and a signal wave $E_s(\rho, z)$ collinear travel along +z and − z direction respectively. Here ρ is a variable in a plane perpendicular to z. Omit time factors $exp(j\omega_c t)$ and $exp(j\omega_s t)$, we get

$$E_c(\rho, z) = \varepsilon_c(\rho, z)exp(-jk_c z) \tag{8.67}$$

and

$$E_s(\rho, z) = \varepsilon_s(\rho, z)exp(-jk_s z) \tag{8.68}$$

They satisfy

$$\frac{\partial \varepsilon_c}{\partial z} - \frac{j}{2k_c}\nabla_\rho^2 \varepsilon_c = 0 \tag{8.69}$$

and

$$\nabla^2 E_s(\rho, z) + k_s^2 E_s(\rho, z) = \mu \frac{\partial^2}{\partial t^2} P_{NL} \tag{8.70}$$

respectively.

Here, ∇_ρ^2 is two dimension Laplace operator.

Assume the complex amplitude of signal satisfy the slow varying approximations, that is

$$\left|\frac{\partial^2 \varepsilon_s}{\partial z}\right| \ll \left|k_s \frac{\partial \varepsilon_s}{\partial z}\right| \ll |k_s^2 \varepsilon_s|$$

then

$$\frac{\partial \varepsilon_s}{\partial z} - \frac{j}{2k_s}\nabla_\rho^2 \varepsilon_s + \frac{1}{2}g(\rho, z)\varepsilon_s = 0 \tag{8.71}$$

Here, $g(\rho, z)$ is a local gain generated by Brillouin interaction. It can be proved that it is proportional to stimulated strength, so it can be written

$$g(\rho, z) = A|E_c(\rho, z)|^2 \tag{8.72}$$

Coefficient A is

$$A = \frac{\pi p^2 \varepsilon_0 n^8}{\alpha \rho_0 v^2 \lambda^2} \tag{8.73}$$

Here

p —photoelastic coefficient

α —sound absorption coefficient of medium

ρ_0 —density of material

v —sound speed in medium

λ —laser wave length

As k_s and k_c have little difference, in a typical situation, $(k_s - k_c)/k_c$ is in the order of 10^{-5}, so we treat $k_s \approx k_c = k$ in the following discussion.

Introduce a set of function $f_\xi(r), \xi = 0, 1, 2, \ldots$, let them satisfy the equation

$$\frac{\partial f_\xi}{\partial z} + \frac{j}{2k_s}\nabla_\rho^2 f_\xi = 0 \tag{8.74}$$

and orthonormal condition

$$\int f_\xi^*(\rho, z) f_\eta^*(\rho, z) d\eta = \delta_{\xi\eta} \tag{8.75}$$

Furthermore choose the difference between its complex conjugate of the first member of this function set is a constant B, that is

$$\varepsilon_c(\rho, z) = B f_0^*(\rho, z) \tag{8.76}$$

Then other members of this function set can be found from equation (8.74) and equation (8.75).

Substituting equation (8,76) into equation (8.72), leads to

$$g(\rho, z) = A|B|^2|f_0|^2$$

Then equation (8.71) can be written as

$$\frac{\partial \varepsilon_s}{\partial z} + \frac{j}{2k}\nabla_\rho^2 \varepsilon_s + \frac{1}{2}A|B|^2|f_0|^2 \varepsilon_s = 0 \tag{8.77}$$

Assume signal field can be expanded by function f as

$$\varepsilon_s(\rho, z) = \sum_{\eta=0}^{\infty} c_\eta(z) f_\eta(\rho, z)$$

Substituting it into equation (8.77), we get

$$\sum_{\eta=0}^{\infty} \frac{dc_\eta}{dz} f_\eta + c_\eta\left(\frac{\partial f_\eta}{\partial z} + \frac{j}{2k}\nabla_\rho^2 f_\eta\right) + \frac{1}{2}A|B|^2|f_0|^2 c_\eta f_\eta = 0$$

Apply equation (8.74), the above equation becomes

$$\sum_{\eta=0}^{\infty}\left[\frac{dc_\eta}{dz} f_\eta + \frac{1}{2}A|B|^2|f_0|^2 c_\eta f_\eta\right] = 0 \tag{8.78}$$

Multiply equation (8.78) both sides by $f_\xi^*(\rho, z)$, and do integration respected to ρ, from the orthonormal property of f we get,

$$\frac{dc_\varepsilon}{dz} + \frac{1}{2}\sum_{\eta=0}^{\infty} g_{\xi\eta}(z)c_\eta(z) = 0 \tag{8.79}$$

here

$$g_{\xi\eta}(z) = A|B|^2 \int d\rho |f_0(\rho, z)|^2 f_\xi^*(\rho, z) f_\eta(\rho, z) \tag{8.80}$$

If the intensity $|f_0(\rho, z)|^2$ of laser light field as a function of ρ has a large fluctuation, then the maximum and minimum values of $|f_0(\rho, z)|^2$ and $f_\xi^*(\rho, z) \times f_\eta(\rho, z)$ generally will have some overlap that will make $g_{\xi\eta}$ being a small number. The only exception is when

$$g_{00}(z) = A|B|^2 \int |f_0(\rho, z)|^4 d\rho$$

Under this conditions, c_0 increases quicker than other c, so after long enough distance, the signal field complex envelop becomes

$$\varepsilon_s(\rho, z) = \sum_{\eta=0}^{\infty} c_\eta(z) f_\eta(\rho, z) \approx c_0(z) f_0(\rho, z) = \frac{c_0(z)}{B^*} \varepsilon_c^*(z) \tag{8.81}$$

That means, the apace signal generated by backward SBS is complex conjugate of incident space field. Therefore it can correct the phase distortion of laser field.

It is noticed that if $f_0(\rho, z)$ is not a strong fluctuation function of ρ, then $g_{\xi\xi}$ and g_{00} have the same order, as a result the rapid increase of c_0 which leads to a phase conjugate won't be happen. Therefore, introducing phase distortion in front of Brillouin box can improve phase conjugate characteristic.

8.9 Photorefractive effect and associated materials

From the discussion of above two sections we know, phase conjugate reproduction of a light wave can be obtained through four-wave mixing or stimulated scattering. For stimulating scattering, it only requires one light wave, but the intensity of this light must be higher than a high threshold. For example, as for SBS, this threshold is about 106 W/cm^2. Although four-wave mixing allows very weak probe wave, but it requires additional pump light. So since four-wave mixing phenomenon has been discovered, people are working hard to find new material. Especially those materials that have response for very weak light signal. Up to now, a large amount materials have been found belong to this category. These materials can change their refraction indexes when interact with light especially weak light. Among them the most sensitive materials which can make wavefront inversion are called photorefractive material.

These materials can produce phase conjugate wave reacted by light which intensity is in the order of micro watts.

Photorefractive effect has different physical mechanism of changing refraction index than other light induced refraction index changes. For instance, refraction index of ordinary material changes as the intensity of incident light changes. While photorefractive coefficient of photorefractive material has nothing to do with the total intensity of incident light, it dependents on the ratio of intensity of incident lights. The total intensity can only affect the speed of photorefractive. Therefore, in general, any weak light, as far as it has enough time to react with medium, the photorefractive can always happen. However, the limited coherence time in which light can be recorded (writing time) sets a maximum value for useful writing time. Hence, there is a restriction of minimum intensity of light. While this limited intensity is small enough it can be in the order of 6–10 W/cm^2.

In this section, we will discuss three problems. We first discuss basic properties of photorefractive effect quantitatively and qualitatively. Then we will talk about some typical photorefractive materials. Finally, we will look at self-pumped phase conjugate phenomenon.

8.9.1 Photorefractive effect

At beginning photorefractive effect was recognized as a harmful phenomenon. In early experiment in which pulsed laser was used to produce second harmonic, people found the producing efficiency of second harmonic of $LiNbO_3$ reduced tremendously after several pulses. This happens because the incident laser causes a semi-permanent change of material's refraction index, hence the phase match condition on which the second harmonic generated is destroyed. Soon after finding this phenomenon, people found this optical damage can be used to stored high quality holographic images in LiNbO3. So the photorefractive effect started to attract attention. In this section, we will talk about the physical mechanism of this phenomenon and then give some qualitative results.

1. The physical interpretation of photorefractive

Usually people use electron hopping model and diffusion model to descript photorefractive. The method we used here is basically the same as the electron hopping model.

Assume there existing some electrons in crystal material, the origin of these electrons is not too clear. Further assume these electrons are in the low potential wells formed by doped crystal and crystal dislocation. Without light, these electrons are frozen in their potential wells. When light shining on the crystal, these electrons can move between the wells. Consequently, a static electric field is generated. It is this electric field that induced a change of refraction indices of crystals that lack of inversion symmetry through linear electro-optic effect (Pockels effect). Light-induced electrostatic field can be very strong, some can be in the order of 10^5 V/m. If the linear electro-optic system is also large, the refraction index will have a big change Δn. For example Δn for $BaTiO_3$ crystal can be in the order of 10^{-3}.

The above theory about the formation of light induced electrostatic field and its effect has a good agreement with the experimental data. It has also successfully explained and predicted some new phenomena.

2. Production of light-induced electrostatic field and its dependence on some parameters

Assume two uniform light waves have same frequency and slow varying amplitude ε_1 and ε_2 incident on to a crystal with an angle 2θ to each other (Fig. 8.22a). The total spatial intensity distribution produced by these two lights is

$$I(\boldsymbol{x}) = I_0(1 + m cos \boldsymbol{k} \cdot \boldsymbol{x}) \tag{8.82}$$

which is a periodic function. Here I_0 is the total intensity of incident light, and unitless m is

$$m = 2(I_1 I_2)^{1/2} \frac{COS2\theta}{I_0}$$

m is called grating modulation index and

$$\boldsymbol{k} = \boldsymbol{k}_1 - \boldsymbol{k}_2$$

is grating wave vector. Since

$$|\boldsymbol{k}_1| = |\boldsymbol{k}_2|$$

so

$$|\boldsymbol{k}|^2 = |\boldsymbol{k}_1|^2 + |\boldsymbol{k}_2|^2 - 2|\boldsymbol{k}_1||\boldsymbol{k}_2| cos 2\theta$$

$$= 4|\boldsymbol{k}_1|^2 sin^2\theta$$

that is

$$|\boldsymbol{k}| = 2|\boldsymbol{k}_1| sin\theta$$

The direction of $\boldsymbol{k}$ is indicated in Fig. 8.22(b).

Assume at beginning there are N movable electric charges, if the action light wave is spatial periodic then the steady state distribution of charge will have a component with the same period. These charges can produce a steady electric field.

$$\boldsymbol{E}(\boldsymbol{x}) = Re\{\boldsymbol{E}\boldsymbol{k}^0 e^{j\boldsymbol{k}\cdot\boldsymbol{x}}\}$$

The direction of electric field is parallel to grating wave vector k. Neglecting induced current, the electric field is

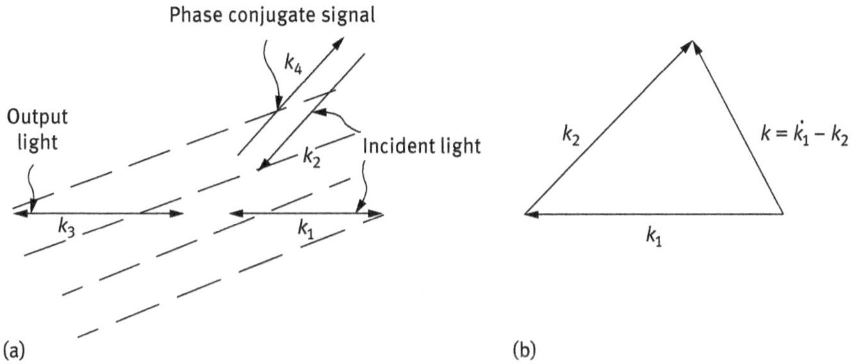

Fig. 8.22: Schematic graph of generating phase conjugate from photorefractive crystal. (a) Formation of periodic grating; (b) Wave vector graph.

$$\mathbf{E}(\mathbf{x}) = Re\left\{ jam \frac{\frac{K_B T}{q}\frac{|\mathbf{k}|}{a} + j\mathbf{f}\cdot\mathbf{k}^0}{1 + \frac{|\mathbf{k}|^2}{a^2}\frac{\mathbf{k}\cdot\mathbf{f}}{a}} \times \mathbf{k}^0 exp(j\mathbf{k}\cdot\mathbf{x}) \right\} \tag{8.83}$$

Here $\mathbf{k}^0$ is unit vector of $\mathbf{k}$

$K_B T$ is thermal energy of lattice

q is total charge of movable electric charge

$a = \left(\frac{Nq^2}{\varepsilon\varepsilon_0 k_B T}\right)^{1/2}$ is material's constant, ε_0 and ε are permittivity of free space and medium respectively. Last,

$$\boldsymbol{f} = \frac{\mathbf{E}_0 q}{k_B T a}$$

Here $\boldsymbol{E}_0$ is steady and uniform electric field in crystal, it can be external or inherent within the crystal. Especially, when $E = 0$, equation (8.83) simplified as

$$\mathbf{E}(\mathbf{x}) = -m\frac{k_B T}{q}\frac{|k|}{1+\frac{|k|^2}{a^2}} sin(\mathbf{k}\cdot\mathbf{x}) \tag{8.84}$$

Now we will briefly analysis the dependence of induced electric field E(x) to light intensity and grating wave vector.

(a) Dependence of E to light intensity

Dependence of E to light intensity is embodied by the dependence of unitless grating modulation index m to light intensity. As

$$m \propto \frac{\sqrt{I_1 I_2}}{I_1 + I_2} = \left[\sqrt{\frac{I_1}{I_2}} + \sqrt{\frac{I_1}{I_2}}^{-1}\right]^{-1}$$

So is m is known, from equation (8.84), we know that E is only dependent on ratio of $\mathbf{I}_1/\mathbf{I}_2$ and has nothing to do with the absolute values of $\mathbf{I}_1$ and $\mathbf{I}_2$. Especially when $\mathbf{I}_1 = \mathbf{I}_2$, m reaches its maximum value. That means when two incident lights have same intensity, amplitude of light induced electric field in the medium will reach its maximum

(b) Dependence of E to grating wave vector

First we notice that **E** and **k** are always in the same direction. So, here, we focus our discussion in the relationship of amplitude of **E** and magnitude of **k**. In situations of $\mathbf{E}_0 = 0$ and $\mathbf{E}_0 \neq 0$, the dependence of **E** to **k** are different, it needs to be discussed separately.

(i) When $E_0 = 0$

When there is no electric field, the dependence of amplitude of **E** to k is reflected through

$$g(k) = \frac{k}{1 + \frac{k^2}{a^2}}$$

From equation (8.84) we can tell, when $k \ll a$, E linearly increases at a rate mk_BT/q with k. When k getting bigger, increasing rate of E becomes smaller. When k \mequala, E gets its maximum $mk_BTa/2q$. As k continues to increase, E reduces slowly. In this situation, to get bigger E, k has to be big enough, it is better to have k has same order of a.

(ii) When $E_0 \neq 0$

When induced electric field exists inside crystal, from equation (8.83) we know, the amplitude of light induced electric field

$$E = am\frac{k_BT}{q}\frac{\frac{k}{a} + j\boldsymbol{f}\cdot\boldsymbol{k^o}}{1 + \frac{k^2}{a^2} + j\frac{k\cdot f}{a}} \tag{8.85}$$

when k is very small,

$$E \approx am\frac{k_BT}{q}\boldsymbol{f}\cdot\boldsymbol{k^o}$$

That means, even if grating wave vector is very small, E can reach expectation value provided a proper value of E_0 can be chosen.

3. Reflectivity of grating and phase conjugate mirror

(a) Grating reflectivity η

As we discussed earlier, when two writing lights form a periodical interference pattern in a refractive material, it will generate a periodical light-induced electric field $E(x)$. This will leads to periodical change of magnetic susceptibility of the material. The action of this change on late coming incident light just like a phase grating. Reading light with intensity I_3 scattered by this grating becomes output light with intensity of I_4. The ratio $\eta = I_4/I_3$ is then defined as grating reflectivity.

Grating reflectivity is first dependent on material's properties, mainly the material's reflectivity, electro-optic coefficient tensor and physical size. It also depends on the polarization status of readout light. And its relationship with writing light is reflected by light induced electric field.

For $BaTiO_3$ crystal, assume the angle between light-induced electric field and C axis (001) of crystal is β. The geometry relationship of these light beams is illustrated in Fig. 8.23. If readout light is an ordinary light,

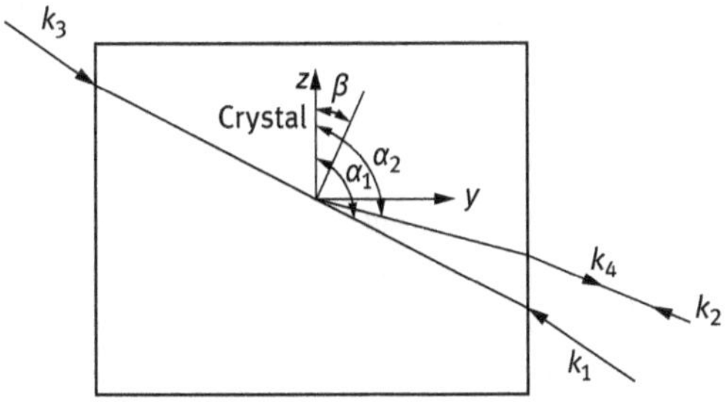

Fig. 8.23: Geometry relationship between light beams in phase grating.

$$\eta_0 = \left| \left(\frac{\omega L}{4c} \right) n_0^3 r_{13} E(\beta) cos\beta \right|^2 \tag{8.86}$$

If readout light is extraordinary light,

$$\eta_e = \left| \frac{\omega L}{4cn_3} E(\beta) cos\beta \left(n_e^4 r_{33} sin\alpha_1 sin\alpha_2 + 2n_e^2 n_0^2 r_{42} sin^2\beta + n_0^4 r_{13} cos\alpha_1 cos\alpha_2 \right) \right|^2 \tag{8.87}$$

For a $BaTiO_3$ crystal, its electro-optic coefficients are $r_{13} = 8$, $r_{33} = 88$, $r_{42} = 820$ (in 10–12 m/V) respectively. So generally, use e light as readout light can easily get higher reflectivity. In Fig. 8.23, L is the length of crystal, α_1 and α_2 as indicated in the figure.

(b) Conjugate reflectivity R

Phase conjugate reflectivity R defines as intensity ratio of a pair of phase conjugate waves. That is

$$R = \frac{I_4}{I_2} = \frac{\eta I_3}{I_2}$$

So, for ordinary light, the readout is

$$R=\frac{I_3}{I_2}\left|\frac{\omega L}{4c}n_0^3r_{13}E(\beta)\cos\beta\right| \tag{8.88}$$

and for extraordinary light, the readout is

$$R=\frac{I_3}{I_4}\left|\frac{\omega L}{4cn_3}\cos\beta\left(n_e^4r_{33}\sin\alpha_1\sin\alpha_2+2r_{42}n_e^2n_0^2sin^2\beta+n_0^2r_{13}\cos\alpha_1\cos\alpha_2\right)\right|^2|E(\beta)|^2 \tag{8.89}$$

equation (8.86) to equation (8.89) shows η and R are all proportional to $|E|^2$, and

$$|E|^2\propto\frac{k^2+a^2f^2}{\left(1+\frac{k^2}{a^2}\right)^2+\frac{k^2f^2}{a^2}}$$

From the relationship of f and E_0 , we know that η and R are related to electric field. It is also easy to see that η and R increase with increase of E_0.

4. The occurring speed of photorefractive phenomenon

As we mentioned before, the strength of photorefractive effect has nothing to do with the strength of incident light. The rate of formation and disappearance of photorefractive index grating varies with the intensity of incident light. And also depends on the characteristics of charge migration of materials. For instance, uniform grating erases phase grating with a reducing rate

$$\mathrm{A}=2I_0Dd^2\left(a^2+k^2\right) \tag{8.90}$$

Here

D—diffusion constant,

d—average jumping length of electron.

These two parameters are usually unknown. While the light conduction

$$\sigma=\varepsilon\varepsilon_0I_0Dd^2d$$

can be detected as a whole, and it leads to

$$A=\frac{2}{\varepsilon\varepsilon_0}\left(1+\frac{k^2}{a^2}\right)\sigma \tag{8.91}$$

Equation (8.90) or (8.91) shows the phase grating reducing speed is proportional to light intensity. To $BaTiO_3$ crystal, in a typical situation, the time constant is in the order of second.

5. Sensitivity of photorefractive

To characterize photorefractive sensitivity, two concepts are usually applied. The first one is, the light energy needed to generate the required change of refractive index. The second one is the light energy of getting required diffraction efficiency (for example 1%). In the first situation, photorefractive sensitivity defined as during the starting stage of recording, refraction change caused by unit energy absorbed by unit volume. That is

$$S_{n_1} = \frac{1}{\alpha}\frac{dn}{d\omega_0}$$

Here α is absorption coefficient, w_0 is incident energy per unit volume. Photorefractive sensitivity can also be defined as for a unit incident energy the variation of refraction index per unit volume. That is

$$S_{n_2} = \frac{dn}{d\omega_0}$$

Obviously,

$$S_{n2} = \alpha S_{n1}$$

For holographic material, more practical definition of photorefractive sensitivity is to use diffraction coefficient η instead of refraction index n . Doing so, the photorefractive is

$$S_{\eta_1} = \frac{1}{\alpha}\frac{d(\eta^{1/2})}{d\omega_0}\frac{1}{L}$$

and

$$S_{\eta_2} = \frac{d(\eta^{1/2})}{d\omega_0}\frac{1}{L}$$

again,

$$S_{\eta_2} = \alpha S_{\eta_1}$$

6. Response time

Refraction index change is due to the electro-optic effect driven by space charge field. The time required to form record grating depends on the speed at which static electric field established by producing electric charge and their transmission. Inertia of nonlinear responds of photorefractive media is an important difference to distinguish themselves from other nonlinear media. The refraction index changes of later originated from electrons and happens instantaneously. The time evolution of forming grating has been analyzed in detail in many literatures. Here, we only give a main result. In a circumstance in which there is continuous light illumination and electron moving distance is not obviously less than grating, the crystal response time is

$$\tau_{eff} = \tau_d \frac{\left(1+\frac{\tau_R}{\tau_u}\right)+\left(\frac{\tau_R}{\tau_E}\right)^2}{\left(1+\frac{\tau_R\tau_d}{\tau_u\tau_I}\right)\left(1+\frac{\tau_R}{\tau_u}\right)+\left(\frac{\tau_R}{\tau_E}\right)^2\frac{\tau_d}{\tau_I}} \tag{8.92}$$

Here

τ_d – dielectric relaxation time of the crystal, τ_R – charge recombination time, τ_E – charge drift time, τ_u – charge diffusion time; τ_I – inverse of light generation and ion combination rate.

8.9.2 Some photorefractive materials

Using photorefractive materials to obtain phase conjugate has many potential applications in modern science and technology frontier. To make these applications more successfully, material performance has to be improved in the following two main aspects. One is to increase gain of phase conjugate, and another one is to increase responding speed. Methods used now to reach these goals are: dosing the existing photorefractive crystals, applying additional electric field, and looking for new materials. In this section, we will briefly discuss progress in these areas.

1. Dosed photorefractive crystal

Dose photorefractive crystal is a common way to improve material performance. Barium titanate crystal is the earliest material that was discovered has photorefractive property and is known as a good all-around material. Figure 8.24 shows beam coupling gain coefficient vs. refraction grating slit spacing for three different Barium titanate crystals. The solid lines represent theoretical values and dot lines represents

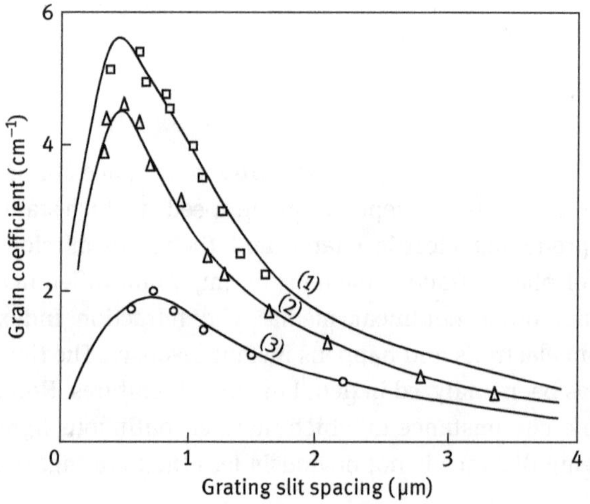

Fig. 8.24: Relationship between beam coupling gain and grating slit spacing.

experimental data. Curves (1) and (2) corresponding to 1 ppm and 50 ppm dosing respectively, whereas curve (3) is for un-dosed crystal. The coherent light used in the experiment was hydrogen ion laser, its wavelength is 514.5 nm, and intensity of pump light is 3 W/cm^2, the illumination uniformity is better than 20% in the sample area. Light polarity direction is perpendicular to crystal's axis. Therefore the only useful electro-optical coefficient is r_{13}, and intensity ratio between pump light and signal light is 800:1. Angle between pump light and signal light is adjustable.

Table 8.1 Beam coupling gain and response time for heat treated dosed and un-dosed crystals. The data show for certain oxidation-reduction treated crystals, gain coefficients and response time strongly depended on doping concentration.

Table 8.1: Beam coupling gain and response time of Barium titanate crystal[1]

Dosing elements	Dosing concentration ($\times 10^{-6}$)	Annealing	Gain (cm^{-1})	Response time (ms)
Co	50	Same length as dosing	3.7	1400
Co	100	Same length as dosing	5.0	730
Cr	50	Same length as dosing	3.2	550
Mn	50	Same length as dosing	2.1	1200
Co	50	CO_2:CO(99:1)	2.0	55
Co	100	CO_2:CO(99:1)	2.7	610
Mn	50	CO_2:CO(99:1)	2.2	200
No-doping			1.9	420

[1] Experimental condition: Wavelength of light 515 nm,light intensity 1 W/cm^2, grating slit spacing 0.7 μm.

2. Photorefractive under external electric field

Apply DC electric field in direction of electro-optical axis can change photorefractive crystal performance. Figure 8.25 is a two-beam coupling experimental apparatus for showing material's photorefractive characteristic. Weak signal wave (intensity $I_2(0)$) is coupling with strong pump wave (intensity $I_1(0)$) in a photorefractive medium. The ratio of two beams' intensity is $I_2(0)/I_1(0) = M < 0.01$. Produced lights have intensities.

$$I_1(z) = I_1(0)exp[-(\Gamma + \alpha)z]$$

$$I_2(z) = I_2(0)exp[(\Gamma - \alpha)z] \tag{8.93}$$

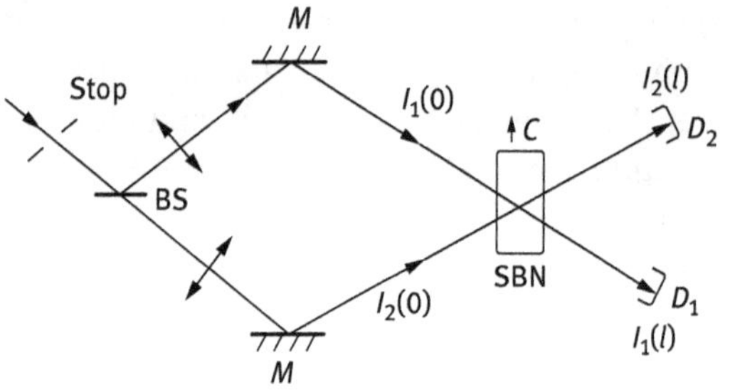

Fig. 8.25: Apparatus of two-beam coupling. *BS* — beam splitter; *M* — reflected mirror; *SBN* — rhodium dosed crystal.

Here α is absorption coefficient and Γ is beam coupling constant. Material response time is given by

$$\tau = t_0 \frac{E_0 + i(E_d + E_\mu)}{E_0 + i(E_d + E_N)} \tag{8.94}$$

Here

$$t_0 = \frac{h_v N_A}{SI_0(N_D - N_A)}$$

is time required for generating N_t photoionization ions in unit volume of material. s is photoionization cross section. E_0 is external DC electric field, the internal characteristics field are defined by

$$E_N = \frac{eN_t}{\varepsilon k}\left(1 - \frac{N_t}{N_D}\right) \approx \frac{eN_t}{\varepsilon k}\text{; for } N_t \ll N_D,$$

$$E_d = \frac{k_B TK}{e}$$

$$E_\mu = \frac{\gamma N_t}{\mu k}$$

Here $k = 2\pi/\lambda_g$ is wave number of corresponding grating period; γ is electron recombination rate; μ is electron drift speed; e is electron charge; k_B is Boltzmann constant; N_t is trap density; N_D is donor's density; and ε is permittivity of material.

The virtual part of τ is oscillation factor of coupling coefficient Γ. Its modulus

$$|\tau| = t_0 \sqrt{\frac{E_0^2 + (E_d + E_\mu)^2}{E_0^2 + (E_d + E_N)^2}} \tag{8.95}$$

is response time.

From equation (8.95), we can see, when external field

$$E_0 = 0$$

we get,

$$|\tau| = t_0 \frac{E_d + E_\mu}{E_d + E_N}$$

Usually as

$$E_\mu > E_N$$

so

$$\tau > t_0$$

As E_0 increases, τ reduces, when E_0 reaches

$$E_0 \gg E_d, E_N, E_\mu,$$

τ gets its minimum value t_0. This is the time required for generating N excitation photons in unit volume.

For SBN(Sr0.6Ba0.4Nb2O6) crystal dosed with rhodium, in a typical situation, $N_A = 2 \times 10^{22} M^{-3}$, $\Lambda_g = 1.5 \mu m$, $E_N = 1.55 kVcm^{-1}$, $E_d = 1.06 kVcm^{-1}$, and $E_\mu = 10 kVcm^{-1}$. Plug these data into equation (8.95), we get a graph of response time varies as external DC electric field. The result is plotted in Fig. 8.26.

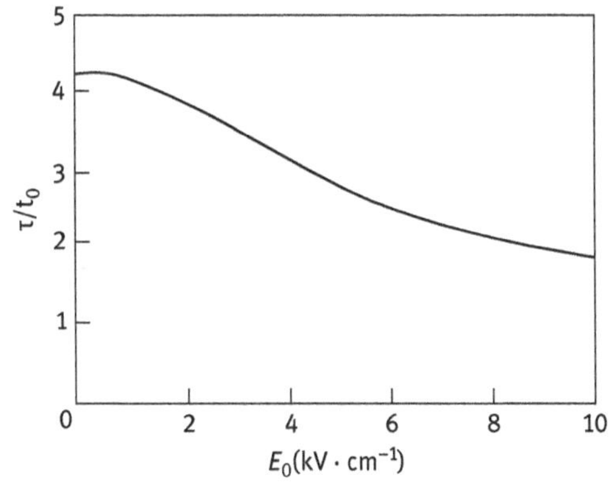

Fig. 8.26: SNB-Rh response time vs. external electric field.

Figures 8.27 and 8.28 show the variations of photorefractive gain coefficient and response time of material without an external electric field corresponding to the change of grating slit space. While Figs. 8.29 and 8.30 are the response of these two parameters when an external electric field changes. In Figs. 8.29 and 8.30, grating split space Λ_g is 1.51 μm, electric field varies from 0 to 10 kV/cm and is applied in the c axis of crystal. For a given grating slit spacing, gain from 10 cm^{-1} when $E_0 = 0$ reduces to about 8 cm^{-1} when $E_0 = 10$ kV/cm.

The coherence light source used in this experiment is an argon ion laser with wave length 514.5 nm, and

$$I_1(0) + I_2(0) \approx \frac{0.25W}{cm^2}, \frac{I_2(0)}{I_1(0)} < 0.01 .$$

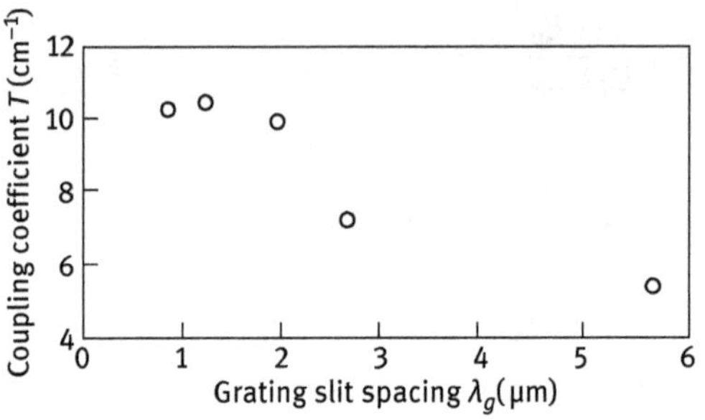

Fig. 8.27: Relationship between coupling gain coefficient and grating slit spacing ($E_0 = 0$).

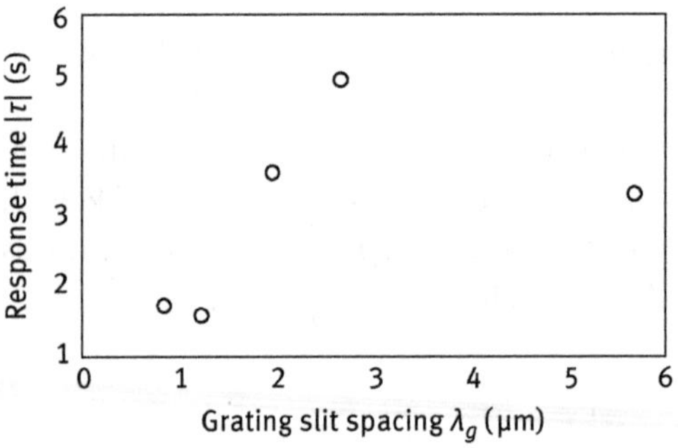

Fig. 8.28: Relationship between response time and grating slit spacing ($E_0 = 0$).

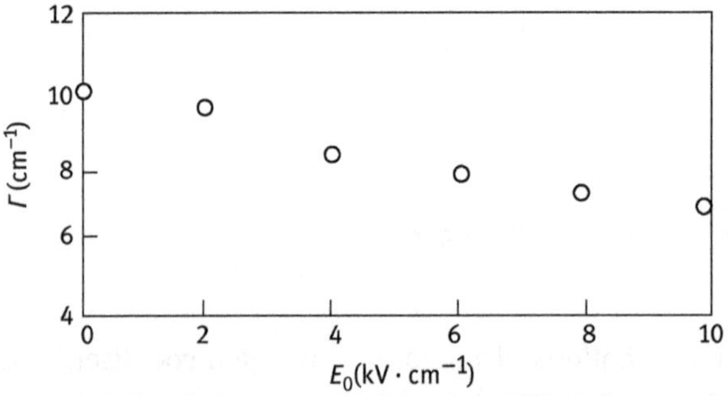

Fig. 8.29: Relationship between gain coefficient and E_0 ($\lambda_g = 1.51 \mu m$).

We can tell from the experimental results that an external electric field E=10 kV/cm can reduce the response time 20 times, and the coupling gain only reduce 0.20, it is worth to do so for those photorefractive devices which requires real-time response.

There are some reports talking about using AC electric field to change the performance of photorefractive materials. We are not going to discuss it, readers who are interested in it can read some references.

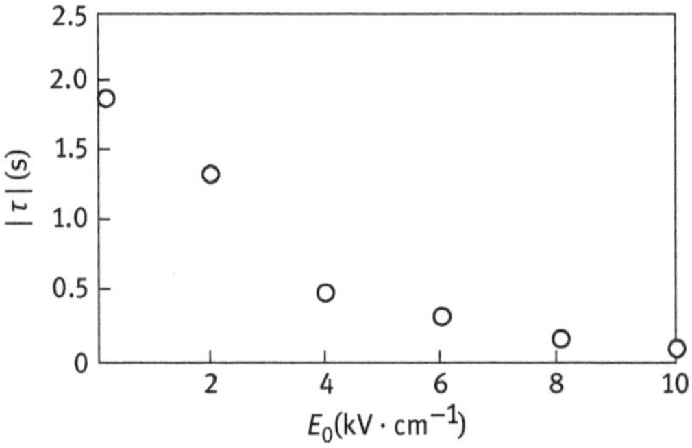

Fig. 8.30: Relationship between response time and E_0 ($\lambda_g = 1.51\mu m$).

3. Other photorefractive materials

In previous sections we discussed mainly on photorefractive crystals. There are other photorefractive materials which can be used to get phase conjugate light such as semi-conductors and liquid crystals. We will end up this book by brief introducing these materials.

(a) Semi-conductor materials

Many semi-conductor materials show photorefractive effect under a certain conditions. CdTe, InP, GaP, and GaAs are common ones. Among them GaP is very useful. It is because its response wavelength is around 0.8 μm which is exactly the center wavelength of GaAlAs diode laser. Most photorefractive materials have no response to this wavelength, only BaTiO3 crystal can be applied in this wavelength, but its response time is as long as several ten seconds.

After study GaP, people found in the case without external electric field nor photo electric field, the gain is

$$\Gamma = \frac{2\pi n^3 \gamma_{41}}{\lambda \cos\theta} \frac{2\pi k_B \gamma_{41}}{e\Lambda\left[1 + \left(\frac{\Lambda_D}{\Lambda}\right)^2\right]} \frac{1}{\left(1 + \frac{\sigma_d}{\sigma_{ph}}\right)} \tag{8.96}$$

Here $n = 3.45$ is material's refractive index, $\gamma_{41} = 1.07$ pV/m is a component of electro-optical coefficient, λ is wavelength, θ is Bragg angle inside crystal, k_B is Boltzmann's constant, $T = 300$ K is temperature, e is electron charge, Λ is grating period, and Λ_D is a Λ value that makes Γ the maximum, σ_d is dark conductance and σ_{ph} is optical conductance. Figures 8.31 and 8.32 are theoretical date (solid line) for wavelength of 633 nm. Experimental data were showed in the same figure (circle ling).
Response time is

$$\tau = \tau_{d_i} \frac{1 + D\tau_r k^2}{1 + \zeta D\tau_r k^2} \tag{8.97}$$

here

$$\zeta = \tau_{d_i}(sI_0 + \beta + \gamma n_0)$$

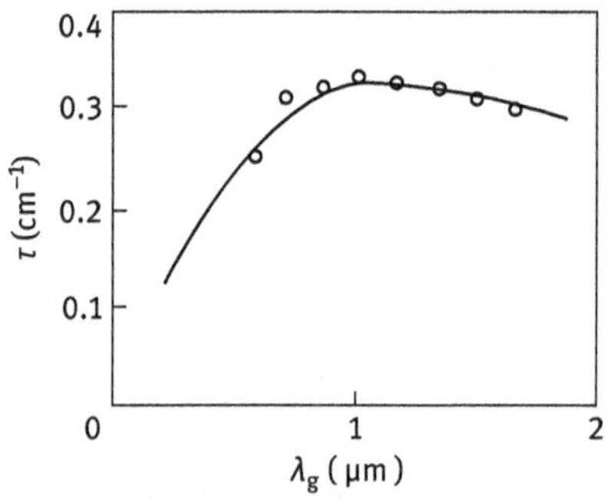

Fig. 8.31: Relationship between GaP gain λ_g.

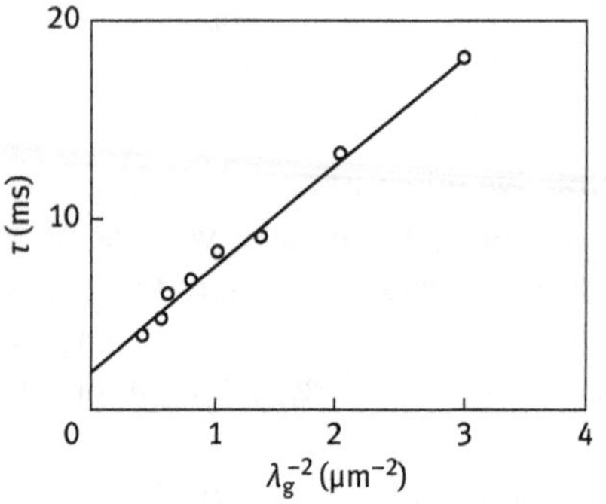

Fig. 8.32: Response time varies vs. λ_g^{-2}.

$\tau_{d_i} = \varepsilon/\sigma$ is dielectric time constant, $k = 2\pi/\Lambda$ is grating wave number, $D = \mu K_B T/e$ is diffusion constant, μ is electron migrate rate, τ_r is recombination time constant. s is absorption cross section divided by photon energy. I_0 is light intensity, β is under-current production rate, γ is recombination coefficient, n is electron density. Straight line in Fig. 8.32 is $\tau = 2.94 + 5.08\lambda_g^{-2}$. And is a special case of equation 8.97. Again circle line is experimental date.

(b) Liquid crystal and organic film

Johnson and others used hydrogenated amorphous silicon (a-Si:H) optical sensor and surface solidified ferroelectric liquid crystal (FLC) modulator to form a so called optical addressing spatial light modulator (OASLM). And used it to achieve a low power fast response phase conjugate. The cross-section of this sandwich kind of device is shown in Fig. 8.33. A uniform indium tin oxide (ITO) electrode is coated on the right side base glass surface; FLC filled $1.75\mu m$ space between two base plates. To

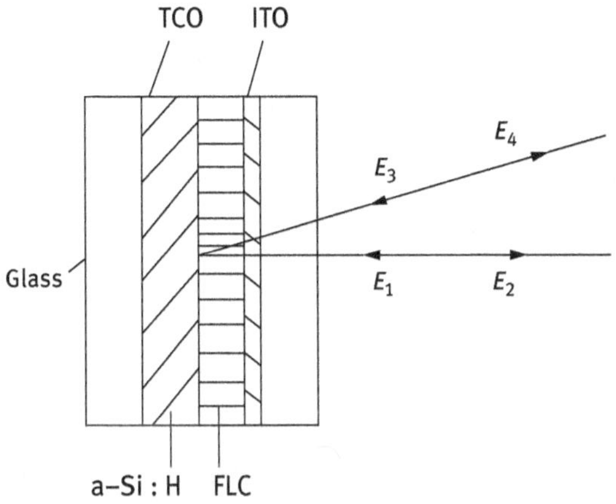

Fig. 8.33: Cross-sectional schematic diagram of OASLM. a – Si:H– hydrogenated amorphous silicon, FLC–ferroelectric liquid crystal; TCO–conductor electrode, ITO–indium tin oxide.

make OASLM work, a 30 V peak to peak square clock voltage and a 5 V bias DC power are applied to the electrodes. Under these arrangement, in an ideal case, when clock voltage is positive and photo diode is in forward bias, there is a 20 V applied to FLC. FLC works like a switch. When clock voltage is -10 V and photo diode is in reverse bias, FLC stays open. If at this moment light incident to the sensor, make the voltage of FLC reduces to a certain value, switch is then closed. As optical axes of FLC in these two situations are in different direction, it forms a refraction index type which dependents on the intensity of incident light. Therefore, read light incidents from FLC side will be modulated by refraction index grating, this will obtain optics phase conjugate. In Fig. 8.33, E_1 is reference light, E_3 is incident signal light, E_2 is reflective reference light, and E_4 is phase conjugate signal light.

Organic polymer film is also a good material to obtain OPC beam. It also can be easily obtained high optical quality and large area samples. But its weak point is to easy to get damaged. Some literatures reported using cooling technique can overcome this shortcoming. Using this method does not affect the efficiency of phase conjugate.

Among photorefractive mediums introduced in this section, wavelength response rang for crystal materials is in the center part of visible light, for most semi-conductor materials is longer than 0.9 μm, very little falls in 0.6–0.9 μm. Barium titanate crystal can response this wave range, but its time constant becomes too big. Reference [8–20] reported GaP can compensate for this deficiency.

8.10 Self-pumped phase conjugate

In section 8.4–8.6 we discussed generating phase conjugate from four-wave mixing. In that case, two light beams travel toward to each other is needed to pump nonlinear medium to form phase grating, probing wave incident in the form of reflection to obtain phase conjugate signal.

Now we are going to introduce other devices which are different from the above ones. In these devices, only one incident light beam is required. That is a probe light which wavefront is waiting to be inversed. The pump light is also comes from the same light. Therefore this is called self-pumped. This can be done by adding two or one ordinary reflector. It can also be done by inner reflection of crystal without any reflector. These will be discussed separately in the following.

8.10.1 Two reflectors

The schematic set up of two reflectors is shown in Fig. 8.34. M_1 and M_2 are two simple ordinary plane mirrors. The C axis of photorefractive crystal is set in such way so that beam 2 will be amplified by beam 1 by two beams coupling. Upon reach stable condition, beam 2 and 3 pump crystal to form PCM. Crystal then gives phase conjugate beam 4 of beam 1. Now the problem is how to find out the reflectivity. That is the intensities ratio $R = I_4(0)/I_1(0)$ at z = 0 of phase conjugate wave 4 to the initial incident beam. Since pump beams 2 and 3 are initiated from incident beam 1, therefore no attenuation pump approximation cannot be applied here. Consider the boundary condition in the situation, apply method used in reference, we can find the reflectivity

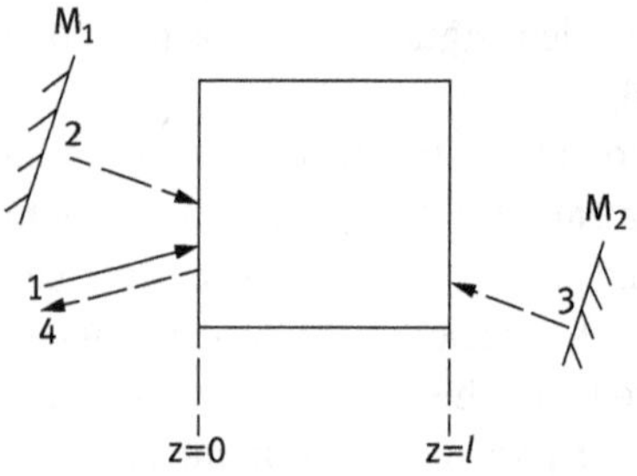

Fig. 8.34: Schematic of double reflectors self-pumping phase conjugate.

$$R = \frac{(\Delta+1)|T|^2}{M_2|T\Delta+a|^2} \tag{8.98}$$

Here

$$\Delta = I_3(l) - I_1(0) - I_2(0)$$

$$T = \tanh\left(\frac{\gamma la}{2}\right) \tag{8.99}$$

$$a = \left[\Delta^2 + \frac{(\Delta+1)^2}{M_2}\right]^{1/2} \tag{8.100}$$

And γl represents coupling strength, l is crystal effective length and γ is coupling constant per unit length. In above equations, $I_0 = I_1(z) + I_2(z) + I_3(z) + I_4(z)$ has been normalized. Δ Can be obtained from the root of equation

$$M_1 M_2 = \left| \frac{T}{T\Delta + \frac{(\Delta+1)M}{M_2}} \right|^2 \tag{8.101}$$

Coupling strength of refractive material has nothing to do with average light intensity I_0. In other medium, such as atomic vapor, γl is proportional to I_0. However, as I_0 is conserved, so as far as other assumptions are true, the discussion here can be applied to other materials.

Now let's consider the threshold for vibration established from zero. This equivalent to take $I_2(0) = I_3(0) = 0$ .So the total intensity is the intensity of incident beam 1. As I_0 is normalized, so $I_1(0) = 0$, therefore,

$$\Delta = -I_1(0) = -1$$

$$a=1$$

$$T = \tanh\left(\frac{\gamma l}{2}\right)$$

From equation (8.99) we have

$$M_1 M_2 = \left| \frac{1 + \tanh\left(\frac{\gamma l}{2}\right)}{1 - \tanh\left(\frac{\gamma l}{2}\right)} \right|^2 = |exp(\gamma l)|^2 = exp[2Re(\gamma l)] \tag{8.102}$$

This shows that the threshold can be defined as the gain in crystal can compensate the loss due to the two reflectors.

8.10.2 Single reflector

For qualitative analysis and without loss of generality. Assume reflector 1 is absent, that is $M_1 = 0$, from equation (8.99), we get,

$$T = -a \quad or \quad \tanh\left(\frac{-\gamma l a}{2}\right) = a \tag{8.103}$$

Substitute into equation (8.98) get

$$R = \frac{1}{M_2}\left(\frac{1+\Delta}{1-\Delta}\right)$$

and from equation (8.100), we find

$$\Delta = \frac{-1 \pm M_2^{1/2}[a^2(1+M_2)-1]^{1/2}}{(1+M_2)}$$

Substitute back to expression of R we get reflectivity

$$R = \left\{\frac{M_2^{1/2} \pm [a^2(1+M_2)-1]^{1/2}]}{M_2 + 2 \mp M_2^{1/2}[a^2(1+M_2)-1]^{1/2}}\right\}^2$$

It can be proved that for above two expressions of R only one of them which has same sign with above can be a steady solution, we then get the final reflectivity of PCM

$$R = \left\{\frac{M_2^{1/2} + [a^2(1+M_2)-1]^{1/2}]}{M_2 + 2 - M_2^{1/2}[a^2(1+M_2)-1]^{1/2}}\right\}^2 \tag{8.104}$$

The threshold of this device is

$$a^2(1+M_2)-1=0$$

That is

$$a_{th}^2 = \frac{1}{1+M_2}$$

The reflectivity at threshold is

$$R_{th} = \frac{M_2}{(M_2+2)^2} \tag{8.105}$$

When $a^2 = 1(\gamma l \to -\infty)$, $R = M_2$, that means reflectivity of PCM is equal to mirror reflectivity. When $M_2 = 1$, $a_{th}^2 = 1/2$ that means when reflectivity of mirror 2 is one, threshold of a^2 is 1/2. Relationship equation (8.103) says $\gamma l \approx 2.5$, this is the threshold of coupling strength without M_1.

Figure 8.35 shows variation of one mirror self-pump phase conjugate reflectivity *R* for a given a^2 vs. reflectivity M_2. Up section of each curve expresses the steady solution of R. We can tell from the figure, for a fixed a^2, steady R increases linearly as M_2 increases; for a given M_2, R increases obviously as a^2 increases.

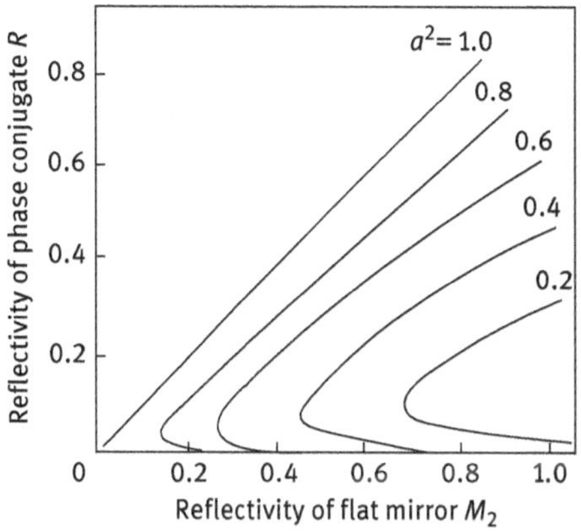

Fig. 8.35: Variation of reflectivity R vs. M_2 for single reflector self-pump PCM.

8.10.3 No external mirror

Figure 8.36 is a schematic beam lines for a no external mirror self-pump phase conjugate set up. Beam 1 incidents crystal with an angle α_1 with respect to C axis, after reaches stable state, pump beams 2 and 3 as well as beams $2'$ and $3'$ generated by beam 1 will travel toward to each other. It forms two action regions (dashed circles). In each action region, there are four beams participate in the action. That is two pump beams, incident beam and its phase conjugate. The pump beam from one action region reflected from crystal surface will enter to another action region, meantime associated with L% energy loss. If the beam reflected point at which the dihedral of two adjacent crystal planes is not a right angle ($\alpha_2' \neq \alpha_2$), the two action regions have different coupling strength, but the device still works.

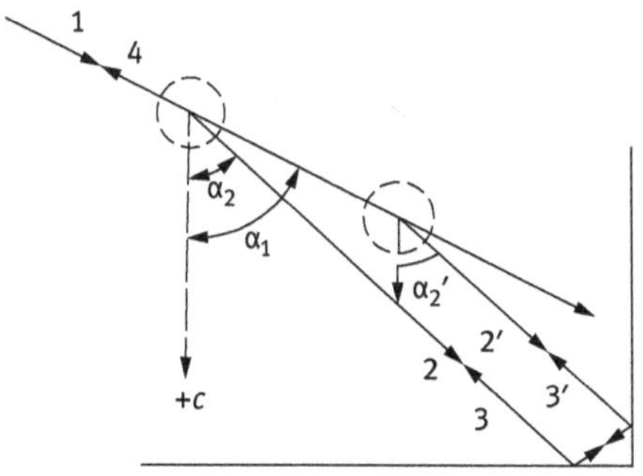

Fig. 8.36: Self-pump phase conjugate produced by inner reflection of crystal.

Assume one action region expands from l_1 to l_2 , the reflectivity is (see reference [8–12])

$$R \equiv \left|\frac{A_4(l_1)}{A^*(l_1)}\right|^2 = \frac{a(r)b(r)}{4(\gamma^2-1)\tanh^4\left(\frac{\gamma lr}{2}\right)} \tag{8.106}$$

here,

$$a(r) = r^2\left[1 - tanh^2\left(\frac{\gamma lr}{2}\right)\right]$$

$$b(r) = \left[r - atanh\left(\frac{\gamma lr}{2}\right)\right]^2 - r^2 tanh^2\left(\frac{\gamma lr}{2}\right)$$

$$l = l_2 - l_1$$

$$r = \left(\Delta^2 + 4|c|^2\right)^{1/2}$$

and

$$\Delta = I_3 + I_4 - I_1 - I_2$$

$$c = A_2 A_3 + A_4 A_1$$

Variation of conjugate wave reflectivity vs. coupling strength is shown in Fig. 8.37. When deriving equation (8.106) and plotting Fig. 8.37, zero energy loss (L\mequal0) was assumed for reflected from one action to another action. Figure 8.37 shows in this situation, threshold for coupling strength $(\gamma l)_{th}$ is approximately 2.3. As L increases, threshold increases slowly at beginning, but as L becomes quite large, threshold increases rapidly as L increases (see Fig. 8.38).

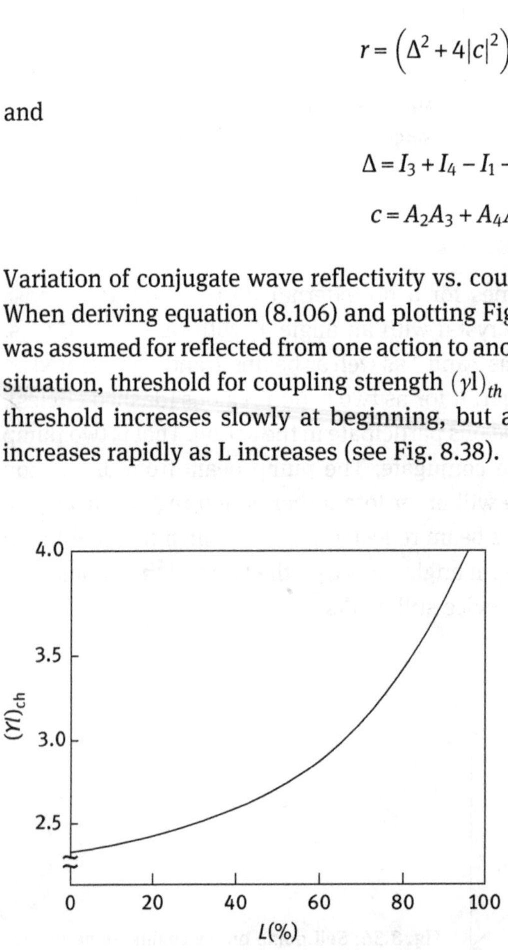

Fig. 8.37: Variation of *R* vs. coupling strength.

In steady state, crystal coupling constant γ [8–13] is

$$\gamma = \frac{\omega}{2nc} \frac{r_{eff}E}{\cos\left(\frac{\alpha_1 - \alpha_2}{2}\right)}$$

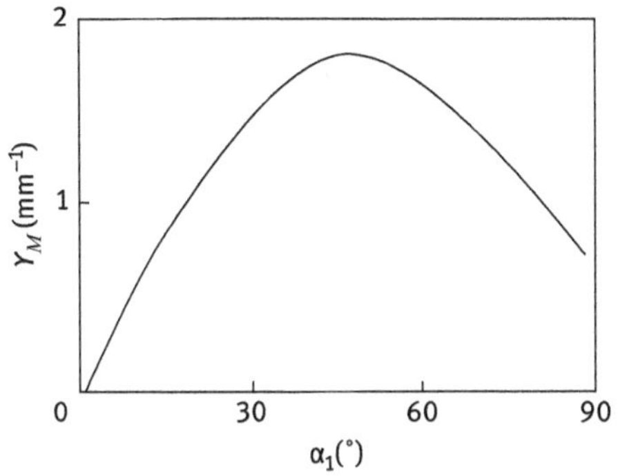

Fig. 8.38: Variation of coupling strength threshold vs. *L*.

Here E is the amplitude of light-induced electric field, and f_{eff} is effective Planck constant. For BaTiO3 kinds of crystal which point group is 4 mm, when use ordinary light as incident light [8–14],

$$r_{eff} = n_0^4 r_{13} sin\left(\frac{\alpha_1 + \alpha_2}{2}\right)$$

For extraordinary light incident,

$$r_{eff} = \left[n_0^4 r_{13} cos\alpha_1 cos\alpha_2 + 2n_e^2 n_0^2 r_{42} \times cos^2\left(\frac{\alpha_1 + \alpha_2}{2}\right) + n_e^4 sinsin\alpha_2\right] \times sin\left(\frac{\alpha_1 + \alpha_2}{2}\right)$$

The above equation shows γ is close dependent on the incident angle, the changes of its maximum with α_1 is shown in Fig. 8.39.

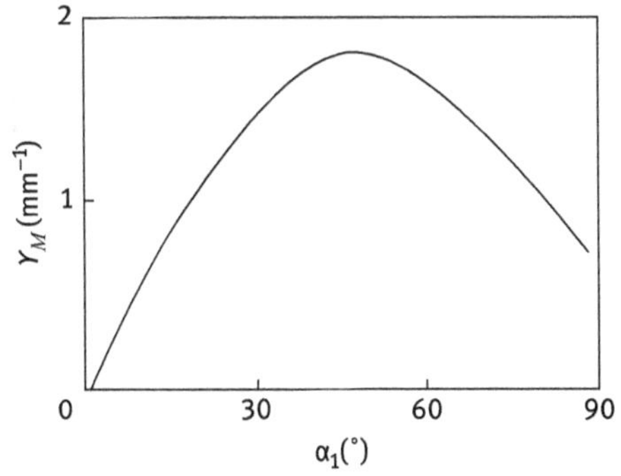

Fig. 8.39: Variation of maximum coupling constant γ_M vs. incident angle α_1.

References

[1] Omatsu, T. (2003). Wavefront correction by self-pumped phase conjugate mirror: application to laser and imaging systems. Ite Technical Report, 27, 17–20.

[2] Pepper, D. M. (1996). Real-time compensated imaging system and method using a double-pumped phase-conjugate mirror. US, US5557431.

[3] Valley, G. C., & Klein, M. B. (1988). Self-pumped phase conjugate mirror and method using AC-field enhanced photorefractive effect. US, US4773739.

[4] Mager, L., Pauliat, G., Garrett, M. H., & Rytz, D. (1994). Nanosecond pulses wavefront correction using photorefractive self-pumped phase conjugate mirror. Lasers and Electro-Optics Europe, 1994 Conference on(pp.20–24). IEEE.

[5] Betin, A. A., Reeder, R. A., & Byren, R. W. (2006). Phase conjugate laser and method with improved fidelity. US, US 7133427 B2.

[6] Byren, R. W., & Filgas, D. (2005). Phase conjugate relay mirror apparatus for high energy laser system and method. US, US 6961171 B2.

[7] Dane, B. C., Hackel, L.,A. & Harris, F.B (2004). Self-seeded single-frequency solid-state ring laser, and single-frequency laser peening method and system using same. US, US 20040228376 A1.

[8] Dane, B. C., Hackel, L. A., & Harris, F. B. (2010). Self-seeded single-frequency solid-state ring laser, laser peening method and system using same. EP, EP1478062.

[9] Dane, B. C., Hackel, L., & Harris, F. B. (2007). Stimulated Brillouin scattering mirror system, high power laser and laser peening method and system using same. US, US7209500.

[10] Dane, B. C., Harris, F. B., Taranowski, J. T., & Brown, S. B. (2010). Active beam delivery system for laser peening and laser peening method.US, US 7750266 B2.

[11] Clauer, A. H., Toller, S. M., & Dulaney, J. L. (2002). Beam path clearing for laser peening. US, US 6359257 B1.

[12] Dane, C. B., Zapata, L. E., Neuman, W. A., & Norton, M. A. (1995). Design and operation of a 150 w near diffraction-limited laser amplifier with SBS wavefront correction. IEEE Journal of Quantum Electronics,31(1), 148–163.

[13] Bischel, W. K., Reed, M. K., Negus, D. K., & Frangineas, G. (1996). System for minimizing the depolarization of a laser beam due to thermally induced birefringence. US, US5504763.

[14] Kewitsch, A. S., & Rakuljic, G. A. (2010). Electronically phase-locked laser systems. US, US7848370.

[15] Klein, M. B., Pepper, D. M., Stephens, R. R., O'Meara, T. R., Welch, D., & Lang, R. J., et al. (1998). Hybrid laser power combining and beam cleanup system using nonlinear and adaptive optical wavefront compensation. US, US5717516.

[16] Li, J., & Chen, X. (2013). Aberration compensation of laser mode unging a novel intra-cavity adaptive optical system. Optik-International Journal for Light and Electron Optics, 124(3),272–275.

[17] Billman, K. W. (2006). Method and system for wavefront compensation.US, US 700212.

[18] Rigamonti, Luca (April 2010). "Schiff base metal complexes for second order nonlinear optics" (PDF). La Chimica l'Industria (Società Chimica Italiana) (3): 118–122.

[19] Bloembergen, Nicolaas (1965). Nonlinear Optics. ISBN 9810225997.

[20] Kouzov, N.I. Egorova, M. Chrysos, F. Rachet 2012,Non-linear optical channels of the polarizability induction in a pair of interacting molecules, NANOSYSTEMS: PHYSICS, CHEMISTRY, MATHEMATICS, 3 (2), P 55.

[21] Abolghasem, Payam; Junbo Han; Bhavin J. Bijlani; Amr S. Helmy (2010). "Type-0 second order nonlinear interaction in monolithic waveguides of isotropic semiconductors". Optics Express 18 (12): 12681–12689.

https://doi.org/10.1515/9783110500608-009

[22] Strauss, CEM; Funk, DJ (1991). "Broadly tunable difference-frequency generation of VUV using two-photon resonances in H2 and Kr". Optics Lett. 16 (15): 1192.
[23] Vladimir Shkunov and Boris Zel'dovich. 1985 "Phase Conjugation, in Scientific American, December 1985.
[24] David M. Pepper. 1986, "Applications of Optical Phase Conjugation," in Scientific American, January 1986.
[25] David M. Pepper, Jack Feinberg, and Nicolai V. Kukhtarev. 1990, "The Photorefractive Effect," in Scientific American, October 1990.
[26] A.Yu. Okulov, (2008) "Angular momentum of photons and phase conjugation", J. Phys. B: At. Mol. Opt. Phys. v. 41, 101001.
[27] A.Y. Okulov, (2008)" Optical and Sound Helical structures in a Mandelstam-Brillouin mirror". JETP Lett, v.88, n.8, pp. 561–566.
[28] Strauss, CEM; Funk, DJ (1991). "Broadly tunable difference-frequency generation of VUV using two-photon resonances in H2 and Kr".Optics Lett. 16 (15): 1192.
[29] Vladimir Shkunov and Boris Zel'dovich. 1985 "Phase Conjugation, in Scientific American", December 1985.
[30] David M. Pepper. 1986, "Applications of Optical Phase Conjugation," in Scientific American, January 1986.
[31] David M. Pepper, Jack Feinberg, and Nicolai V. Kukhtarev. 1990, "The Photorefractive Effect," in Scientific American, October 1990.
[32] A.Yu. Okulov, (2008) "Angular momentum of photons and phase conjugation", J. Phys. B: At. Mol. Opt. Phys. v. 41, 101001.
[33] A.Y. Okulov, (2008)"Optical and Sound Helical structures in a Mandelstam-Brillouin mirror". JETP Lett, v.88, n.8, pp. 561–566.

Subject Index

https://doi.org/10.1515/9783110500608-010

www.ingramcontent.com/pod-product-compliance
Lightning Source LLC
Chambersburg PA
CBHW080657220326
41598CB00033B/5237